LOGNORMAL DISTRIBUTIONS

STATISTICS: Textbooks and Monographs

A Series Edited by

Vol. 1: The Generalized Jackknife Statistic, *H. L. Gray and W. R. Schucany*
Vol. 2: Multivariate Analysis, *Anant M. Kshirsagar*
Vol. 3: Statistics and Society, *Walter T. Federer*
Vol. 4: Multivariate Analysis: A Selected and Abstracted Bibliography, 1957-1972, *Kocherlakota Subrahmaniam and Kathleen Subrahmaniam* (out of print)
Vol. 5: Design of Experiments: A Realistic Approach, *Virgil L. Anderson and Robert A. McLean*
Vol. 6: Statistical and Mathematical Aspects of Pollution Problems, *John W. Pratt*
Vol. 7: Introduction to Probability and Statistics (in two parts), Part I: Probability; Part II: Statistics, *Narayan C. Giri*
Vol. 8: Statistical Theory of the Analysis of Experimental Designs, *J. Ogawa*
Vol. 9: Statistical Techniques in Simulation (in two parts), *Jack P. C. Kleijnen*
Vol. 10: Data Quality Control and Editing, *Joseph I. Naus* (out of print)
Vol. 11: Cost of Living Index Numbers: Practice, Precision, and Theory, *Kali S. Banerjee*
Vol. 12: Weighing Designs: For Chemistry, Medicine, Economics, Operations Research, Statistics, *Kali S. Banerjee*
Vol. 13: The Search for Oil: Some Statistical Methods and Techniques, *edited by D. B. Owen*
Vol. 14: Sample Size Choice: Charts for Experiments with Linear Models, *Robert E. Odeh and Martin Fox*
Vol. 15: Statistical Methods for Engineers and Scientists, *Robert M. Bethea, Benjamin S. Duran, and Thomas L. Boullion*
Vol. 16: Statistical Quality Control Methods, *Irving W. Burr*
Vol. 17: On the History of Statistics and Probability, *edited by D. B. Owen*
Vol. 18: Econometrics, *Peter Schmidt*
Vol. 19: Sufficient Statistics: Selected Contributions, *Vasant S. Huzurbazar (edited by Anant M. Kshirsagar)*
Vol. 20: Handbook of Statistical Distributions, *Jagdish K. Patel, C. H. Kapadia, and D. B. Owen*
Vol. 21: Case Studies in Sample Design, *A. C. Rosander*
Vol. 22: Pocket Book of Statistical Tables, *compiled by R. E. Odeh, D. B. Owen, Z. W. Birnbaum, and L. Fisher*
Vol. 23: The Information in Contingency Tables, *D. V. Gokhale and Solomon Kullback*

Vol. 24: Statistical Analysis of Reliability and Life-Testing Models: Theory and Methods, *Lee J. Bain*
Vol. 25: Elementary Statistical Quality Control, *Irving W. Burr*
Vol. 26: An Introduction to Probability and Statistics Using BASIC, *Richard A. Groeneveld*
Vol. 27: Basic Applied Statistics, *B. L. Raktoe and J. J. Hubert*
Vol. 28: A Primer in Probability, *Kathleen Subrahmaniam*
Vol. 29: Random Processes: A First Look, *R. Syski*
Vol. 30: Regression Methods: A Tool for Data Analysis, *Rudolf J. Freund and Paul D. Minton*
Vol. 31: Randomization Tests, *Eugene S. Edgington*
Vol. 32: Tables for Normal Tolerance Limits, Sampling Plans, and Screening, *Robert E. Odeh and D. B. Owen*
Vol. 33: Statistical Computing, *William J. Kennedy, Jr. and James E. Gentle*
Vol. 34: Regression Analysis and Its Application: A Data-Oriented Approach, *Richard F. Gunst and Robert L. Mason*
Vol. 35: Scientific Strategies to Save Your Life, *I. D. J. Bross*
Vol. 36: Statistics in the Pharmaceutical Industry, *edited by C. Ralph Buncher and Jia-Yeong Tsay*
Vol. 37: Sampling from a Finite Population, *J. Hajek*
Vol. 38: Statistical Modeling Techniques, *S. S. Shapiro*
Vol. 39: Statistical Theory and Inference in Research, *T. A. Bancroft and C.-P. Han*
Vol. 40: Handbook of the Normal Distribution, *Jagdish K. Patel and Campbell B. Read*
Vol. 41: Recent Advances in Regression Methods, *Hrishikesh D. Vinod and Aman Ullah*
Vol. 42: Acceptance Sampling in Quality Control, *Edward G. Schilling*
Vol. 43: The Randomized Clinical Trial and Therapeutic Decisions, *edited by Niels Tygstrup, John M. Lachin, and Erik Juhl*
Vol. 44: Regression Analysis of Survival Data in Cancer Chemotherapy, *Walter H. Carter, Jr., Galen L. Wampler, and Donald M. Stablein*
Vol. 45: A Course in Linear Models, *Anant M. Kshirsagar*
Vol. 46: Clinical Trials: Issues and Approaches, *edited by Stanley H. Shapiro and Thomas H. Louis*
Vol. 47: Statistical Analysis of DNA Sequence Data, *edited by B. S. Weir*
Vol. 48: Nonlinear Regression Modeling: A Unified Practical Approach, *David A. Ratkowsky*
Vol. 49: Attribute Sampling Plans, Tables of Tests and Confidence Limits for Proportions, *Robert E. Odeh and D. B. Owen*
Vol. 50: Experimental Design, Statistical Models, and Genetic Statistics, *edited by Klaus Hinkelmann*
Vol. 51: Statistical Methods for Cancer Studies, *edited by Richard G. Cornell*
Vol. 52: Practical Statistical Sampling for Auditors, *Arthur J. Wilburn*
Vol. 53: Statistical Signal Processing, *edited by Edward J. Wegman and James G. Smith*
Vol. 54: Self-Organizing Methods in Modeling: GMDH Type Algorithms, *edited by Stanley J. Farlow*
Vol. 55: Applied Factorial and Fractional Designs, *Robert A. McLean and Virgil L. Anderson*
Vol. 56: Design of Experiments: Ranking and Selection, *edited by Thomas J. Santner and Ajit C. Tamhane*
Vol. 57: Statistical Methods for Engineers and Scientists. Second Edition, Revised and Expanded, *Robert M. Bethea, Benjamin S. Duran, and Thomas L. Boullion*
Vol. 58: Ensemble Modeling: Inference from Small-Scale Properties to Large-Scale Systems, *Alan E. Gelfand and Crayton C. Walker*
Vol. 59: Computer Modeling for Business and Industry, *Bruce L. Bowerman and Richard T. O'Connell*
Vol. 60: Bayesian Analysis of Linear Models, *Lyle D. Broemeling*

Vol. 61: Methodological Issues for Health Care Surveys, *Brenda Cox and Steven Cohen*

Vol. 62: Applied Regression Analysis and Experimental Design, *Richard J. Brook and Gregory C. Arnold*

Vol. 63: Statpal: A Statistical Package for Microcomputers – PC-DOS Version for the IBM PC and Compatibles, *Bruce J. Chalmer and David G. Whitmore*

Vol. 64: Statpal: A Statistical Package for Microcomputers – Apple Version for the II, II+, and IIe, *David G. Whitmore and Bruce J. Chalmer*

Vol. 65: Nonparametric Statistical Inference, Second Edition, Revised and Expanded, *Jean Dickinson Gibbons*

Vol. 66: Design and Analysis of Experiments, *Roger G. Petersen*

Vol. 67: Statistical Methods for Pharmaceutical Research Planning, *Sten W. Bergman and John C. Gittins*

Vol. 68: Goodness-of-Fit Techniques, *edited by Ralph B. D'Agostino and Michael A. Stephens*

Vol. 69: Statistical Methods in Discrimination Litigation, *edited by D. H. Kaye and Mikel Aickin*

Vol. 70: Truncated and Censored Samples from Normal Populations, *Helmut Schneider*

Vol. 71: Robust Inference, *M. L. Tiku, W. Y. Tan, and N. Balakrishnan*

Vol. 72: Statistical Image Processing and Graphics, *edited by Edward J. Wegman and Douglas J. DePriest*

Vol. 73: Assignment Methods in Combinatorial Data Analysis, *Lawrence J. Hubert*

Vol. 74: Econometrics and Structural Change, *Lyle D. Broemeling and Hiroki Tsurumi*

Vol. 75: Multivariate Interpretation of Clinical Laboratory Data, *Adelin Albert and Eugene K. Harris*

Vol. 76: Statistical Tools for Simulation Practitioners, *Jack P. C. Kleijnen*

Vol. 77: Randomization Tests, Second Edition, *Eugene S. Edgington*

Vol. 78: A Folio of Distributions: A Collection of Theoretical Quantile-Quantile Plots, *Edward B. Fowlkes*

Vol. 79: Applied Categorical Data Analysis, *Daniel H. Freeman, Jr.*

Vol. 80: Seemingly Unrelated Regression Equations Models : Estimation and Inference, *Virendra K. Srivastava and David E. A. Giles*

Vol. 81: Response Surfaces: Designs and Analyses, *Andre I. Khuri and John A. Cornell*

Vol. 82: Nonlinear Parameter Estimation: An Integrated System in BASIC, *John C. Nash and Mary Walker-Smith*

Vol. 83: Cancer Modeling, *edited by James R. Thompson and Barry W. Brown*

Vol. 84: Mixture Models: Inference and Applications to Clustering, *Geoffrey J. McLachlan and Kaye E. Basford*

Vol. 85: Randomized Response: Theory and Techniques, *Arijit Chaudhuri and Rahul Mukerjee*

Vol. 86: Biopharmaceutical Statistics for Drug Development, *edited by Karl E. Peace*

Vol. 87: Parts per Million Values for Estimating Quality Levels, *Robert E. Odeh and D. B. Owen*

Vol. 88: Lognormal Distributions: Theory and Applications, *edited by Edwin L. Crow and Kunio Shimizu*

Vol. 89: Properties of Estimators for the Gamma Distribution, *K. O. Bowman and L. R. Shenton*

Vol. 90: Spline Smoothing and Nonparametric Regression, *Randall L. Eubank*

Vol. 91: Linear Least Squares Computations, *R. W. Farebrother*

ADDITIONAL VOLUMES IN PREPARATION

LOGNORMAL DISTRIBUTIONS

THEORY AND APPLICATIONS

edited by

Edwin L. Crow
Institute for Telecommunication Sciences
National Telecommunications and
Information Administration
U.S. Department of Commerce
Boulder, Colorado

Kunio Shimizu
Department of Information Sciences
Faculty of Science and Technology
Science University of Tokyo
Noda City, Chiba, Japan

MARCEL DEKKER, INC. New York and Basel

Library of Congress Cataloging-in-Publication Data

Lognormal distributions : theory and applications / edited by Edwin L.
Crow, Kunio Shimizu.
p. cm. — (Statistics, textbooks and monographs ; vol. 88)
Includes indexes.
ISBN 0-8247-7803-0
1. Lognormal distribution. I. Crow, Edwin L. II. Shimizu,
Kunio, [date]. III. Series: Statistics, textbooks and monographs ;
v. 88.
QA273.6.L64 1988
519.2—dc19 87-24375
CIP

MARCEL DEKKER, INC.
270 Madison Avenue, New York, New York 10016

Current printing (last digit):
10 9 8 7 6 5 4 3 2 1

PRINTED IN THE UNITED STATES OF AMERICA

Preface

The lognormal distribution may be defined as the distribution of a random variable whose logarithm is normally distributed. It is a typical example of positively skewed frequency curves. It has two parameters but may be generalized by a translation parameter, by truncation and censoring, by adjoining a point probability mass, by extension to two or more dimensions, and by transformation. These generalizations, as well as the two-parameter distribution, are treated in this book.

Aitchison and Brown's 1957 monograph presented a unified treatment of the theory, beginning with a discussion of the genesis of the lognormal distribution from elementary random processes, passing through problems of point and interval estimation of parameters, and ending with a review of applications. Johnson and Kotz, in 1970, concisely summarized the history of theory and application of lognormal distributions and the development of estimation for the two- and three-parameter lognormal and related distributions.

Since the publication of the books by Aitchison and Brown and by Johnson and Kotz, the theory of lognormal distributions has steadily progressed and fields of application have greatly increased. (Johnson and Kotz had stated [1970, page 128], "It is quite likely that the lognormal distribution will be one of the most widely applied distributions in practical statistical work in the near future.") The present book emphasizes, but is not limited to, the more recent developments in the genesis, application, and properties of lognormal distributions, estimation and test theories, and some results for related distributions. The first seven chapters are primarily theory, the last seven primarily application.

This book is directed primarily to applied statisticians and graduate students in statistics, but will also be useful to researchers in a variety of subject-matter fields, including economics, biology, ecology, geology, atmospheric sciences, business, and industry. Subject-matter specialists have often not applied, or even had available, the recent more precise methods in the first seven chapters, while statisticians may find impetus for further development of methods in the discussions and references of the last seven chapters.

The book presupposes an introductory course in mathematical statistics and a knowledge of calculus. Thus the book requires some knowledge of basic results for the normal distribution, for many of the properties of the lognormal distribution may immediately be derived from those of the normal. As far as possible, the development is self-contained. A unified treatment is attempted despite the substantial number of authors, but the chapters can be read essentially independently. Each chapter has its own references; in addition, there is a complete author index with the location of full references italicized, as well as a complete subject index.

Chapter 1 gives a brief history (a more extensive history, including 217 references, is available in Aitchison and Brown [1957]), the genesis of the two-parameter distribution from qualitative assumptions (law of proportionate effect and breakage process), and basic properties, including moments, order statistics, and distributions of products and sums. Chapter 2 brings up to date the theory of point estimation of the two-parameter distribution, its multivariate analog, and the delta generalizations, in which a point probability mass is adjoined to the distribution. The formulas for estimates and their variances are often rendered compact and systematized by the use of generalized hypergeometric functions.

Chapter 3 presents the theory of interval estimation and testing of hypotheses for the two-parameter distribution. Chapters 4 and 5 deal with estimation for the three-parameter distribution and for data that have been censored, truncated, or grouped. The Bayesian approach to estimation has been particularly useful in life testing and is presented in Chapter 6.

Chapter 7 on the Poisson-lognormal distribution is unique in this book because that distribution is not strictly a lognormal distribution; it is a mixture of Poisson distributions, the Poisson means being lognormally distributed. It has been found useful in describing the abundance of species in biology.

The last seven chapters describe in detail the extensive applications in economics, business, and human affairs (8 and 9), industry (10), biology, especially growth models (11), ecology (12), atmospheric sciences (13), and geology (14).

The contributions of the ten other authors were invited and reviewed by us. We are grateful to them for their dedication and cooperation and to the chairman and editors of Marcel Dekker, Inc., Maurits Dekker, Vickie Kearn, John K. Cook, and Lila Harris, for their ready and invaluable professional assistance. The book's typesetters, The Bartlett Press, Inc., deserve a hand for dealing elegantly with often unwieldy equations.

While our employers have no responsibility for this work, we are grateful to the Institute for Telecommunication Sciences (ITS), U.S. Department of Commerce (DoC), and the Science University of Tokyo for the use of facilities and the assistance of individuals. In addition, both of us are indebted to the National Center for Atmospheric Research (NCAR) and its sponsor, the National Science Foundation, for positions as visitors in the Convective Storms Division, directed by Edward J. Zipser, during the preparation of the volume. Jean M. Bankhead (DoC), Jane L. Watterson (DoC), and Gayl H. Gray (NCAR) expertly and cheerfully provided bibliographical aid, and Kathy E. Mayeda (ITS), Carmen du Bouchet (ITS), and Sudie J. Kelly (NCAR) similarly provided word-processing aid. We are pleased to express our thanks to all.

Edwin L. Crow
Kunio Shimizu

Contents

Contributors

GREGORY CAMPBELL, Ph.D. Senior Staff Fellow, Laboratory of Statistical and Mathematical Methodology, Division of Computer Research and Technology, National Institutes of Health, Bethesda, Maryland

A. CLIFFORD COHEN, Ph.D. Professor Emeritus of Statistics, Department of Statistics, University of Georgia, Athens, Georgia

EDWIN L. CROW, Ph.D. Mathematical Statistician, Institute for Telecommunication Sciences, National Telecommunications and Information Administration, U.S. Department of Commerce, Boulder, Colorado

BRIAN DENNIS, Ph.D. Associate Professor, College of Forestry, Wildlife, and Range Sciences, University of Idaho, Moscow, Idaho

CHARLES E. LAND, Ph.D. Health Statistician, Radiation Epidemiology Branch, National Cancer Institute, National Institutes of Health, Bethesda, Maryland

RAYMOND J. LAWRENCE, M.A. Professor of Marketing, Department of Marketing, University of Lancaster, Bailrigg, Lancaster, England

JAMES E. MOSIMANN, Ph.D. Chief, Laboratory of Statistical and Mathematical Methodology, Division of Computer Research and Technology, National Institutes of Health, Bethesda, Maryland

WILLIAM J. PADGETT, Ph.D. Professor and Chairman, Department of Statistics, University of South Carolina, Columbia, South Carolina

G. P. PATIL, Ph.D., D.Sc. Professor of Mathematical Statistics, Director, Center for Statistical Ecology and Environmental Statistics, Department of Statistics, The Pennsylvania State University, University Park, Pennsylvania; Visiting Professor of Biostatistics, Department of Biostatistics, Harvard School of Public Health and Dana-Farber Cancer Institute, Harvard University, Boston, Massachusetts

JEAN-MICHEL M. RENDU, Eng. Sc.D. Director, Technical and Scientific Systems Group, Newmont Mining Corporation, Danbury, Connecticut

S. A. SHABAN, Ph.D. Associate Professor, Faculty of Commerce, Economics and Political Science, Department of Insurance and Statistics, Kuwait University, Kuwait

KUNIO SHIMIZU, D.Sc. Assistant Professor, Department of Information Sciences, Faculty of Science and Technology, Science University of Tokyo, Noda City, Chiba, Japan

LOGNORMAL DISTRIBUTIONS

1

History, Genesis, and Properties

KUNIO SHIMIZU Department of Information Sciences, Faculty of Science and Technology, Science University of Tokyo, Noda City, Chiba, Japan

EDWIN L. CROW Institute for Telecommunication Sciences, National Telecommunications and Information Administration, U.S. Department of Commerce, Boulder, Colorado

1. INTRODUCTION

The lognormal distribution (with two parameters) may be defined as the distribution of a random variable whose logarithm is normally distributed. Such a variable is necessarily positive. Since many variables in real life, from the sizes of organisms and the numbers of species in biology to rainfalls in meteorology and sizes of incomes in economics, are inherently positive, the lognormal distribution has been widely applied in an empirical way for fitting data. In addition, it has been derived theoretically from qualitative assumptions; Gibrat (1930, 1931) did this in 1930, calling it the law of proportionate effect, but Kapteyn (1903) had described a machine that was the mechanical equivalent. Kolmogoroff (1941) derived the distribution as the asymptotic result of an iterative process of successive breakage of a particle into two randomly sized particles.

Thus there is a theoretical basis as well as empirical application of lognormal distributions, but why is there much to say about them if the data analysis can be referred to the intensively studied normal distribution by taking the logarithm? There are several reasons:

(1) The parameter estimates resulting from the inverse transformation are biased.
(2) The two-parameter distribution is often not a sufficient description; a third parameter, the threshold or location parameter, is needed, for example, for the distribution of ages at first marriage.
(3) The distribution may be censored or truncated (e.g., low income data are often missing), or the data may be classified into groups, so that special methods are needed. (See Chapter 5.)

Aitchison and Brown (1957) and Johnson and Kotz (1970) have described the early history of lognormal distributions, but a brief summary is desirable here. Galton (1879) and McAlister (1879) initiated the study of the distribution in papers published together, relating it to the use of the geometric mean as an estimate of location. Much later Kapteyn (1903) discussed the genesis of the distribution, and Kapteyn and Van Uven (1916) gave a graphical method for estimating the parameters. Wicksell (1917) used the method of moments for three-parameter estimation, introducing a third parameter, the threshold, to fit the distribution of ages at first marriage. Nydell (1919) obtained asymptotic standard errors for the moment estimates. The distribution appeared in papers of the 1930s that developed probit analysis in bioassay. Yuan (1933) introduced the bivariate lognormal distribution.

Later work on genesis of the lognormal distribution is most appropriately recorded in connection with exposition of the basic methods of genesis in Section 3. Similarly, later developments of properties of lognormal distributions, including distributions of products, quotients, and sums, are discussed in Section 4. The two- and three-parameter univariate (i.e., one-dimensional) lognormal distributions and the multivariate lognormal distribution are defined in Section 2. The later developments in inference and applications of lognormal distributions are considered in the later chapters.

2. DEFINITION AND NOTATION

A positive random variable X is said to be lognormally distributed with two parameters μ and σ^2 if $Y = \ln X$ is normally distributed with mean μ and variance σ^2. The two-parameter lognormal distribution is denoted by $\Lambda(\mu, \sigma^2)$; the corresponding normal distribution is denoted by $N(\mu, \sigma^2)$. The probability density function of X having $\Lambda(\mu, \sigma^2)$ is

$$f(x) = \begin{cases} \dfrac{1}{\sqrt{2\pi}\sigma x} \exp\left\{ -\dfrac{(\ln x - \mu)^2}{2\sigma^2} \right\} & x > 0 \\ 0 & x \leq 0 \end{cases} \tag{2.1}$$

Figure 1 illustrates the probability density function of $\Lambda(0, \sigma^2)$ for $\sigma = 0.1, 0.3, 1.0$. The probability density function of $\Lambda(\mu, \sigma^2)$ can be written as $f(x \mid \mu, \sigma^2) = e^{-\mu} f(xe^{-\mu} \mid 0, \sigma^2)$, so it has the same shape as that of $\Lambda(0, \sigma^2)$. The distribution is unimodal and positively skew.

In addition, a random variable X which can take any value exceeding a fixed value τ is said to be lognormally distributed with three parameters τ, μ, and σ^2 if $Y = \ln(X - \tau)$ is $N(\mu, \sigma^2)$. The three-parameter lognormal distribution is denoted by $\Lambda(\tau, \mu, \sigma^2)$. The parameter τ is called the *threshold* parameter. Thus the two-parameter lognormal distribution $\Lambda(\mu, \sigma^2)$ is a special case of the three-parameter lognormal distribution $\Lambda(\tau, \mu, \sigma^2)$ for which $\tau = 0$. But estimation procedures developed for the two-parameter case are not directly applicable to the three-parameter case. (See Chapter 4).

Further consider a vector $(X_1, \ldots, X_n)'$ of positive random variables such that $(Y_1, \ldots, Y_n)' = (\ln X_1 \ldots, \ln X_n)'$ has an n-dimensional normal distribution with mean vector $\mu = (\mu_1, \ldots, \mu_n)'$ and variance-

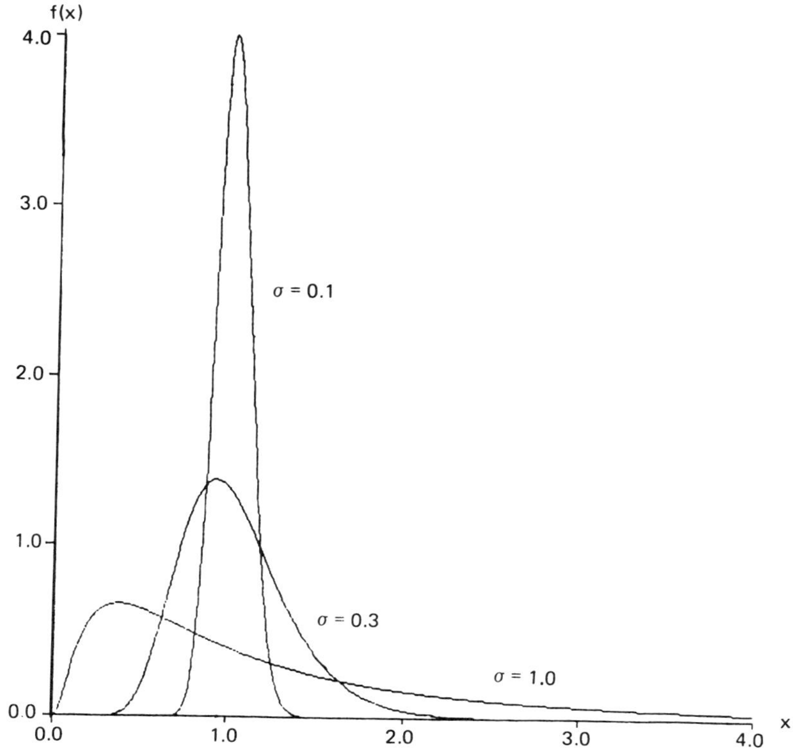

Figure 1 Probability density function of $\Lambda(0, \sigma^2)$.

covariance matrix $\Sigma = (\sigma_{ij})$, $i, j = 1, \ldots, n$, with $\sigma_{ij} = \sigma_{ji}$ and $\sigma_{ii} = \sigma_i^2$. The distribution of $(X_1, \ldots, X_n)'$ is said to be an n-dimensional lognormal distribution with parameters μ and Σ and denoted by $\Lambda_n(\mu, \Sigma)$. The corresponding n-dimensional normal distribution is denoted by $N_n(\mu, \Sigma)$. The probability density function of $(X_1, \ldots, X_n)'$ having $\Lambda_n(\mu, \Sigma)$ is

$$f(x_1, \ldots, x_n) = \begin{cases} \dfrac{1}{(2\pi)^{n/2}\sqrt{|\Sigma|}x_1 \cdots x_n} \exp\left\{-\dfrac{1}{2}(\ln x - \mu)'\Sigma^{-1}(\ln x - \mu)\right\} \\ \qquad x \in R^{n+} \\ 0, \qquad x \in (R^{n+})^c \end{cases}$$

where $x = (x_1, \ldots, x_n)', \ln x = (\ln x_1, \ldots, \ln x_n)'$, and $R^{n+} = \{x \mid x_i > 0$ for all $i = 1, \ldots, n\}$.

3. GENESIS

3.1 The Law of Proportionate Effect

Numerous processes have been devised for generating the lognormal distribution. One of them is the *law of proportionate effect*, so called by Gibrat (1930, 1931).

Suppose that an initial variable X_0 is positive. The equation considered by Kapteyn (1903), which successively calculates the jth step variable X_j, is

$$X_j - X_{j-1} = \varepsilon_j \phi(X_{j-1}) \tag{3.1}$$

where $\{\varepsilon_j\}$ is a set of mutually independent and identically distributed random variables and is also statistically independent of $\{X_j\}$, and ϕ is a certain function. We consider here the important special case $\phi(X) = X$; then the process $\{X_j\}$ is said to obey the law of proportionate effect. Thus (3.1) reduces to

$$X_j = X_{j-1}(1 + \varepsilon_j) \tag{3.2}$$

The generation of a lognormal distribution is roughly explained as follows. The relation (3.2) leads to

$$X_n = X_0 \prod_{j=1}^{n} (1 + \varepsilon_j)$$

Assuming that the absolute value of ε_j is small compared with 1, we approximately have from the Taylor expansion of $\ln(1 + x)$ that

$$\ln X_n = \ln X_0 + \sum_{j=1}^{n} \varepsilon_j$$

By the additive central limit theorem $\ln X_n$ is asymptotically normally distributed and hence X_n is asymptotically lognormally distributed in a two-parameter form.

The result rigorously follows from the multiplicative analogue for the additive central limit theorem: If $\{Z_j\}$ is a sequence of positive, independent, and identically distributed random variables such that

$$E(\ln Z_j) = \mu < \infty$$

$$\text{Var}(\ln Z_j) = \sigma^2 < \infty$$

then the product $\prod_{j=1}^{n} Z_j$ is asymptotically distributed as $\Lambda(n\mu, n\sigma^2)$.

Kapteyn's analogue machine (a photograph is in Aitchison and Brown (1957)) is based on the generating model (3.2), where the distribution of the random variable ε_j is

$$Pr(\varepsilon_j = a) = \tfrac{1}{2} \qquad Pr(\varepsilon_j = -a) = \tfrac{1}{2}$$

for all j, where a is a positive constant. For $0 < a < 1$, this situation satisfies the above multiplicative central limit theorem. Hence, the final step variable X_n is asymptotically lognormally distributed in a two-parameter form. Sand poured at the top of the machine forms a skew histogram approximating that given by a two-parameter lognormal distribution.

Kalecki (1945) states that Gibrat's argument is formally correct but its implications may be unrealistic. Consequently, Kalecki reasoned that there must be something wrong in the underlying assumptions. The first case he considered, which appears in the case of many economic phenomena, is that the variance M of $\ln X_j$ remains constant through time. Since

$$\begin{aligned}\operatorname{Var}(\ln X_j) = {} & \operatorname{Var}(\ln X_{j-1}) + \operatorname{Var}(\ln(1+\varepsilon_j)) \\ & + 2\operatorname{Cov}(\ln X_{j-1}, \ln(1+\varepsilon_j))\end{aligned}$$

the constancy of the variance of $\ln X_j$ implies a negative correlation between $\ln X_{j-1}$ and $\ln(1+\varepsilon_j)$. A further assumption by Kalecki is that the regression of $\ln(1+\varepsilon_j)$ on $\ln X_{j-1}$ is linear,

$$\ln(1+\varepsilon_j) = -\alpha_j \ln X_{j-1} + \eta_j$$

where α_j is a constant and η_j is statistically independent of $\ln X_{j-1}$. The equation

$$X_j = X_{j-1} \exp(-\alpha_j \ln X_{j-1} + \eta_j)$$

leads to

$$\begin{aligned}&\ln X_0 + \sum_{j=1}^{n} \ln(1+\varepsilon_j) \\ &\quad = \ln X_0 \prod_{j=1}^{n} (1-\alpha_j) + \eta_1 \prod_{j=2}^{n} (1-\alpha_j) + \cdots + \eta_{n-1}(1-\alpha_n) + \eta_n\end{aligned}$$

Under the conditions that the variance of $\ln(1+\varepsilon_j)$ is small compared with M (thus $0 < 1 - \alpha_j < 1$), that α_j does not tend to zero as n tends to infinity, and that the absolute value of η_j is small compared with M, the final distribution of X_n is again approximately lognormal. More general conditions, for instance, the case where the second moment of $\ln X_j$ changes through time, were also considered.

The three-parameter lognormal distribution is generated from the modified model

$$X_j - X_{j-1} = \varepsilon_j (X_{j-1} - \tau)$$

where it is assumed that the variables can take any value exceeding a fixed value τ. The equation may be rewritten as

$$(X_j - \tau) - (X_{j-1} - \tau) = \varepsilon_j (X_{j-1} - \tau)$$

Since $X_n - \tau$ is asymptotically lognormallly distributed in a two-parameter form, it is clear that X_n is asymptotically lognormally distributed in a three-parameter form.

3.2 The Theory of Breakage

Consider the distribution of small particles such as crushed stones. The histogram of the weights or masses is often skew and is similar to that given by a lognormal distribution. Such distributions are called mass-size distribution rather than weight-size.

A theory of breakage explains the occurrence of a lognormal distribution. Suppose that the mass of a particle is initially X_0 and that after j random splittings it is X_j. The breakage process is continued as follows. The $(j-1)$th generation particle is split in two and one of two particles is neglected. Thus a remaining particle forms the jth generation particle and the mass of the particle is X_j. The problem is to find the distribution of the final mass X_n.

Kolmogoroff (1941) showed that the distribution of X_n is asymptotically lognormal. The theory is an inverse application of the law of proportionate effect. He carried it out in terms of distribution functions.

The above-mentioned model can be varied as follows. Let X_0 be the mass of an initial particle. The particle is split in two with interdependence Z_1 and $1 - Z_1$ and the second-generation masses are represented by Z_1X_0 and $(1 - Z_1)X_0$. Both of these particles are split in two. This process is continued until 2^n particles are formed after n steps. The problem is to find the distribution of X_n representing the mass-size of all particles. Schultz (1975) discussed some models of the mass-size distributions. Epstein (1947) presented a different approach to a breakage mechanism which asymptotically generates a lognormal distribution.

The fundamental idea of the breakage process is very closely analogous to that of a theory of classification. Aitchison and Brown (1957) have written (p. 27): "It is a curious fact that when a large number of items is classified on some homogeneity principle, the variate defined as the number of items in a class is often approximately lognormal." Brown and Sanders (1981) provided a mathematical justification of the above observation. This idea of classification has guided the development of various types of application in many fields. Examples of applications (Aitchison and Brown, 1957) are the number of persons in a census occupation class and the distribution of incomes in econometrics, the distribution of stars in the universe in astronomy, and the distribution of the radical component of Chinese characters in philology.

3.3 Other Geneses

Haldane (1942) dealt with moments of the distributions of powers and products of normal variables. We treat here the asymptotic distribution of the nth power of a normal variable. Let X be a random variable having $N(\mu, \sigma^2)$. Then

$$X^n = (\mu + \sigma U)^n = \mu^n \left(1 + n\gamma \frac{U}{n}\right)^n$$

where U is $N(0,1)$ and γ is the coefficient of variation of X. Assuming that n is very large and γ is very small, but $n\gamma < \infty$, we approximately have

$$\ln X^n = n \ln \mu + n\gamma U$$

Thus X^n is asymptotically lognormally distributed.

Broadbent (1956) dealt with lognormal approximation to products and quotients. Consider a combination of the X_i of the form

$$Q = \frac{X_1 X_2 \ldots X_j}{X_{j+1} \ldots X_n} \qquad 1 \leq j \leq n$$

The moments of $\ln Q$ were found and the method of fitting by moments was used.

Koch (1966, 1969) discussed mechanisms which generate the lognormal distribution under a variety of biological and pharmacological circumstances. Consider the classical equation of the metabolic turnover. The concentration C of a substance at any time is given by

$$\frac{dC}{dt} = -kC$$

where $k(> 0)$ is the rate constant. When the concentration at time $t = 0$ is C_0, the solution of the differential equation is

$$C = C_0 e^{-kt}$$

The minimal lethal dose calculated from this equation is lognormally distributed, provided, for instance, k is normally distributed in different individuals of the population. Mathematically analogous circumstances also arise from the Beer-Lambert law in optics and from an autocatalytic growth process.

Many applications in biochemistry including mechanisms generating the lognormal distribution can be seen in Masuyama (1984) and its references.

4. PROPERTIES

4.1 Moments and Other Characteristics

Let X be a random variable having $\Lambda(\mu, \sigma^2)$ with density function (2.1). The rth moment of X about the origin is

$$\mu'_r = E(X^r) = \exp\left(r\mu + \tfrac{1}{2}r^2\sigma^2\right) \tag{4.1}$$

from the properties of the moment generating function $E(e^{rY})$ of $N(\mu, \sigma^2)$, where $Y = \ln X$. From (4.1), the mean is

$$\mu'_1 = \exp(\mu + \tfrac{1}{2}\sigma^2) \tag{4.2}$$

and the variance is

$$\mu_2 = \exp(2\mu + \sigma^2)\left\{\exp(\sigma^2) - 1\right\} \tag{4.3}$$

The median is

$$\text{Med}(X) = \exp(\mu) \tag{4.4}$$

and the mode is

$$\text{Mode}(X) = \exp(\mu - \sigma^2) \tag{4.5}$$

The relation between the mean (4.2), the median (4.4), and the mode (4.5) is

$$\text{Mode}(X) < \text{Med}(X) < E(X)$$

Since

$$\frac{X - \mu'_1}{\sqrt{\mu_2}} = U + \tfrac{1}{2}(U^2 - 1)\sigma + \frac{1}{12}(2U^3 - 9U)\sigma^2 + O(\sigma^3)$$

where U is $N(0,1)$, the standardized lognormal distribution tends to $N(0,1)$ as σ tends to zero.

The quantile (or $100q$ percentage point) is

$$\theta_q = \exp(\mu + t_q \sigma) \tag{4.6}$$

where q is a given lower fraction and t_q is a value satisfying

$$q = Pr(U \le t_q)$$

where U is $N(0,1)$. It is clear that (4.6) reduces to (4.4) when $q = 0.5$. The closeness in agreement between the standardized percentage points of the lognormal and the corresponding Pearson Type IV distributions was shown by Pearson, Johnson, and Burr (1979).

The reliability function is

$$R(t) = Pr(X \ge t) = 1 - \Phi\left(\frac{\ln t - \mu}{\sigma}\right) \qquad t > 0 \tag{4.7}$$

where Φ is the distribution function of $N(0,1)$.

The shape factors (the square of skewness and the kurtosis) are

$$\beta_1 = \frac{\mu_3^2}{\mu_2^3} = (\omega - 1)(\omega + 2)^2 \tag{4.8}$$

$$\beta_2 = \frac{\mu_4}{\mu_2^2} = \omega^4 + 2\omega^3 + 3\omega^2 - 3 \tag{4.9}$$

where

$$\mu_r = E[(X - \mu_1')^r] \qquad \omega = \exp(\sigma^2)$$

Thus σ is commonly called the *shape* parameter of the lognormal distribution. Johnson's general family of distributions (Johnson, 1949; Johnson and Kotz, 1970, p. 22) which can be transformed to the normal distribution includes the lognormal distribution as one of three types; the lognormal line is the boundary between the regions of the other two types in the (β_1, β_2) plane. Several different measures of skewness were considered by Nichols and Gibbons (1979).

The square of the coefficient of variation is

$$\frac{\mu_2}{(\mu_1')^2} = \omega - 1 \tag{4.10}$$

The Theil coefficient is

$$E\left[\frac{X}{E(X)}\ln\left(\frac{X}{E(X)}\right)\right] = \tfrac{1}{2}\sigma^2 \tag{4.11}$$

and the modified Theil coefficients are

$$E\left[\frac{X}{\text{Med}(X)}\ln\left(\frac{X}{\text{Med}(X)}\right)\right] = \sigma^2\exp\left(\tfrac{1}{2}\sigma^2\right) \tag{4.12}$$

$$E\left[\frac{X}{\text{Mode}(X)}\ln\left(\frac{X}{\text{Mode}(X)}\right)\right] = 2\sigma^2\exp\left(\tfrac{3}{2}\sigma^2\right) \tag{4.13}$$

Gini's coefficient of mean difference is

$$G = 2\mu_1'\left\{2\Phi\left(\frac{\sigma}{\sqrt{2}}\right) - 1\right\} \tag{4.14}$$

Gini's concentration coefficient is

$$L = 2\Phi\left(\frac{\sigma}{\sqrt{2}}\right) - 1 \tag{4.15}$$

which is a measure relating to the Lorenz diagram. The Lorenz curves of $\Lambda(\mu,\sigma^2)$ are symmetrical with respect to the diagonal.

All of (4.1) through (4.15) are variously used characteristics of the two-parameter lognormal distribution.

The location characteristics of $\Lambda(\tau,\mu,\sigma^2)$ are greater by τ than those of $\Lambda(\mu,\sigma^2)$: The mean is at $\tau + \exp(\mu + \sigma^2/2)$; the median at $\tau + \exp(\mu)$; and the mode at $\tau + \exp(\mu - \sigma^2)$. The quantiles are displaced from θ_q to $\tau + \theta_q$. The rth moment about τ is

$$E\{(X-\tau)^r\} = \exp\left(r\mu + \frac{r^2\sigma^2}{2}\right)$$

so that the moments about the mean and hence the shape factors remain unchanged. A lognormal distribution with unit shape parameter, $\Lambda(\tau, \ln\xi, 1)$, was considered by Gibbons (1978). Thus τ is the threshold parameter and ξ is a *scale* parameter.

It is known (Thorin, 1977) that the lognormal distribution is infinitely divisible; that is, for any integer n it can be represented as the distribution of a sum of n independent random variables with a common distribution.

Let $(X_1, \ldots, X_n)'$ be a random vector having $\Lambda_n(\mu, \Sigma)$ with $\mu = (\mu_1, \ldots, \mu_n)'$ and $\Sigma = (\sigma_{ij})$, $i, j = 1, \ldots, n$. The density function is given by (2.2). Then we have from the form of the moment generating function for $N_n(\mu, \Sigma)$ that

$$E(X_1^{s_1} \cdots X_n^{s_n}) = \exp(s'\mu + \tfrac{1}{2}s'\Sigma s) \tag{4.16}$$

where $s = (s_1, \ldots, s_n)'$. It follows from (4.16) that, for any $i = 1, \ldots, n$,

$$E(X_i^r) = \exp(r\mu_i + \tfrac{1}{2}r^2\sigma_{ii})$$

and for any $i, j = 1, \ldots, n$,

$$\mathrm{Cov}(X_i, X_j) = \exp\left\{\mu_i + \mu_j + \tfrac{1}{2}(\sigma_{ii} + \sigma_{jj})\right\}\left\{\exp(\sigma_{ij}) - 1\right\} \tag{4.17}$$

It is clear that (4.17) reduces to (4.3) when $i = j$.

Nalbach-Leniewska (1979) discussed four measures of dependence in the multivariate lognormal distribution: the multiple correlation coefficient and ratio, and the screening and monotone dependence functions.

4.2 Non-Uniqueness of the Moment Problem

The lognormal density function is not uniquely determined by its moments. To show this, Heyde (1963) gave an example of the distribution with exactly the same moments. Let X be a random variable having $\Lambda(\tau, \mu, \sigma^2)$. This distribution possesses the same moments

$$\mu'_p = \sum_{r=0}^{p} \binom{p}{r} \tau^{p-r} \exp(r\mu + \tfrac{1}{2}r^2\sigma^2) \qquad p = 0, 1, 2, \ldots$$

as the distribution with probability density function

$$g(x) = \begin{cases} \dfrac{1}{\sqrt{2\pi}\sigma(x-\tau)} \exp\left[-\dfrac{1}{2\sigma^2}\{\ln(x-\tau) - \mu\}^2\right] & \\ \quad \times \left(1 + \varepsilon \sin\left[\dfrac{2\pi k}{\sigma^2}\{\ln(x-\tau) - \mu\}\right]\right) & \tau < x < \infty \\ 0 \qquad x \le \tau & \end{cases}$$

where $0 < \varepsilon < 1$ and k is a positive integer.

Leipnik (1981) dealt with the same problem. Let F be a distribution function on $(-\infty, \infty)$, with finite moments of all integer orders. A necessary and sufficient condition for the moments of F to determine F uniquely, which holds for an absolute continuous F with density function f, is

$$\int_{-\infty}^{\infty} \frac{\ln f(x)}{1+x^2}\,dx = -\infty \tag{4.18}$$

The density function of $\Lambda(0,\sigma^2)$ fails to satisfy (4.18), since

$$\int_0^{\infty} \frac{|\ln x|^k}{1+x^2}\,dx = \int_{-\infty}^{\infty} \frac{|y|^k e^y}{1+e^{2y}}\,dy \le \int_{-\infty}^{0} |y|^k e^y\,dy + \int_0^{\infty} |y|^k e^{-y}\,dy < \infty$$

for $k = 0, 1, 2$

Leipnik (1981) also showed that it is possible to construct a one-parameter family of distributions, each of which has the same moments as $\Lambda(0,\sigma^2)$, using characteristic functions resembling elliptic theta functions and also resembling, in a way, the non-differentiable functions of Weierstrass.

For $a > 0$, let

$$\Phi_a(t) = \sum_{n=-\infty}^{\infty} a^{-n} L^{-n^2/2} \exp(iaL^n t)$$

where t is real and $L > 1$. Clearly $\Phi_a(t)$ is rapidly convergent. Since the complex-valued function $\Phi_a(t)$ of the real variable t is non-negative definite, i.e.,

(1) $\Phi_a(t)$ is continuous as a dominated sum of continuous functions,

(2) $\sum_{j,k} \beta_j \beta_k^* \Phi_a(t_j - t_k) = \sum_n a^{-n} L^{-n^2/2} |\sum_j \beta_j \exp(iaL^n t_j)|^2 \ge 0$,
Bochner's theorem shows that

$$\phi_a(t) = \frac{\Phi_a(t)}{\Phi_a(0)}$$

is a characteristic function. Note that $\Phi_a(0)$ is expressed by an elliptic theta function. Let X_a be a random variable whose characteristic function is $\phi_a(t)$. Since

$$\left.\frac{d^r}{dt^r}\Phi_a(t)\right|_{t=0} = \sum_{n=-\infty}^{\infty} a^{-n}(iaL^n)^r L^{-n^2/2}$$

we see that the rth moment of X_a about the origin is

$$\begin{aligned}\mu_r'(X_a) &= i^{-r}\frac{d^r}{dt^r}\phi_a(t)\mid_{t=0}\\ &= \frac{i^{-r}}{\Phi_a(0)}\sum_n a^{-n}(iaL^n)^r L^{-n^2/2}\\ &= \frac{L^{r^2/2}}{\Phi_a(0)}\sum_n a^{-(n-r)}L^{-(n-r)^2/2}\\ &= L^{r^2/2}\end{aligned}$$

If we take $L = \exp(\sigma^2)$, then the rth moment of the distribution

$$Pr(X_a = aL^n) = \frac{a^{-n}L^{-n^2/2}}{\Phi_a(0)}, \qquad n = 0, \pm 1, \ldots$$

is equal to the rth moment of $\Lambda(0,\sigma^2)$. Thus we have a concrete example of a discrete distribution quite different from $\Lambda(0,\sigma^2)$ but with the same moments.

4.3 Distributions of the Products and the Quotients

It is well known that the normal distribution possesses additive reproductive properties. From the characteristic property of the logarithmic function $\ln X_1 + \ln X_2 = \ln X_1X_2$, the two-parameter lognormal distribution will have multiplicative reproductive properties.

The simple reproductive property is as follows. If X_1 and X_2 are independent random variables having $\Lambda(\mu_1,\sigma_1^2)$ and $\Lambda(\mu_2,\sigma_2^2)$, then the product X_1X_2 is $\Lambda(\mu_1+\mu_2,\sigma_1^2+\sigma_2^2)$. A more general form is as follows. If $\{X_i\}$, $i = 1,\ldots,n$, are independent random variables having $\Lambda(\mu_i,\sigma_i^2)$, $\{b_i\}$ $i = 1,\ldots,n$, constants, and $c = \exp(a)$ a positive constant, then the product $c\prod_{i=1}^n X_i^{b_i}$ is $\Lambda(a+\sum_{i=1}^n b_i\mu_i, \sum_{i=1}^n b_i^2\sigma_i^2)$. In particular, if X_1 and X_2 are independent and distributed as $\Lambda(\mu_1,\sigma_1^2)$ and $\Lambda(\mu_2,\sigma_2^2)$, then the quotient X_1/X_2 is $(\Lambda(\mu_1-\mu_2,\sigma_1^2+\sigma_2^2)$ and if $X_i, i = 1,\ldots,n$, are independent and identically distributed as $\Lambda(\mu,\sigma^2)$, then their geometric mean $(\prod_{i=1}^n X_i)^{1/n}$ is $\Lambda(\mu,\sigma^2/n)$.

A similar result holds if $(X_1,\ldots,X_n)'$ has an n-dimensional lognormal distribution. If $(X_1,\ldots,X_n)'$ is a random vector having $\Lambda_n(\mu,\Sigma)$, $b = (b_1,\ldots,b_n)'$ a vector of constants, and $c = \exp(a)$ a positive constant, then the product $c\prod_{i=1}^n X_i^{b_i}$ is $\Lambda(a+b'\mu, b'\Sigma b)$. In particular, if $(X_1,X_2)'$ is

$\Lambda_2(\mu, \Sigma)$, where $\mu = (\mu_1, \mu_2)'$ and

$$\Sigma = \begin{bmatrix} \sigma_1^2 & \sigma_1\sigma_2\rho \\ \sigma_1\sigma_2\rho & \sigma_2^2 \end{bmatrix}$$

then the product X_1X_2 is $\Lambda(\mu_1 + \mu_2, \sigma_1^2 + 2\sigma_1\sigma_2\rho + \sigma_2^2)$ and the quotient X_1/X_2 is $\Lambda(\mu_1 - \mu_2, \sigma_1^2 - 2\sigma_1\sigma_2\rho + \sigma_2^2)$.

It is also noted that the multiplicative analogue for the additive central limit theorem is constructed. A simple form appeared in Section 3.1.

4.4 Distribution of the Sums

Consider the sum

$$X = \sum_{i=1}^{n} X_i$$

where X_i are independent random variables each having a lognormal distribution. The problem is to find the distributions of X and $\ln X$. Applications arise in multihop scatter systems (Fenton, 1960), in statistical detection problems (Schleher, 1977), and in other problems.

By the central limit theorem, the distribution of the sum of independent lognormally distributed random variables tends to a normal distribution as n tends to infinity. The distribution has also been approached by other methods. Fenton (1960) approximated the distribution by a lognormal distribution which has the same moments as the exact sum distribution. Barakat (1976) used the inverse Fourier transform of the characteristic function. Schleher (1977) had an approach by a generalized form of the Gram-Charlier series and compared with FFT numerical techniques.

The distribution of the logarithm of the sum of two lognormal variables is of importance in many areas of communications. The problem arises when the logarithms of the powers (in watts) from two sources are normally distributed, and it is required to find the distribution of the logarithm of the sum of the two sources. Naus (1969) gave the moment generating function and the mean and variance when the two sources are independent and the logarithms of the powers have zero means and equal variances. Hamdan (1971) gave the like results when the logarithms of the powers have a bivariate normal distribution with zero means, unequal variances, and correlation coefficient. Schwartz and Yeh (1982) gave the mean and the variance when the two sources are independent and the logarithms of the powers have unequal means and variances. They also presented simulation results, validating the normal approximation for the power sum.

Here we extend Naus' approach to the case of a bivariate normal distribution with means μ_1 and μ_2, variances σ_1^2 and σ_2^2, and correlation coefficient ρ.

Let $(Y_1, Y_2)'$ have $N_2(\mu, \Sigma)$, where $\mu = (\mu_1, \mu_2)'$ and

$$\Sigma = \begin{bmatrix} \sigma_1^2 & \sigma_1\sigma_2\rho \\ \sigma_1\sigma_2\rho & \sigma_2^2 \end{bmatrix}$$

It is required to find the moment generating function $M_V(s)$ of $V = \ln(e^{Y_1} + e^{Y_2})$. Define

$$R = \tfrac{1}{2}(Y_1 + Y_2) \qquad W = \tfrac{1}{2}(Y_2 - Y_1)$$

Then

$$M_V(s) = E(e^{Y_1} + e^{Y_2})^s = E\left\{e^{Rs}(e^{-W} + e^{W})^s\right\}$$

Since R and W have a bivariate normal distribution with means m_1 and m_2, variances a_1^2 and a_2^2, and correlation coefficient d, where

$$m_1 = \tfrac{1}{2}(\mu_1 + \mu_2) \qquad m_2 = \tfrac{1}{2}(\mu_2 - \mu_1) \qquad a_1 a_2 d = \tfrac{1}{4}(\sigma_2^2 - \sigma_1^2)$$
$$a_1^2 = \tfrac{1}{4}(\sigma_1^2 + 2\sigma_1\sigma_2\rho + \sigma_2^2) \qquad a_2^2 = \tfrac{1}{4}(\sigma_1^2 - 2\sigma_1\sigma_2\rho + \sigma_2^2)$$

the moment generating function is given by

$$\begin{aligned} M_V(s) = e^{m_1 s} \sum_{k=0}^{\infty} \binom{s}{k} e^{2a_2^2k^2} \Bigg[& e^{m_2(2k-s)+ks(\sigma_1\sigma_2\rho-\sigma_1^2)+\sigma_1^2 s^2/2} \\ & \times \Phi\left(-\frac{m_2}{a_2} - 2a_2k + (a_2 - a_1 d)s\right) \\ & + e^{m_2(s-2k)+ks(\sigma_1\sigma_2\rho-\sigma_2^2)+\sigma_2^2 s^2/2} \\ & \times \Phi\left(\frac{m_2}{a_2} - 2a_2k + (a_2 + a_1 d)s\right)\Bigg] \end{aligned} \tag{4.19}$$

where Φ is the distribution function of $N(0, 1)$. The proof is similar to that of Naus (1969) and Hamdan (1971).

From (4.19) the mean of V is

$$E_V = \mu_2 - 2m_2\Phi\left(-\frac{m_2}{a_2}\right) + 2a_2\phi\left(\frac{m_2}{a_2}\right)$$

$$+\sum_{k=1}^{\infty}\frac{(-1)^{k-1}}{k}e^{2a_2^2k^2}$$

$$\times\left[e^{2m_2k}\Phi\left(-\frac{m_2}{a_2}-2a_2k\right)+e^{-2m_2k}\Phi\left(\frac{m_2}{a_2}-2a_2k\right)\right]$$

and the second moment of V is

$$E(V^2) = \mu_2^2 + \sigma_2^2 - (4m_1m_2 + \sigma_2^2 - \sigma_1^2)\Phi\left(-\frac{m_2}{a_2}\right)$$

$$+ 4(a_2 \ln 2 + m_1a_2)\phi\left(\frac{m_2}{a_2}\right)$$

$$+2\sum_{k=1}^{\infty}\frac{(-1)^{k-1}}{k}e^{2a_2^2k^2}\left[e^{2m_2k}\{\mu_1 + k(\sigma_1\sigma_2\rho - \sigma_1^2)\}\right.$$

$$\times\Phi\left(-\frac{m_2}{a_2}-2a_2k\right)$$

$$\left.+e^{-2m_2k}\{\mu_2 + k(\sigma_1\sigma_2\rho - \sigma_2^2)\}\Phi\left(\frac{m_2}{a_2}-2a_2k\right)\right]$$

$$+2\sum_{k=2}^{\infty}(-1)^{k-2}\frac{1}{k}\left(1+\tfrac{1}{2}+\cdots+\frac{1}{k-1}\right)e^{2a_2^2k^2}$$

$$\times\left\{e^{2m_2k}\Phi\left(-\frac{m_2}{a_2}-2a_2k\right)+e^{-2m_2k}\Phi\left(\frac{m_2}{a_2}-2a_2k\right)\right\}$$

where ϕ is the density function of $N(0,1)$.

The distribution of power sums of n independent random variables was considered by Marlow (1967) and Nasell (1967). A generalization to the case of n jointly distributed random variables was treated by Naus (1973).

4.5 Order Statistics

Let $X_1, X_2, \ldots, X_n$ be independent random variables each having a standard lognormal distribution $\Lambda(0,1)$, and let $X_{(k)}$ denote the kth order statistic; i.e., $X_{(1)} \leq X_{(2)} \leq \cdots \leq X_{(n)}$.

The rth moment of $X_{(k)}$ about the origin is given by

$$\mu_r'(k,n) = k\binom{n}{k}\int_0^\infty x^r[F(x)]^{k-1}[1-F(x)]^{n-k}f(x)\,dx$$

$$= k\binom{n}{k}\int_{-\infty}^\infty e^{rx}[\Phi(x)]^{k-1}[1-\Phi(x)]^{n-k}\phi(x)\,dx$$

where $F(x)$ and $f(x)$ denote the distribution and density functions of X_1, and $\Phi(x)$ and $\phi(x)$ denote the standard normal distribution and density functions. It follows that

$$\sum_{k=1}^n \mu_j'(k,n) = \sum_{k=1}^n E(X_k^j) = ne^{j^2/2}$$

Let $1 \leq i < j \leq n$. Then the product moments of $X_{(i)}$ and $X_{(j)}$ are given by

$$E(X_{(i)}X_{(j)}) = C\int_{-\infty}^\infty q(x)\,dx$$

where

$$C = C(i,j,n) = \frac{n!}{(i-1)!(j-i-1)!(n-j)!}$$

and

$$q(x) = \int_{-\infty}^x e^{x+y}[\Phi(y)]^{i-1}[\Phi(x)-\Phi(y)]^{j-i-1} \times [1-\Phi(x)]^{n-j}\phi(x)\phi(y)dy$$

It follows that

$$\sum_{i=1}^{n}\sum_{j=i+1}^{n} E(X_{(i)}X_{(j)}) = \tfrac{1}{2}n(n-1)\{E(X_1)\}^2$$
$$= \tfrac{1}{2}n(n-1)e$$

Given a lower fraction q, the $100q$ percentage point ξ_q of the kth order statistic $X_{(k)}$ can be obtained as follows. It follows that

$$\begin{aligned} q &= Pr(X_{(k)} \leq \xi_q) \\ &= k\binom{n}{k}\int_0^{F(\xi_q)} u^{k-1}(1-u)^{n-k}\,du \\ &\equiv I_{F(\xi_\alpha)}(k, n-k+1) \end{aligned} \tag{4.20}$$

where $I_x(a,b)$ is the incomplete beta integral

$$I_x(a,b) = \frac{B_x(a,b)}{B_1(a,b)}$$

$$B_x(a,b) = \int_0^x u^{a-1}(1-u)^{b-1}\,du$$

Thus for a given $q, F(\xi_q)$ can be determined from (4.20) and $F(\xi_q) = \Phi(\ln \xi_q)$ yields

$$\xi_q = \exp[\Phi^{-1}(F(\xi_q))]$$

The moments, the product moments and the percentage points of the various order statistics were computed and tabulated for all samples of size 20 or less by Gupta, McDonald, and Galarneau (1974). They discussed the best unbiased estimators of the threshold and scale parameters of $\Lambda(\tau, \ln\xi, 1)$ as an application of the tables. The moments of order statistics for $\Lambda(\mu, 1)$ are obtainable from their tables but not those for $\Lambda(0, \sigma^2)$ or $\Lambda(\mu, \sigma^2)$ for general μ and σ.

The concentration of air pollutants (carbon monoxide, nitrogen dioxide, sulfur dioxide, etc.) is a measure for expressing the degree of pollution. Singpurwalla (1972) made an attempt to show how certain empirical relationships observed in an analysis of air pollution data can be interpreted by use of extreme value theory applied to the maximum value from the lognormal random variables.

Cramér (1946) has given the limiting distribution of extreme values from a distribution of the continuous type. This was used in finding the limiting distribution of the maximum and minimum values from a lognormal distribution by Singpurwalla (1972) and by Bury (1975), respectively.

Let $X_{(n)}$ and $X_{(1)}$ denote the maximum and minumum values of n independent random variables having $\Lambda(\mu, \sigma^2)$. Then the maximum value $X_{(n)}$ satisfies

$$\lim_{n\to\infty} Pr\left\{X_{(n)} \le \sigma_n\left(1+\frac{k}{\lambda_n}\right)\right\} = e^{-e^{\Gamma k}} \tag{4.21}$$

for $-\infty < k < \infty$, where

$$\sigma_n = \exp(\mu + \sigma\alpha_n)$$

$$\lambda_n = \frac{\sqrt{2\ln n}}{\sigma}$$

$$\alpha_n = \sqrt{2\ln n} - \frac{\ln\ln n + \ln 4\pi}{2\sqrt{2\ln n}}$$

The above form of the distribution function, (4.21), is known as a reduced Type I asymptotic distribution of largest values. The limiting distribution of the minimum value $X_{(1)}$ is the Weibull distribution whose density function is given by

$$f(x) = \frac{\lambda_n}{\tau_n}\left(\frac{x}{\tau_n}\right)^{\lambda_n - 1}\exp\left\{-\left(\frac{x}{\tau_n}\right)^{\lambda_n}\right\} \qquad x > 0$$

where

$$\tau_n = \exp(\mu - \sigma\alpha_n)$$

The limiting distribution of the maximum value from a nonstationary stochastic process involving a logarithmic transformation was treated by Horowitz and Barakat (1979) and Horowitz (1980).

4.6 Size and Shape Variables

This section provides a brief discussion of size and shape variables, a unified treatment of which is given in Chapter 11.

Let $X = (X_1, X_2, \ldots, X_n)'$, $n \geq 2$, be a vector of positive random variables and $G(X)$ a random variable from R^{n+} into R^{1+}, where $R^{n+} = \{x = (x_1, \ldots, x_n)' \mid x_i > 0 \text{ for all } i = 1, \ldots, n\}$. The random variable $G(X)$ is said to be a *size variable* (Mosimann, 1970, 1975a,b) if G satisfies the homogeneity condition $G(ax) = aG(x)$ for all $a \in R^{1+}$ and $x \in R^{n+}$. The vector variable $Z(X) = X/G(X)$ is called the *shape vector associated with* G.

Examples of size variables are $\sum_{i=1}^{n} X_i$, X_i, $(\sum_{i=1}^{n} X_i^2)^{1/2}$, $\prod_{i=1}^{n} X_i^{1/n}$, and $\max_{1 \leq i \leq n}(X_i)$. Corresponding shape vectors are proportions $X/\sum_{i=1}^{n} X_i$, ratios X/X_i, direction cosines $X/(\sum_{i=1}^{n} X_i^2)^{1/2}$, and the unnamed vectors $X/\prod_{i=1}^{n} X_i^{1/n}$ and $X/\max_{1 \leq i \leq n}(X_i)$.

The main result given by Mosimann (1970) is as follows. Let $G_1(X)$ be a size variable and $Z_1(X)$ a nondegenerate (at a point) shape vector. (G_1 is not necessarily the size variable associated with Z_1.) If $Z_1(X)$ is independent of $G_1(X)$, then any other shape vector $Z_2(X)$ must be independent of $G_1(X)$, and no shape can be independent of any other size variable $G_2(X)$ unless $G_2(X)/G_1(X)$ is a degenerate random variable.

Mosimann (1975a) considered a population of adult humans to illustrate the above result. Let $X = (X_1, X_2, X_3)'$ be a vector of three measurements, the length, width, and depth of the femur for each individual. The length $M_1(X) = X_1$ is a size variable and $Z_1(X) = X/M_1$ is the shape vector associated with M_1. The product of width and depth, X_2X_3, is an approximation to the area of the bone shaft in cross-section, and $M_2(X) = (X_2X_3)^{1/2}$ is a size variable. In addition, a measurement indicative of volume $M_3(X) = (X_1X_2X_3)^{1/3}$ is also a size variable.

Let $X = (X_1, X_2, X_3)'$ be a trivariate lognormal distribution $\Lambda_3(\mu, \Sigma)$ with parameters $\mu = (\mu_1, \mu_2, \mu_3)'$ and $\Sigma = (\sigma_{ij})$, and let $Y_i = \ln X_i$, $i = 1, 2, 3$. Then we shall show the fact that the above two size variables M_1 and M_3 cannot be simultaneously independent of shape Z unless the ratio M_1/M_3 is degenerate.

First, we consider the possible independence of the shape vector $Z_1 = X/M_1$ and size variable M_1. Let $V = (0, Y_2 - Y_1, Y_3 - Y_1)'$. Then V and Y_1 are independent (which is equivalent to the statement that Z_1 is independent of M_1) if and only if

$$\text{Cov}(Y_1, Y_i - Y_1) = 0, \qquad i = 2, 3,$$

i.e.,

$$\Sigma = \begin{bmatrix} \sigma_{11} & \sigma_{11} & \sigma_{11} \\ \sigma_{11} & \sigma_{22} & \sigma_{23} \\ \sigma_{11} & \sigma_{32} & \sigma_{33} \end{bmatrix} \tag{4.22}$$

Such independence implies the independence of M_2/M_1 and M_1 since

$$\text{Cov}(Y_1, \ln M_2 - Y_1) = \frac{\sigma_{12} + \sigma_{13}}{2} - \sigma_{11}$$

Next, we consider the possible independence of the shape vector Z_1 and size variable M_3. Let $S = Y_1 + Y_2 + Y_3$. Then Z_1 and M_3 are independent if and only if Z_1 and S are independent, which is equivalent to the fact that

$$\text{Cov}(Y_i, S) = \text{Cov}(Y_1, S) \qquad i = 2, 3,$$

i.e., each column of Σ has the same total t, where $t = \text{Cov}(Y_1, S)$.

Now, if M_3 is independent of Z_1, then $(1, 1, 1)'$ is an eigenvector of Σ with eigenvalue t. In addition, if M_1 is independent of Z_1, then (4.22) implies that $t = 3\sigma_{11}$, $\sigma_{22} + \sigma_{32} = 2\sigma_{11}$, and $\sigma_{23} + \sigma_{33} = 2\sigma_{11}$. Hence, $(2, -1, -1)'$ must also be an eigenvector of Σ with eigenvalue zero. Thus

$$\text{Var}(2Y_i - Y_2 - Y_3) = 0$$

and M_1/M_3 is degenerate, i.e., $Pr(M_1 = aM_3) = 1$ for some constant $a > 0$.

REFERENCES

Aitchison, J. and Brown, J. A. C. (1957). *The Lognormal Distribution*, Cambridge University Press, Cambridge.

Barakat, R. (1976). Sums of independent lognormally distributed random variables, *J. Opt. Soc. Am.*, *66*, 211–216.

Broadbent, S. R. (1956). Lognormal approximation to products and quotients, *Biometrika*, *43*, 404–417.

Brown, G. and Sanders, J. W. (1981). Lognormal genesis, *J. Appl. Prob.*, *18*, 542–547.

Bury, K. V. (1975). Distribution of smallest log-normal and Gamma extremes, *Statistische Hefte*, *16*, 105–114.

Cramér, H. (1946). *Mathematical Methods of Statistics*, Princeton University Press, Princeton.

Epstein, B. (1947). The mathematical description of certain breakage mechanisms leading to the logarithmico-normal distribution, *J. Franklin Institute*, *244*, 471–477.

Fenton, L. F. (1960). The sum of log-normal probability distributions in scatter transmission systems, *IRE Trans. Commun. Systems*, *CS-8*, 57–67.

Galton, F. (1879). The geometric mean in vital and social statistics, *Proc. Roy. Soc.*, *29*, 365–367.

Gibbons, D. I. (1978). An evaluation of two model specification techniques for a lognormal distribution, *IEEE Trans. Reliability*, *R-27*, 60–63.

Gibrat, R. (1930). Une loi des répartitions économiques: l'effet proportionnel, *Bull. Statist. Gén. Fr.*, *19*, 469ff.

Gibrat, R. (1931). *Les Inégalités Économiques*, Libraire du Recueil Sirey, Paris.

Gupta, S. S., McDonald, G. C. and Galarneau, D. I. (1974). Moments, product moments and percentage points of the order statistics from the lognormal distribution for samples of size twenty and less, *Sankhyā*, *36*, 230–260.

Haldane, J. B. S. (1942). Moments of the distributions of powers and products of normal variates, *Biometrika*, *32*, 226–242.

Hamdan, M. A. (1971), The logarithm of the sum of two correlated lognormal variates, *J. Amer. Statist. Assoc.*, *66*, 105–106.

Heyde, C. C. (1963). On a property of the lognormal distribution, *J. R. Statist. Soc.*, *25*, 392–393.

Horowitz, J. (1980). Extreme values from a nonstationary stochastic process: an application to air quality analysis, *Technometrics*, *22*, 469–478.

Horowitz, J. and Barakat, S. (1979). Statistical analysis of the maximum concentration of air pollutant: effects of autocorrelation and non-stationarity, *Atmospheric Environment*, *13*, 811–818.

Johnson, N. L. (1949). Systems of frequency curves generated by methods of translation, *Biometrika*, *36*, 149–176.

Johnson, N. L. and Kotz, S. (1970). *Distributions in Statistics: Continuous Univariate Distributions-1*, Houghton Mifflin Company, Boston.

Kalecki, M. (1945). On the Gibrat distribution, *Econometrica*, *13*, 161–170.

Kapteyn, J. C. (1903). *Skew Frequency Curves in Biology and Statistics*, Astronomical Laboratory, Noordhoff, Groningen.

Kapteyn, J. C. and Van Uven, M. J. (1916). *Skew Frequency Curves in Biology and Statistics*, Hoitsema Brothers, Inc., Groningen.

Koch, A. L. (1966). The logarithm in biology. I. Mechanisms generating the lognormal distribution exactly, *J. Theoret. Biol.*, *12*, 276–290.

Koch, A. L. (1969). The logarithm in biology. II. Distributions simulating the log-normal, *J. Theoret. Biol.*, *23*, 251–268.

Kolmogoroff, A. N. (1941). Über das logarithmisch Normale Verteilungsgesetz der Dimensionen der Teilchen bei Zerstückelung, *C. R. Acad. Sci. (Doklady) URSS*, *XXXI*, 99–101.

Leipnik, R. (1981). The lognormal distribution and strong non-uniqueness of the moment problem, *Theor. Probability Appl.*, *26*, 850–852.

Marlow, N. A. (1967). A normal limit theorem for power sums of independent random variables, *B. S. T. J.*, *46*, 2081–2089.
Masuyama, M. (1984). A measure of biochemical individual variability, *Biom. J.*, *26*, 337–346.
McAlister, D. (1879). The law of the geometric mean, *Proc. Roy. Soc.*, *29*, 367–375.
Mosimann, J. E. (1970). Size allometry: Size and shape variables with characterizations of the lognormal and generalized gamma distributions, *J. Amer. Statist. Assoc.*, *65*, 930–945.
Mosimann, J. E. (1975a). Statistical problems of size and shape. I. Biological applications and basic theorems, *Statistical Distributions in Scientific Work*, *2* (G. P. Patil, S. Kotz, and J. K. Ord, eds.), D. Reidel Publishing Company, Dordrecht, pp. 187–217.
Mosimann, J. E. (1975b). Statistical problems of size and shape. II. Characterizations of the lognormal, gamma and Dirichlet distributions, *Statistical Distruibutions in Scientific Work*, *2* (G. P. Patil, S. Kotz, and J. K. Ord, eds.), D. Reidel Publishing Company, Dordrecht, pp. 219–239.
Nalbach-Leniewska, A. (1979). Measures of dependence of the multivariate lognormal distribution, *Math. Operationsforsh. Statist.*, *10*, 381–387.
Nasell, I. (1967). Some properties of power sums of truncated normal random variables. *B. S. T. J.*, *46*, 2091–2110.
Naus, J. I. (1969). The distribution of the logarithm of the sum of two log-normal variates, *J. Amer. Statist. Assoc.*, *64*, 655–659.
Naus, J. I. (1973). Power sum distributions, *J. Amer. Statist. Assoc.*, *68*, 740–742.
Nichols, W. G. and Gibbons, J. D. (1979). Parameter measures of skewness, *Commun. Statist.-Simula. Computa.*, *B8*, 161–167.
Nydell, S. (1919). The mean errors of the characteristics in logarithmic-normal distribution, *Skand. Aktuar. Tidskr.*, *2*, 134–144.
Pearson, E. S., Johnson, N. L., and Burr, I. W. (1979). Comparisons of the percentage points of distributions with the same first four moments, chosen from eight different systems of frequency curves, *Commun. Statist.-Simula. Computa.*, *B8*, 191–229.
Schleher, D. C. (1977). Generalized Gram-Charlier series with application to the sum of log-normal variates, *IEEE Trans. Inform. Theory*, *IT-23*, 275–280.
Schultz, D. M. (1975). Mass-size distributions: a review and a proposed new model, *Statistical Distributions in Scientific Work*, *2* (G. P. Patil, S. Kotz, and J. K. Ord, eds.), D. Reidel Publishing Company, Dordrecht, pp. 275–288.
Schwartz, S. C. and Yeh, Y. S. (1982). On the distribution function and moments of power sums with log-normal components, *B. S. T. J.*, *61*, 1441–1462.
Singpurwalla, N. D. (1972). Extreme values from a lognormal law with applications to air pollution problems, *Technometrics*, *14*, 703–711.
Thorin, O. (1977). On the infinite divisibility of the lognormal distribution, *Scand. Actuarial J.*, *3*, 121–148.

Wicksell, S. D. (1917). On logarithmic correlation with an application to the distribution of ages at first marriage, *Medd. Lunds Astr. Obs.*, *84*, 1–21.
Yuan, P. T. (1933). On the logarithmic frequency distribution and the semilogarithmic correlation surface, *Ann. Math. Statist.*, *4*, 30–74.

2
Point Estimation

KUNIO SHIMIZU Department of Information Sciences, Faculty of Science and Technology, Science University of Tokyo, Noda City, Chiba, Japan

1. INTRODUCTION

Aitchison and Brown (1957) summarized early methods of estimating the mean and variance of a two-parameter lognormal distribution. These methods included the method of moments, quantiles, a graphical method (lognormal probability paper), and Finney's (1941) uniformly minimum variance unbiased (UMVU) estimators. Aitchison and Brown studied the efficiencies of the first three methods empirically using 65 pseudorandom samples varying in size from 32 to 512 and also gave formulas for asymptotic efficiencies.

Since the appearance of Aitchison and Brown's book and even Johnson and Kotz's (1970) summary, many authors have contributed to the theory of point estimation for the lognormal distribution, as indicated by the extensive list of references for this chapter. They have considered more general parametric functions, obtained UMVU estimators not only of these functions but also of the variances of the estimators, and considered multivariate distributions. In most cases the formulas have been expressed more concisely by use of generalized hypergeometric functions. While these functions are infinite series in most cases, they converge rapidly and are thus readily calculated.

This chapter is restricted to lognormal distributions which have their lefthand limit (threshold) at the origin, nonzero thresholds being considered in Chapter 4. Section 2 considers estimation in the univariate distribution, whereas generalizations to multivariate distributions and distributions with a nonzero probability mass at the origin (delta-lognormal distributions) are considered in Section 3. A related topic in Section 3 is the UMVU estimation of the mean and variance of variables transformed from a normal variable by general transformations beyond the logarithmic, a topic introduced by Neyman and Scott (1960).

Some work on sequential estimation of the mean of a lognormal distribution and on prediction of lognormal processes is summarized in Section 4.

2. UNIVARIATE LOGNORMAL DISTRIBUTIONS

2.1 Parametric Functions

Let X be a random variable having a univariate lognormal distribution with two parameters μ and σ^2, $\Lambda(\mu,\sigma^2)$. Several population characteristics of $\Lambda(\mu,\sigma^2)$ are given by (4.1) through (4.15) in Chapter 1.

Consider a function of the parameters μ and σ^2,

$$\theta_{a,b,c} = \sigma^{2c} \exp(a\mu + b\sigma^2) \tag{2.1}$$

where a, b, and c are real constants. Then, for example, the mean is represented by $\theta_{1,1/2,0}$, and the rth moment by $\theta_{r,r^2/2,0}$. The median, the mode, and the (modified) Theil coefficient are also represented by use of (2.1).

The variance is $\theta_{2,2,0} - \theta_{2,1,0}$ and the square of skewness is $\theta_{0,3,0} + 3\theta_{0,2,0} - 4$. Thus, some of the population characteristics are represented by the finite linear combination of (2.1),

$$\theta = \sum_{(a,b,c)} C_{a,b,c}\theta_{a,b,c} \tag{2.2}$$

where $C_{a,b,c}$ are real constants. The parametric function (2.2) also contains the kurtosis and the square of coefficient of variation.

The reliability function, the quantile except for the median, Gini's coefficient of mean difference, and Gini's concentration coefficient are not contained in (2.1) and (2.2).

2.2 Methods of Estimation

Let X_1, X_2, ..., X_n, $n \geq 2$, be independent and identically distributed (iid) random variables having $\Lambda(\mu, \sigma^2)$. We assume that the parameters μ and σ^2 are unknown. If σ^2 is known, the situation will be easier and omitted.

The sample mean and the sample variance are

$$\overline{X} = \frac{1}{n}\sum_{i=1}^{n} X_i$$

and $S_X/(n-1)$, where

$$S_X = \sum_{i=1}^{n}(X_i - \overline{X})^2$$

Let $Y_i = \ln X_i$, $i = 1, 2, \ldots, n$. Then the variables Y_i are $N(\mu, \sigma^2)$. And let

$$\overline{Y} = \frac{1}{n}\sum_{i=1}^{n} Y_i \qquad \text{and} \qquad S_Y = \sum_{i=1}^{n}(Y_i - \overline{Y})^2$$

The variable $\overline{Y}$ follows $N(\mu, \sigma^2/n)$ and the variable S_Y divided by σ^2 follows the chi-square distribution with $n-1$ degrees of freedom, denoted by $\chi^2(n-1)$. Statistics $\overline{Y}$ and S_Y are jointly complete and sufficient for $N(\mu, \sigma^2)$ and mutually independent.

According to the Lehmann-Scheffé theorem, if a function $g(\mu, \sigma^2)$ admits an unbiased estimator based on $(\overline{Y}, S_Y)$, then it has a uniformly minimum variance unbiased (UMVU) estimator. The maximum likelihood (ML) estimator of $g(\mu, \sigma^2)$ is $g(\overline{Y}, S_Y/n)$ because the ML estimators of μ and σ^2 are $\overline{Y}$ and S_Y/n.

Parametric functions of the lognormal distribution $\Lambda(\mu, \sigma^2)$ can also be estimated by equating several sample moments, for instance $\overline{X}$ and S_X/n, to population moments, by using sample quantiles, and by using lognormal probability paper.

2.3 UMVU Estimators

We shall derive the UMVU estimator of (2.1) by use of the method due to Finney (1941). This method was used by Laurent (1963), by Oldham

(1965), by Bradu and Mundlak (1970), and by Gleit (1982) in several situations.

Notice that for an arbitrary real number p,

$$E\left(e^{p\overline{Y}}\right) = \exp\left[p\mu + \frac{p^2}{2n}\sigma^2\right]$$

In finding the UMVU estimator of (2.1), it is sufficient to find a function h of S_Y satisfying

$$E[h(S_Y)] = \sigma^{2c} \exp\left[\left(b - \frac{a^2}{2n}\right)\sigma^2\right]$$

$$= \sum_{j=0}^{\infty} \frac{1}{j!}\left(b - \frac{a^2}{2n}\right)^j \sigma^{2(j+c)}$$

because the statistics $\overline{Y}$ and S_Y are mutually independent.

Generally, if a variable S divided by σ^2 follows $\chi^2(d)$, then for a non-negative integer j,

$$E(S^j) = \frac{2^j\Gamma(d/2+j)}{\Gamma(d/2)}\sigma^{2j} \tag{2.3}$$

where Γ stands for the Gamma function.

Therefore, the function h must be

$$h(S_Y) = \sum_{j=0}^{\infty} \frac{1}{j!}\left(b - \frac{a^2}{2n}\right)^j \frac{\Gamma\left((n-1)/2\right)}{2^{j+c}\Gamma\left((n-1)/2+j+c\right)} S_Y^{j+c}$$

$$= \frac{\Gamma\left((n-1)/2\right)}{\Gamma\left((n-1)/2+c\right)}\left(\frac{S_Y}{2}\right)^c {}_0F_1\left(\frac{n-1}{2}+c; \frac{2bn-a^2}{4n}S_Y\right)$$

where ${}_0F_1$ stands for a particular member of the generalized hypergeometric functions, i.e.,

$${}_0F_1(\alpha; z) = \sum_{j=0}^{\infty} \frac{z^j}{(\alpha)_j j!}$$

$$(\alpha)_j = \begin{cases} \alpha(\alpha+1)\dots(\alpha+j-1) & j \geq 1 \\ 1 & j = 0, \end{cases}$$

where α is a complex number and z is a complex variable. Hence, for $(n-1)/2+c>0$ the UMVU estimator of (2.1) is

$$\hat{\theta}_{a,b,c} = \frac{\Gamma((n-1)/2)}{\Gamma((n-1)/2+c)} e^{a\overline{Y}} \left(\frac{S_Y}{2}\right)^c {}_0F_1\left(\frac{n-1}{2}+c; \frac{2bn-a^2}{4n} S_Y\right) \tag{2.4}$$

and the UMVU estimator of (2.2) is clearly

$$\hat{\theta} = \sum_{(a,b,c)} C_{a,b,c} \hat{\theta}_{a,b,c} \tag{2.5}$$

The use of generalized hypergeometric functions was shown in Shimizu and Iwase (1981).

In spite of the positiveness of (2.1), there exist combinations of a, b, and n in which the value of ${}_0F_1$ is negative. Such an estimator will be inappropriate. Evans and Shaban (1974) may be noted as a reference.

NOTE: The notation of generalized hypergeometric functions is, as stated in Erdélyi et al. (1953), Vol. 1, p. 182,

$${}_pF_q(\alpha_1, \ldots, \alpha_p; \beta_1, \ldots, \beta_q; z) = \sum_{j=0}^{\infty} \frac{(\alpha_1)_j \cdots (\alpha_p)_j}{(\beta_1)_j \cdots (\beta_q)_j} \frac{z^j}{j!}$$

where α_i, β_k are complex numbers and p, q are non-negative integers. If $p < q+1$, the series converges absolutely at every point of the finite complex plane and, therefore, represents an entire function. If $p = q+1$, it converges absolutely for $|z| < 1$. If $p > q+1$, it diverges for all $z \neq 0$ unless it terminates.

Substituting $a = 1$, $b = 1/2$, and $c = 0$ into (2.4), we have the UMVU estimator of the mean $\mu_1' = \exp(\mu + \sigma^2/2)$ as

$$\hat{\mu}_1' = e^{\overline{Y}} {}_0F_1\left(\frac{n-1}{2}; \frac{n-1}{4n} S_Y\right) \tag{2.6}$$

The UMVU estimator of the variance $\mu_2 = \exp(2\mu + 2\sigma^2) - \exp(2\mu + \sigma^2)$ is

$$\hat{\mu}_2 = e^{2\overline{Y}} \left\{ {}_0F_1\left(\frac{n-1}{2}; \frac{n-1}{n} S_Y\right) - {}_0F_1\left(\frac{n-1}{2}; \frac{n-2}{2n} S_Y\right) \right\} \tag{2.7}$$

Aitchison and Brown (1957) and Thöni (1969) have made a table for obtaining values (2.6) and (2.7). Infinite series are involved in (2.4), but the speed of convergence of the summation for the hypergeometric function ${}_0F_1$ is very rapid. It is sufficient to sum up a series until the desired degree of numerical accuracy is achieved.

2.4 Variances of UMVU Estimators

We shall give the variance of (2.4). The method is due to Shimizu (1982). The following formula is useful:

$$\{{}_0F_1(\rho; z)\}^2 = {}_1F_2\left(\frac{1}{2}(2\rho-1); \rho, 2\rho-1; 4z\right)$$

which is a special case of the equation (2) in Erdélyi et al. (1953), Vol. 1, p. 185. Since

$$E\left[\left\{S_Y^c {}_0F_1\left(\frac{n-1}{2}+c; \frac{2bn-a^2}{4n}S_Y\right)\right\}^2\right]$$

$$= E\left[S_Y^{2c} {}_1F_2\left(\frac{1}{2}(n-2+2c); \frac{n-1}{2}+c, n-2+2c; \frac{2bn-a^2}{n}S_Y\right)\right]$$

$$= \frac{\Gamma((n-1)/2+2c)}{\Gamma((n-1)/2)}(2\sigma^2)^{2c}$$

$$\times {}_2F_2\left(\frac{1}{2}(n-2+2c), \frac{n-1}{2}+2c; \frac{n-1}{2}+c, n-2+2c; \frac{2(2bn-a^2)}{n}\sigma^2\right)$$

the variance of $\hat{\theta}_{a,b,c}$ is given by

$$\mathrm{Var}(\hat{\theta}_{a,b,c}) = \theta_{a,b,c}^2\left[\frac{\Gamma((n-1)/2)\,\Gamma((n-1)/2+2c)}{\Gamma^2((n-1)/2+c)}\exp\left\{2\left(\frac{a^2}{n}-b\right)\sigma^2\right\}\right.$$

$$\times {}_2F_2\left(\frac{1}{2}(n-2+2c), \frac{n-1}{2}+2c; \frac{n-1}{2}+c,\right.$$

$$\left.\left. n-2+2c; \frac{2(2bn-a^2)}{n}\sigma^2\right) - 1\right] \qquad (2.8)$$

In particular, if $c = 0$, then (2.8) reduces to

$$\operatorname{Var}(\hat{\theta}_{a,b,0}) = \theta^2_{a,b,0}$$
$$\times \left[\exp\left\{2\left(\frac{a^2}{n} - b\right)\sigma^2\right\} {}_1F_1\left(\frac{1}{2}(n-2); n-2; \frac{2(2bn - a^2)}{n}\sigma^2\right) - 1\right]$$

Moreover, we know that

$${}_1F_1\left(\nu + \frac{1}{2}; 2\nu + 1; 2iz\right) = e^{iz}{}_0F_1\left(\nu + 1; -\frac{z^2}{4}\right)$$

which is given in Erdélyi et al. (1953), Vol. 2, p. 4. Hence, we have

$$\operatorname{Var}(\hat{\theta}_{a,b,0}) = \theta^2_{a,b,0}\left\{e^{a^2\sigma^2/n}{}_0F_1\left(\frac{n-1}{2}; \frac{(2bn - a^2)^2}{4n^2}\sigma^4\right) - 1\right\} \tag{2.9}$$

Bradu and Mundlak (1970) gave an alternative form of (2.9). Relations between the expressions are seen in Evans and Shaban (1974) and Likeš (1977). The large-sample evaluation of certain infinite series which appear was shown by Ebbeler (1973).

The variance of the UMVU estimator of the mean is clearly

$$\operatorname{Var}(\hat{\mu}_1') = {\mu_1'}^2\left\{e^{\sigma^2/n}{}_0F_1\left(\frac{n-1}{2}; \frac{(n-1)^2}{4n^2}\sigma^4\right) - 1\right\}$$

To find the variance of the UMVU estimator of the variance, we use the following. Kummer's transformation is

$${}_1F_1(\gamma; \delta; z) = e^z{}_1F_1(\delta - \gamma; \delta; -z) \qquad \delta \neq \text{negative integer}$$

Laguerre polynomials are

$$L_j^{(\nu)}(x) = \frac{\Gamma(\nu + j + 1)}{\Gamma(\nu + 1)\Gamma(j + 1)}{}_1F_1(-j; \nu + 1; x), \qquad \nu \neq \text{negative integer}$$

Hardy's formula [a combination of Erdélyi et al. (1953), Vol. 2, p. 189, (18) and p. 4, (3)] is

$$\sum_{j=0}^{\infty} \frac{1}{\Gamma(j + \nu + 1)} L_j^{(\nu)}(x) z^j = \frac{e^z}{\Gamma(\nu + 1)}{}_0F_1(\nu + 1; -xz) \qquad \operatorname{Re}\nu > -1$$

We have from these formulas that

$$E\left\{ {}_0F_1\left(\frac{n-1}{2}; \frac{a}{2}S_Y\right) {}_0F_1\left(\frac{n-1}{2}; \frac{b}{2}S_Y\right)\right\}$$

$$= \sum_{j=0}^{\infty}\sum_{k=0}^{\infty} \frac{1}{j!k!\left((n-1)/2\right)_j \left((n-1)/2\right)_k} \left(\frac{a}{2}\right)^j \left(\frac{b}{2}\right)^k$$

$$\times \frac{2^{j+k}\Gamma\left((n-1)/2+j+k\right)}{\Gamma\left((n-1)/2\right)}\sigma^{2(j+k)}$$

$$= \sum_{j=0}^{\infty} \frac{(a\sigma^2)^j \Gamma\left((n-1)/2+j\right)}{j!\left((n-1)/2\right)_j \Gamma\left((n-1)/2\right)} {}_1F_1\left(\frac{n-1}{2}+j; \frac{n-1}{2}; b\sigma^2\right)$$

$$= e^{(a+b)\sigma^2} {}_0F_1\left(\frac{n-1}{2}; ab\sigma^4\right) \tag{2.10}$$

The formula (2.10) plays an essential role for expressing the variance of (2.7) and, moreover, (2.5) with $c = 0$. Also, the method used here is applicable to a bivariate case.

The variance of $\hat{\theta}$ with $c = 0$ is given by

$$\mathrm{Var}(\hat{\theta}) = \sum_{(a,b)} C_{a,b,0}^2 e^{2a\mu+2b\sigma^2}\left\{e^{a^2\sigma^2/n} {}_0F_1\left(\frac{n-1}{2}; \frac{(2bn-a^2)^2}{4n^2}\sigma^4\right) - 1\right\}$$

$$+ 2 \sum_{(a,b)\neq(a',b')} C_{a,b,0}C_{a',b',0} e^{(a+a')\mu+(b+b')\sigma^2}$$

$$\times \left\{e^{aa'\sigma^2/n} {}_0F_1\left(\frac{n-1}{2}; \frac{1}{4n^2}(2bn-a^2)(2b'n-a'^2)\sigma^4\right) - 1\right\}$$

In particular, the variance of $\hat{\mu}_2$ is given by

$$\mathrm{Var}(\hat{\mu}_2) = e^{4\mu+2\sigma^2}\left[e^{2\sigma^2}\left\{e^{4\sigma^2/n} {}_0F_1\left(\frac{n-1}{2}; \frac{4(n-1)^2}{n^2}\sigma^4\right) - 1\right\}\right.$$

$$+ e^{4\sigma^2/n} {}_0F_1\left(\frac{n-1}{2}; \frac{(n-2)^2}{n^2}\sigma^4\right) - 1 - 2e^{\sigma^2}$$

$$\left.\times \left\{e^{4\sigma^2/n} {}_0F_1\left(\frac{n-1}{2}; \frac{2(n-1)(n-2)}{n^2}\sigma^4\right) - 1\right\}\right]$$

It is likewise easy to obtain the UMVU estimators of the rth moment, shape factors, square of coefficient of variation, median, mode, and (modified) Theil coefficient, and their variances.

2.5 Other Derivations of the Variances of UMVU Estimators

Our aim here is to derive the formula (2.10) by use of the Bhattacharyya bounds. This section is due to Shimizu (1984).

Let $Y_1, Y_2, \ldots, Y_n$, $n \geq 1$, be iid random variables having a normal distribution with mean zero and variance θ, $N(0,\theta)$. A complete and sufficient statistic $t = n^{-1}\sum_{i=1}^{n} Y_i^2$ has the density function

$$f(t,\theta) = \left(\frac{n}{2\theta}\right)^{n/2} \frac{t^{n/2-1}e^{-nt/(2\theta)}}{\Gamma(n/2)} \qquad t > 0$$

Let $\tau(\theta) = e^{\beta\theta}$, where β is a real constant. If we use the Taylor expansion of $\tau(\theta)$ and

$$E(t^k) = \left(\frac{2\theta}{n}\right)^k \frac{\Gamma(n/2+k)}{\Gamma(n/2)}$$

then we easily have the UMVU estimator of $\tau(\theta)$ as

$$T(t) = {}_0F_1\left(\frac{n}{2}; \frac{\beta n}{2}t\right)$$

The parametric function $\tau(\theta)$, the density function $f(t,\theta)$, and the estimator $T(t)$ satisfy the regularity conditions of the theorem by Blight and Rao (1974). Hence, the variance of $T(t)$ is given by

$$\text{Var}(T) = \sum_{i=1}^{\infty} \left\{\frac{\tau^{(i)}}{J_i}\right\}^2$$

where

$$J_i^2 = J_i^2(\theta) = E_\theta\left[\left\{\frac{f^{(i)}}{f}\right\}^2\right]$$

and $\tau^{(i)}$ and $f^{(i)}$ stand for the ith derivatives of τ and f with respect to θ. Direct computation leads to

$$J_i^2 = \frac{\Gamma(n/2+i)i!}{\Gamma(n/2)\theta^{2i}}$$

Hence,

$$\mathrm{Var}(T) = e^{2\beta\theta}\left\{{}_0F_1\left(\frac{n}{2};\beta^2\theta^2\right) - 1\right\}$$

so that

$$E\left[\left\{{}_0F_1\left(\frac{n}{2};\frac{\beta n}{2}t\right)\right\}^2\right] = e^{2\beta\theta}{}_0F_1\left(\frac{n}{2};\beta^2\theta^2\right) \tag{2.11}$$

Similarly, let $\tau^*(\theta) = e^{\beta_1\theta} + e^{\beta_2\theta}$, where β_1 and β_2 are real constants. The UMVU estimator of $\tau^*(\theta)$ is

$$T^*(t) = {}_0F_1\left(\frac{n}{2};\frac{\beta_1 n}{2}t\right) + {}_0F_1\left(\frac{n}{2};\frac{\beta_2 n}{2}t\right)$$

and its variance is

$$\begin{aligned}\mathrm{Var}(T^*) &= \sum_{i=1}^{\infty}\left(\beta_1^i e^{\beta_1\theta} + \beta_2^i e^{\beta_2\theta}\right)^2 \frac{\theta^{2i}\Gamma(n/2)}{i!\Gamma(n/2+i)} \\ &= \sum_{j=1}^{2} e^{2\beta_j\theta}\left\{{}_0F_1\left(\frac{n}{2};\beta_j^2\theta^2\right) - 1\right\} \\ &\quad + 2e^{(\beta_1+\beta_2)\theta}\left\{{}_0F_1\left(\frac{n}{2};\beta_1\beta_2\theta^2\right) - 1\right\}\end{aligned} \tag{2.12}$$

Combining (2.11) with (2.12), we have

$$E\left\{{}_0F_1\left(\frac{n}{2};\frac{\beta_1 n}{2}t\right){}_0F_1\left(\frac{n}{2};\frac{\beta_2 n}{2}t\right)\right\} = e^{(\beta_1+\beta_2)\theta}{}_0F_1\left(\frac{n}{2};\beta_1\beta_2\theta^2\right)$$

which is equal to (2.10).

2.6 Further Discussions on the Mean and the Variance

The expansion in powers to n^{-3} of (2.9) is given by

$$\begin{aligned}\operatorname{Var}(\hat{\theta}_{a,b,0}) = \theta^2_{a,b,0} \Bigg[& \frac{1}{n}\sigma^2(a^2+2b^2\sigma^2) + \frac{1}{n^2}\sigma^4 \\ & \times \left(\tfrac{1}{2}a^4 - 2a^2b + 2b^2 + 2a^2b^2\sigma^2 + 2b^4\sigma^4\right) \\ & + \frac{1}{n^3}\sigma^4 \left\{\tfrac{1}{2}(2b-a^2)^2 + \left(\tfrac{1}{6}a^6 + 2a^2b^2 - 2a^4b\right)\sigma^2 \right. \\ & + \left. (a^4b^2 - 4a^2b^3)\sigma^4 + 2a^2b^4\sigma^6 + \tfrac{4}{3}b^6\sigma^8\right\}\Bigg] + O(n^{-4})\end{aligned}$$

And we have

$$\begin{aligned}\operatorname{Var}(\hat{\mu}'_1) = {\mu'_1}^2 & \left\{\frac{1}{n}\sigma^2\left(1+\frac{1}{2}\sigma^2\right) + \frac{1}{2n^2}\sigma^6\left(1+\frac{1}{4}\sigma^2\right)\right. \\ & \left. + \frac{1}{n^3}\sigma^6\left(-\frac{1}{3} - \frac{1}{4}\sigma^2 + \frac{1}{8}\sigma^4 + \frac{1}{48}\sigma^6\right)\right\} + O(n^{-4}) \\ \operatorname{Var}(\hat{\mu}_2) = e^{4\mu+2\sigma^2} & \left[\frac{2}{n}\sigma^2\left\{2\left(e^{\sigma^2}-1\right)^2 + \sigma^2\left(2e^{\sigma^2}-1\right)^2\right\}\right. \\ & \left. + \frac{2}{n^2}\sigma^4\left\{1 + 4\sigma^2\left(2e^{\sigma^2}-1\right)^2 + \sigma^4\left(4e^{\sigma^2}-1\right)^2\right\}\right] + O(n^{-3})\end{aligned}$$

Finney (1941) gave the expansion in powers to the n^{-2} and n^{-1} for the variances of $\hat{\mu}'_1$ and $\hat{\mu}_2$. Mehran (1973) pointed out that Finney's approximation for the variance of $\hat{\mu}'_1$ was very good. This is the reason why the expansion is performed up to the term of n^{-2}. But Finney's approximation for the variance of $\hat{\mu}_2$ is not so good because the expansion is only performed up to the term of n^{-1}. Shimizu and Iwase (1981) pointed out that the approximation up to the term of n^{-2} for the variance of $\hat{\mu}_2$ was fairly good. Likeš (1980) also has a similar result.

We shall compare the UMVU estimators of the mean and the variance with the ML and nonparametric unbiased estimators of them in terms of the mean square errors. We refer to Shimizu, Iwase, and Ushizawa (1979). See also Likeš (1983).

ML estimators of μ'_1 and μ_2 are

$$\begin{aligned}\hat{\mu}'_{1(1)} &= e^{\overline{Y}+S_Y/(2n)} \\ \hat{\mu}_{2(1)} &= e^{2\overline{Y}}\left(e^{2S_Y/n} - e^{S_Y/n}\right)\end{aligned}$$

Since S_Y divided by σ^2 is $\chi^2(n-1)$, the following formula holds:

$$E(S_Y^c e^{bS_Y/n}) = \frac{\Gamma\left((n-1)/2+c\right)}{\Gamma\left((n-1)/2\right)}(2\sigma^2)^c\left(1-\frac{2b}{n}\sigma^2\right)^{-\{(n-1)/2+c\}} \quad (2.13)$$

where it is assumed that $n > 2b\sigma^2$ and c is a non-negative integer. The formula (2.13) is useful for expressing the mathematical expectations and mean square errors of the ML estimators. We have

$$E(\hat{\mu}'_{1(1)}) = e^{\mu+\sigma^2/(2n)}H(1)$$

$$\text{MSE}(\hat{\mu}'_{1(1)}) = e^{2\mu}\left\{e^{2\sigma^2/n}H(2) - 2e^{(1+1/n)\sigma^2/2}H(1) + e^{\sigma^2}\right\}$$

$$E(\hat{\mu}_{2_{(1)}}) = e^{2\mu+2\sigma^2/n}\{H(4) - H(2)\}$$

$$\begin{aligned}\text{MSE}(\hat{\mu}_{2(1)}) &= e^{4\mu+8\sigma^2/n}\{H(8) - 2H(6) + H(4)\} - 2e^{2\mu+\sigma^2}(e^{\sigma^2}-1) \\ &\quad \times e^{2\mu+2\sigma^2/n}\{H(4) - H(2)\} + e^{4\mu+2\sigma^2}(e^{\sigma^2}-1)^2\end{aligned}$$

where

$$H(\alpha) = \left(1-\frac{\alpha}{n}\sigma^2\right)^{-(n-1)/2}$$

and it is assumed that $n > \alpha\sigma^2$.

Nonparametric unbiased estimators of μ'_1 and μ_2 are

$$\hat{\mu}'_{1(2)} = \overline{X}$$

$$\hat{\mu}_{2(2)} = \frac{1}{n-1}S_X$$

and their variances are

$$\text{Var}\left(\hat{\mu}'_{1(2)}\right) = \frac{1}{n}\mu_2$$

$$\begin{aligned}\text{Var}\left(\hat{\mu}_{2(2)}\right) &= \frac{e^{4\mu+2\sigma^2}}{(n-1)^2}\left\{n\left(e^{6\sigma^2} - 4e^{3\sigma^2} - e^{2\sigma^2} + 8e^{\sigma^2} - 4\right)\right. \\ &\quad - 2\left(e^{6\sigma^2} - 4e^{3\sigma^2} - 2e^{2\sigma^2} + 10e^{\sigma^2} - 5\right) \\ &\quad \left. + \frac{1}{n}\left(e^{6\sigma^2} - 4e^{3\sigma^2} - 3e^{2\sigma^2} + 12e^{\sigma^2} - 6\right)\right\}\end{aligned}$$

Table 1 Efficiency of the ML Estimators Relative to the UMVU Estimators

$\text{eff}\left(\hat{\mu}'_{1(1)}, \hat{\mu}'_1\right) = \text{Var}\left(\hat{\mu}'_1\right)/\text{MSE}\left(\hat{\mu}'_{1(1)}\right)$

n \ σ	0.1	0.2	0.3	0.4	0.5	0.6	0.7
10	1.000	1.000	.999	.997	.992	.983	.969
50	1.000	1.000	1.000	.999	.998	.996	.994
100	1.000	1.000	1.000	1.000	.999	.998	.997
150	1.000	1.000	1.000	1.000	.999	.999	.998
200	1.000	1.000	1.000	1.000	1.000	.999	.998

$\text{eff}\left(\hat{\mu}_{2(1)}, \hat{\mu}_2\right) = \text{Var}\left(\hat{\mu}_2\right)/\text{MSE}\left(\hat{\mu}_{2(1)}\right)$

n \ σ	0.1	0.2	0.3	0.4	0.5	0.6	0.7
10	1.161	1.130	1.071	.975	.842	.676	.488
50	1.029	1.024	1.013	.996	.971	.937	.892
100	1.014	1.012	1.006	.998	.986	.969	.946
150	1.010	1.008	1.004	.999	.990	.979	.964
200	1.007	1.006	1.003	.999	.993	.984	.973

Table 1 shows the efficiency of the ML estimators of the mean and the variance relative to the respective UMVU estimators. Figures 1a and 1b show the efficiency of the nonparametric unbiased estimators of the mean and the variance relative to the respective UMVU estimators.

2.7 Estimation of a Quantile

The quantile θ_q of $\Lambda(\mu, \sigma^2)$ is not contained in (2.1) except for the median, where $q = \Pr[X \leq \theta_q]$. Efficiencies of UMVU and ML estimators and ones based on sample quantiles and sample moments of θ_q were treated by Shimizu (1983d). Only the results are given here.

The UMVU estimator of θ_q is

$$\hat{\theta}_q = e^{\overline{Y}} \left[\sum_{k=0}^{\infty} H_k\left(\sqrt{n}t_q\right) \frac{\Gamma\left((n-1)/2\right)}{k!(2n)^{k/2}\Gamma\left((k+n-1)/2\right)} S_Y^{k/2} \right]$$

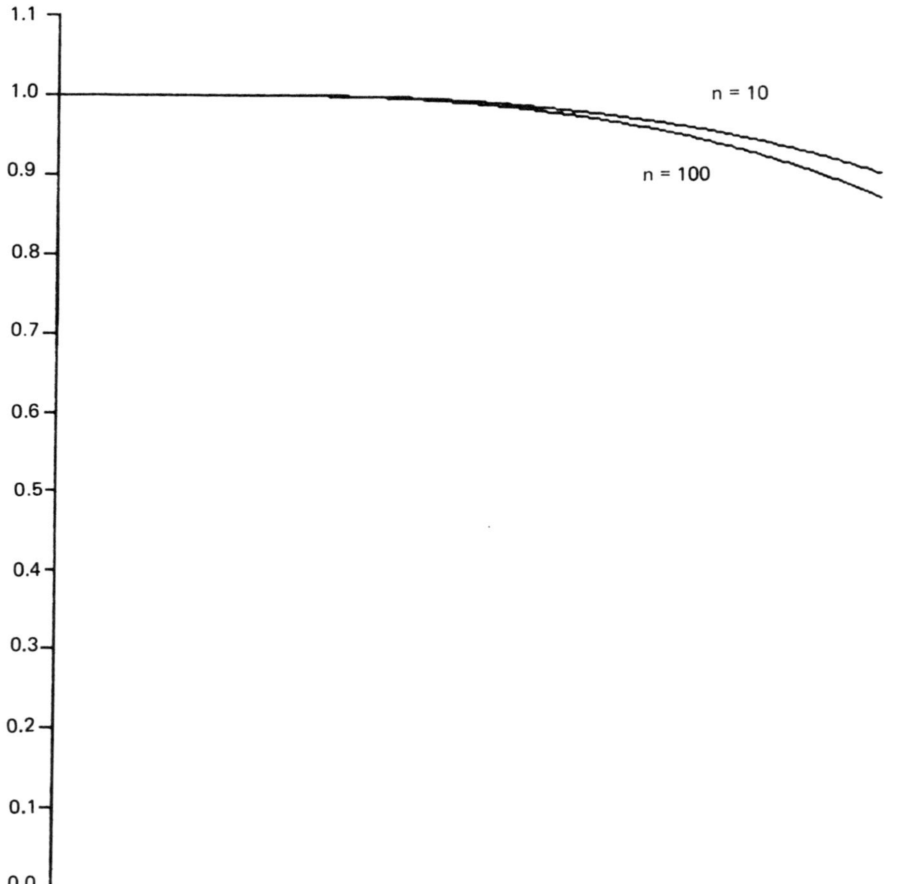

Figure 1a Efficiency of the nonparametric unbiased estimators relative to the UMVU estimators. eff $\left(\hat{\mu}'_{1(2)}, \hat{\mu}'_1\right) = \mathrm{Var}(\hat{\mu}'_1)/\mathrm{Var}\left(\hat{\mu}'_{1(2)}\right)$. (*Continued on opposite page.*)

where H_k are the Hermite polynomials

$$H_k(x) = (-1)^k e^{x^2/2} \left(\frac{d^k}{dx^k} e^{-x^2/2} \right)$$

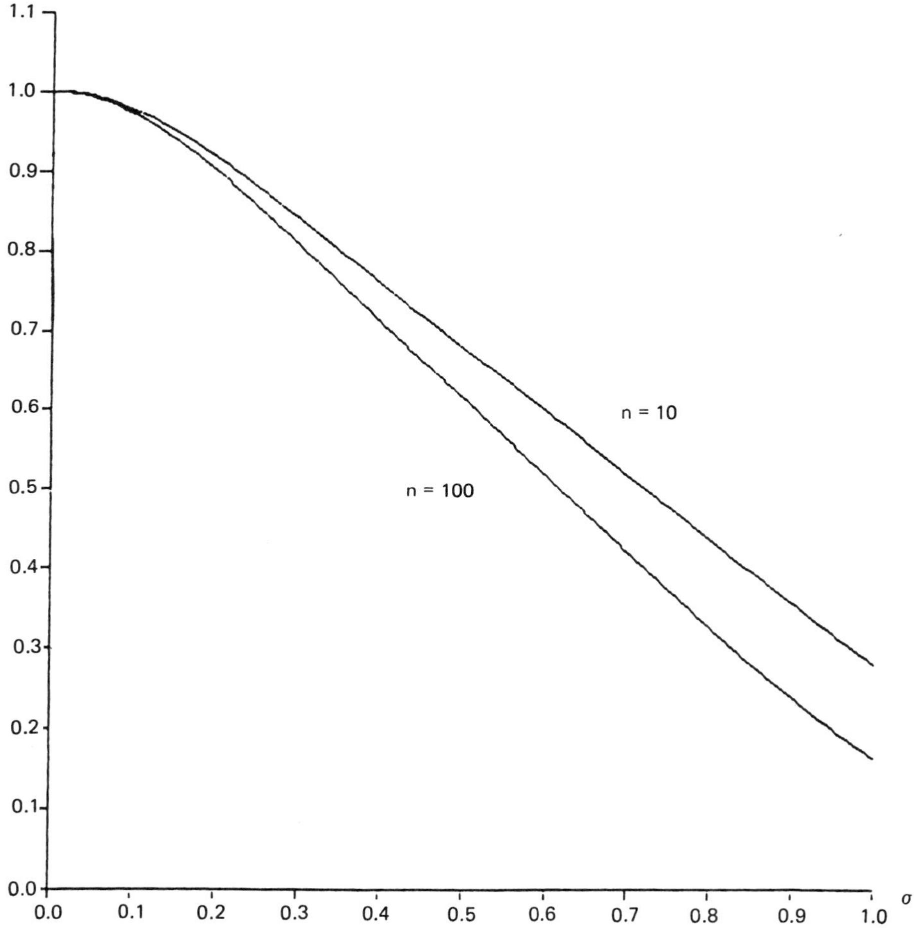

Figure 1b (*Continued*) eff $(\hat{\mu}_{2(2)}, \hat{\mu}_2) = \text{Var}(\hat{\mu}_2)/\text{Var}(\hat{\mu}_{2(2)})$.

The variance of $\hat{\theta}_q$ is

$$\text{Var}(\hat{\theta}_q) = e^{2\mu}\left[e^{2\sigma^2/n}\sum_{k=0}^{\infty} W_k \frac{\Gamma((k+n-1)/2)}{\Gamma((n-1)/2)}\left(\frac{\sigma}{\sqrt{n}}\right)^k - e^{2t_q\sigma}\right]$$

where

$$W_k = \sum_{r+s=k} \frac{\Gamma^2((n-1)/2)}{\Gamma((r+n-1)/2)\,\Gamma((s+n-1)/2)\,r!s!} H_r(\sqrt{n}t_q)\,H_s(\sqrt{n}t_q)$$

and r and s are integers such that $0 \le r,\ s \le k$.

The ML estimator of θ_q is

$$\hat{\theta}_{q(1)} = \exp\left(\overline{Y} + t_q\sqrt{\frac{S_Y}{n}}\right)$$

The mathematical expectation of $\hat{\theta}_{q(1)}$ is

$$E\left(\hat{\theta}_{q(1)}\right) = e^{\mu+\sigma^2/(2n)} \times \left\{ {}_1F_1\left(\frac{n-1}{2};\frac{1}{2};\frac{t_q^2\sigma^2}{2n}\right) + \sqrt{\frac{2}{n}}\frac{\Gamma(n/2)}{\Gamma((n-1)/2)}t_q\sigma\ {}_1F_1\left(\frac{n}{2};\frac{3}{2};\frac{t_q^2\sigma^2}{2n}\right)\right\}$$

and the mean square error of $\hat{\theta}_{1(1)}$ is

$$\begin{aligned}\mathrm{MSE}\left(\hat{\theta}_{q(1)}\right) = e^{2\mu}\Bigg[& e^{2\sigma^2/n}\left\{ {}_1F_1\left(\frac{n-1}{2};\frac{1}{2};\frac{2t_q^2\sigma^2}{n}\right)\right. \\ & \left. + 2\sqrt{\frac{2}{n}}t_q\sigma\frac{\Gamma(n/2)}{\Gamma((n-1)/2)}\,{}_1F_1\left(\frac{n}{2};\frac{3}{2};\frac{2t_q^2\sigma^2}{n}\right)\right\} \\ & - 2e^{\sigma^2/(2n)+t_q\sigma}\left\{ {}_1F_1\left(\frac{n-1}{2};\frac{1}{2};\frac{t_q^2\sigma^2}{2n}\right)\right. \\ & \left. + \sqrt{\frac{2}{n}}t_q\sigma\frac{\Gamma(n/2)}{\Gamma((n-1)/2)}\,{}_1F_1\left(\frac{n}{2};\frac{3}{2};\frac{t_q^2\sigma^2}{2n}\right)\right\} + e^{2t_q\sigma}\Bigg]\end{aligned}$$

An estimator of θ_q based on sample quantiles is

$$\hat{\theta}_{q(2)} = X_{([qn+1])}$$

where $[a]$ denotes the largest integer not greater than a and $X_{(1)} \le \cdots \le X_{(n)}$ are the order statistics of $X_1, \ldots, X_n$. The estimator $\hat{\theta}_{q(2)}$ asymptotically has a normal distribution with mean θ_q and variance

$$\operatorname{Var}\left(\hat{\theta}_{q(2)}\right) = \frac{1}{n}\left\{2q(1-q)\pi\sigma^2 \exp\left(t_q^2 + 2t_q\sigma + 2\mu\right)\right\}$$

An estimator of θ_q based on sample moments is

$$\hat{\theta}_{q(3)} = \overline{Y}\left(1 + \frac{S_X}{n\overline{X}^2}\right)^{-1/2} \exp\left[t_q\left\{\ln\left(1 + \frac{S_X}{n\overline{X}^2}\right)\right\}^{1/2}\right]$$

The estimator $\hat{\theta}_{q(3)}$ asymptotically has a normal distribution with mean θ_q and variance

$$\begin{aligned}\operatorname{Var}\left(\hat{\theta}_{q(3)}\right) = {} & \frac{1}{n} e^{2\mu - 2\sigma^2 + 2t_q\sigma}\left(e^{\sigma^2} - 1\right)\left[\left\{1 + \left(e^{\sigma^2} - 1\right)\left(2 - \frac{t_q}{\sigma}\right)\right\}^2\right. \\ & + \left(e^{\sigma^2} - 1\right)\left(e^{\sigma^2} + 2\right)\left\{1 + \left(e^{\sigma^2} - 1\right)\left(2 - \frac{t_q}{\sigma}\right)\right\}\left(\frac{t_q}{\sigma} - 1\right) \\ & \left. + \frac{1}{4}\left(e^{\sigma^2} - 1\right)\left(e^{4\sigma^2} + 2e^{3\sigma^2} + 3e^{2\sigma^2} - 4\right)\left(\frac{t_q}{\sigma} - 1\right)^2\right]\end{aligned}$$

Numerical computation shows that ML estimates should be practically used because they are easier to calculate than UMVU estimates. Moment estimates may be used when $q = 0.75$, and also used when $q = 0.5$, 0.98 under the prior information that σ is small. The efficiency of the quantile estimator is low.

A simulation was carried out for $n = 50$, $\mu = 0$, and $q = 0.5, 0.75, 0.98$ because the mean square errors of the quantile and the moment estimators are expressed only asymptotically. Ten thousand samples (each of size 50) were obtained for each of the three sets of parameter values. It was found that the value obtained from the asymptotic expressions is close to the value obtained from the simulation.

2.8 Smaller Mean Square Error Estimation

The aim is to derive an estimator having smaller mean square error than the ML and UMVU estimators. This section is due to the work by Evans and Shaban (1976).

Consider the estimation of $\theta_{a,b} = \exp(a\mu + b\sigma^2)$ from independent observations $X_1, X_2, \ldots, X_n$ having $\Lambda(\mu, \sigma^2)$.

First, let us assume that σ^2 is known and μ is unknown. We shall consider the class of estimators

$$\hat{\theta}_C = e^{a\overline{Y}} f(\sigma^2) \tag{2.14}$$

where $\overline{Y} = n^{-1} \sum_{i=1}^{n} \ln X_i$. The problem is to find a function f such that the mean square error is minimum in the class (2.14).

The ML estimator

$$\hat{\theta}_L = \exp(a\overline{Y} + b\sigma^2)$$

and the UMVU estimator

$$\hat{\theta}_U = \exp\left\{a\overline{Y} + b\sigma^2 - \frac{a^2\sigma^2}{2n}\right\}$$

are both in the class (2.14).

The mean square error of (2.14) is

$$\begin{aligned}\mathrm{MSE}(\hat{\theta}_C) &= f^2(\sigma^2)\exp\left\{2a\mu + \frac{2a^2\sigma^2}{n}\right\} \\ &\quad - 2f(\sigma^2)\exp\left\{2a\mu + b\sigma^2 + \frac{a^2\sigma^2}{2n}\right\} + \exp(2a\mu + 2b\sigma^2)\end{aligned}$$

which is readily seen to be minimal when

$$f(\sigma^2) = \exp\left\{b\sigma^2 - \frac{3a^2\sigma^2}{2n}\right\}$$

Hence, the minimum mean square error estimator in the class (2.14) is

$$\hat{\theta}_M = \exp(a\overline{Y} + k\sigma^2) \tag{2.15}$$

where $k = b - 3a^2/(2n)$. We obtain

$$\mathrm{MSE}(\hat{\theta}_L) > \mathrm{MSE}(\hat{\theta}_U) > \mathrm{MSE}(\hat{\theta}_M)$$

Second, let us assume that both μ and σ^2 are unknown. Put $S_Y = \sum_{i=1}^{n}(\ln X_i - \overline{Y})^2$. Replacing σ^2 in (2.15) by $S_Y/(n+1)$, which is the minimum mean square error estimator of σ^2 of the form $S_Y/(n+c)$, we have an estimator of $\theta_{a,b}$

$$\hat{\theta}_1 = \exp\left\{a\overline{Y} + \frac{kS_Y}{n+1}\right\} \tag{2.16}$$

An alternative estimator can be obtained by equating

$$E\{f(S_Y)\} = \exp(k\sigma^2)$$

Such a function must be

$$f(S_Y) = {}_0F_1\left(\frac{n-1}{2}; \frac{k}{2}S_Y\right)$$

as a result of the discussion in Section 2.3. Hence, we have an estimator of $\theta_{a,b}$

$$\hat{\theta}_2 = e^{a\overline{Y}}{}_0F_1\left(\frac{n-1}{2}; \frac{k}{2}S_Y\right) \tag{2.17}$$

The mean square errors of (2.16) and (2.17) are

$$\begin{aligned}\mathrm{MSE}(\hat{\theta}_1) = \theta_{a,b}^2 \Bigg[& e^{(2a^2/n-2b)\sigma^2}\left(1 - \frac{4k}{n+1}\sigma^2\right)^{-(n-1)/2} \\ & - 2e^{\{a^2/(2n)-b\}\sigma^2}\left(1 - \frac{2k}{n+1}\sigma^2\right)^{-(n-1)/2} + 1\Bigg]\end{aligned}$$

by use of (2.13), where it is assumed that $n > 4k\sigma^2 - 1$, and

$$\mathrm{MSE}(\hat{\theta}_2) = \theta_{a,b}^2\left\{e^{-a^2\sigma^2/n}{}_0F_1\left(\frac{n-1}{2}; k^2\sigma^4\right) - 2e^{-a^2\sigma^2/n} + 1\right\}$$

by use of (2.10).

The variance of the UMVU estimator

$$\hat{\theta}_U = e^{a\overline{Y}}{}_0F_1\left(\frac{n-1}{2}; \frac{2bn-a^2}{4n}S_Y\right) \tag{2.18}$$

is

$$\operatorname{Var}(\hat{\theta}_U) = \theta_{a,b}^2 \left\{ e^{a^2\sigma^2/n} {}_0F_1\left(\frac{n-1}{2}; \frac{(2bn-a^2)^2}{4n^2}\sigma^4\right) - 1 \right\}$$

and the mean square error of the ML estimator

$$\hat{\theta}_L = e^{a\overline{Y} + bS_Y/n} \tag{2.19}$$

is

$$\operatorname{MSE}(\hat{\theta}_L) = \theta_{a,b}^2 \left[e^{(2a^2/n - 2b)\sigma^2} \left(1 - \frac{4b}{n}\sigma^2\right)^{-(n-1)/2} - 2e^{\{a^2/(2n) - b\}\sigma^2} \left(1 - \frac{2b}{n}\sigma^2\right)^{-(n-1)/2} + 1 \right]$$

where it is assumed that $n > 4b\sigma^2$.

There is no minimum mean square error estimator among the four estimators (independent of σ^2).

For large enough n, asymptotic expansions to n^{-2} are

$$\operatorname{MSE}(\hat{\theta}_1) = \theta_{a,b}^2 \frac{\sigma^2}{n} \left\{ a^2 + 2b^2\sigma^2 + \frac{\sigma^2}{2n}(14b^4\sigma^4 - 8b^3\sigma^2 - 4a^2b^2\sigma^2 - 4b^2 - 12a^2b - a^4) \right\} + O(n^{-3})$$

$$\operatorname{MSE}(\hat{\theta}_2) = \sigma_{a,b}^2 \frac{\sigma^2}{n} \left\{ a^2 + 2b^2\sigma^2 + \frac{\sigma^2}{2n}(4b^4\sigma^4 - 4a^2b^2\sigma^2 + 4b^2 - 12a^2b - a^4) \right\} + O(n^{-3})$$

$$\operatorname{Var}(\hat{\theta}_U) = \theta_{a,b}^2 \frac{\sigma^2}{n} \left\{ a^2 + 2b^2\sigma^2 + \frac{\sigma^2}{2n}(4b^4\sigma^4 + 4a^2b^2\sigma^2 + 4b^2 - 4a^2b + a^4) \right\} + O(n^{-3})$$

$$\mathrm{MSE}(\hat{\theta}_L) = \theta_{a,b}^2 \frac{\sigma^2}{n} \left\{ a^2 + 2b^2\sigma^2 + \frac{\sigma^2}{4n}(28b^4\sigma^4 + 8b^3\sigma^2 + 28a^2b^2\sigma^2 \right.$$

$$\left. - 4b^2 - 12a^2b + 7a^4) \right\} + O(n^{-3})$$

All four mean square errors are equal to $O(n^{-1})$, so that for very large n there is very little to choose among the four estimators. For values of n greater than 50, according to Evans and Shaban (1976), one would presumably prefer $\hat{\theta}_1$ or $\hat{\theta}_L$, since neither can take negative values and each is expressed completely in terms of the exponential function. However, Rukhin (1986) constructed the estimator of $\theta_{a,b}, b \geq a^2/(2n)$,

$$\hat{\theta}_3 = e^{a\overline{Y}} {}_0F_1\left(\frac{n-1}{2}; \frac{k(n-1)}{2(n+1)} S_Y\right)$$

for positive $k = b - 3a^2/(2n)$, which improves on the estimator (2.17), the UMVU estimator (2.18), and the ML estimator (2.19). In addition, he suggested for practical use a generalized Bayes estimator, which is locally optimal for both small and large values of σ^2.

Improved estimation of the mean when the coefficient of variation is known was treated by Sen (1978).

3. ESTIMATION IN RELATED DISTRIBUTIONS AND MODELS

3.1 Univariate Delta-Lognormal Distributions

Let X^* be a random variable having a delta-lognormal distribution with parameters δ, μ, and σ^2, denoted by $\Delta(\delta, \mu, \sigma^2)$, i.e.,

$$\Pr(X^* < 0) = 0$$

$$\Pr(X^* = 0) = \delta$$

$$\Pr(X^* \leq x) = \delta + (1-\delta)\Lambda(x \mid \mu, \sigma^2) \qquad x > 0$$

where $\Lambda(x|\mu, \sigma^2)$ is the distribution function of $\Lambda(\mu, \sigma^2)$. Treatments of this distribution appeared in Aitchison and Brown (1957). Pennington (1983) applied it to an ichthyoplankton survey.

If $\delta = 0$, then X^* has $\Lambda(\mu, \sigma^2)$. This means that a delta-lognormal distribution is a generalization of a lognormal distribution.

The rth moment of X^* about zero is

$$E\{X^{*r}\} = (1-\delta)\exp\left(r\mu + \frac{r^2\sigma^2}{2}\right)$$

Therefore, the mean is

$$\kappa = (1-\delta)\exp\left(\mu + \frac{\sigma^2}{2}\right)$$

and the variance is

$$\rho^2 = (1-\delta)\left\{\exp(\sigma^2) - (1-\delta)\right\}\exp(2\mu + \sigma^2)$$

Note that the mean and the variance are members of a parametric function

$$\eta_{a,b,c} = (1-\delta)^c \exp(a\mu + b\sigma^2) \tag{3.1}$$

and its linear combination

$$\eta = \sum_{(a,b,c)} C_{a,b,c}\eta_{a,b,c} \tag{3.2}$$

where c is a positive integer, and a, b, and $C_{a,b,c}$ are real constants.

We shall find the UMVU estimators of (3.1) and (3.2) and evaluate their variances.

Let $X_1^*, X_2^*, \ldots, X_n^*$, $n \geq 2$, be iid random variables having $\Delta(\delta, \mu, \sigma^2)$ with unknown parameters δ, μ, and σ^2. Let n_1 be the number of non-zero values among X_i^*, $i = 1, \ldots, n$. Without loss of generality, let $X_1^*, \ldots, X_{n_1}^*$ be non-zero. Put

$$\overline{Y}^* = \begin{cases} \dfrac{1}{n_1}\displaystyle\sum_{i=1}^{n_1} Y_i^* & n_1 > 0 \\ 0 & n_1 = 0 \end{cases}$$

$$S_Y^* = \begin{cases} \displaystyle\sum_{i=1}^{n_1}(Y_i^* - \overline{Y}^*)^2 & n_1 > 1 \\ 0 & n_1 = 0, 1 \end{cases}$$

where $Y_i^* = \ln X_i^*$ for $i = 1, \ldots, n_1$.

From Shimizu and Iwase (1981), the UMVU estimator of (3.1) is

$$\hat{\eta}_{a,b,c}(n_1, \overline{Y}^*, S_Y^*) = \begin{cases} \dfrac{n_{1(c)}}{n_{(c)}} e^{a\overline{Y}^*} {}_0F_1\left(\dfrac{n_1-1}{2}; \dfrac{2bn_1 - a^2}{4n_1} S_Y^*\right) & n_1 \geq c \\ 0 & n_1 < c \end{cases}$$

and its variance is

$$\begin{aligned}\operatorname{Var}(\hat{\eta}_{a,b,c}) = \exp(2a\mu + 2b\sigma^2) \times \Bigg[& \frac{1}{n_{(c)}^2} \sum_{j=c}^{n} \binom{n}{j} (1-\delta)^j \delta^{n-j} j_{(c)}^2 \\ & \times \exp\left(\frac{a^2\sigma^2}{j}\right) {}_0F_1\left(\frac{j-1}{2}; \frac{(2bj-a^2)^2}{4j^2}\sigma^4\right) - (1-\delta)^{2c} \Bigg]\end{aligned}$$

where ${}_0F_1(\cdot;0) = 1$ and $n_{(c)} = n(n-1)\ldots(n-c+1)$.

Also, the UMVU estimator of (3.2) is

$$\hat{\eta} = \sum_{(a,b,c)} C_{a,b,c}\hat{\eta}_{a,b,c}$$

and its variance is

$$\begin{aligned}\operatorname{Var}(\hat{\eta}) = \sum_{(a,b,c)} C_{a,b,c}^2 \exp(2a\mu + 2b\sigma^2) \times \Bigg[& \frac{1}{n_{(c)}^2} \sum_{j=c}^{n} \binom{n}{j} (1-\delta)^j \delta^{n-j} j_{(c)}^2 \\ & \times \exp\left(\frac{a^2\sigma^2}{j}\right) {}_0F_1\left(\frac{j-1}{2}; \frac{(2bj-a^2)^2}{4j^2}\sigma^4\right) - (1-\delta)^{2c} \Bigg] \\ + 2 \sum_{(a,b,c)\neq(a',b'c')} & C_{a,b,c}C_{a',b',c'} \exp\left\{(a+a')\mu + (b+b')\sigma^2\right\} \\ \times \Bigg[& \frac{1}{n_{(c)}n_{(c')}} \sum_{j=\max(c,c')}^{n} \binom{n}{j} (1-\delta)^j \delta^{n-j} j_{(c)} j_{(c')} \exp\left(\frac{aa'\sigma^2}{j}\right) \\ & \times {}_0F_1\left(\frac{j-1}{2}; \frac{1}{4j^2}(2bj-a^2)(2b'j-a'^2)\sigma^4\right) - (1-\delta)^{c+c'} \Bigg]\end{aligned}$$

These are proved by use of (2.10) and

$$E(n_{1(c)}) = n_{(c)}(1-\delta)^c$$

The UMVU estimators of the mean κ and the variance ρ^2 are

$$\hat{\kappa} = \begin{cases} \dfrac{n_1}{n} e^{\overline{Y}^*} {}_0F_1\left(\dfrac{n_1-1}{2}; \dfrac{n_1-1}{4n_1} S_Y^*\right) & n_1 > 1 \\ \dfrac{X_1^*}{n} & n_1 = 1 \\ 0 & n_1 = 0 \end{cases}$$

$$\hat{\rho}^2 = \begin{cases} \dfrac{n_1}{n} e^{2\overline{Y}^*} \left\{ {}_0F_1(\dfrac{n_1-1}{2}; \dfrac{n_1-1}{n_1} S_Y^*) \right. & \\ \qquad \left. - \dfrac{n_1-1}{n-1} {}_0F_1\left(\dfrac{n_1-1}{2}; \dfrac{n_1-2}{2n_1} S_Y^*\right)\right\}, & n_1 > 1 \\ \dfrac{1}{n} {X_1^*}^2 & n_1 = 1 \\ 0 & n_1 = 0 \end{cases}$$

and their variances are

$$\begin{aligned} \mathrm{Var}(\hat{\kappa}) = {} & \exp(2\mu+\sigma^2) \left[\frac{1}{n^2} \sum_{j=1}^{n} \binom{n}{j} (1-\delta)^j \delta^{n-j} j^2 \exp\left(\frac{\sigma^2}{j}\right) \right. \\ & \left. \times {}_0F_1\left(\frac{j-1}{2}; \frac{(j-1)^2}{4j^2}\sigma^4\right) - (1-\delta)^2 \right] \end{aligned}$$

$$\begin{aligned} \mathrm{Var}(\hat{\rho}^2) = {} & \exp(4\mu+4\sigma^2) \left[\frac{1}{n^2} \sum_{j=1}^{n} \binom{n}{j} (1-\delta)^j \delta^{n-j} j^2 \exp\left(\frac{4\sigma^2}{j}\right) \right. \\ & \left. \times {}_0F_1\left(\frac{j-1}{2}; 4\left(1-\frac{1}{j}\right)^2 \sigma^4\right) - (1-\delta)^2 \right] + \exp(4\mu+2\sigma^2) \end{aligned}$$

$$\times \left[\frac{1}{n^2(n-1)^2} \sum_{j=2}^{n} \binom{n}{j} (1-\delta)^j \delta^{n-j} j^2 (j-1)^2 \exp\left(\frac{4\sigma^2}{j}\right) \right.$$

$$\left. \times \, {}_0F_1 \left(\frac{j-1}{2}; \left(1 - \frac{2}{j}\right)^2 \sigma^4 \right) - (1-\delta)^4 \right] - 2\exp(4\mu + 3\sigma^2)$$

$$\times \left[\frac{1}{n^2(n-1)} \sum_{j=2}^{n} \binom{n}{j} (1-\delta)^j \delta^{n-j} j^2 (j-1) \exp\left(\frac{4\sigma^2}{j}\right) \right.$$

$$\left. \times \, {}_0F_1 \left(\frac{j-1}{2}; 2\left(1 - \frac{1}{j}\right)\left(1 - \frac{2}{j}\right)\sigma^4 \right) - (1-\delta)^3 \right]$$

The asymptotic expansions in powers to n^{-1} of the variances of $\hat{\kappa}$ and $\hat{\rho}^2$ are

$$\mathrm{Var}(\hat{\kappa}) = \frac{1}{n}(1-\delta)\left\{\delta + \frac{1}{2}\sigma^2(\sigma^2+2)\right\}\exp(2\mu+\sigma^2) + O(n^{-2})$$

$$\mathrm{Var}(\hat{\rho}^2) = \frac{1}{n}(1-\delta)\left\{\delta\left(e^{\sigma^2} - 2 + 2\delta\right)^2 + 4\sigma^2\left(e^{\sigma^2} - 1 + \delta\right)^2 \right.$$

$$\left. + \, 2\sigma^4\left(2e^{\sigma^2} - 1 + \delta\right)^2\right\}\exp(4\mu + 2\sigma^2) + O(n^{-2})$$

These coincide with the results by Aitchison and Brown (1957).

3.2 Lognormal Linear Models

Consider the model

$$Y = X_1^{\alpha_1} X_2^{\alpha_2} \ldots X_p^{\alpha_p} e^u = \exp(\alpha_1 x_1 + \alpha_2 x_2 + \cdots + \alpha_p x_p + u)$$

where $x_j > 0$, $j = 1 \ldots p$, are explanatory variables, $x_j = \ln X_j$, α_j are parameters, and the error term u has a normal distribution with mean zero and unknown variance σ^2. This model was introduced by Bradu and Mundlak (1970).

For instance, $\exp(s'\alpha)$ is a parametric function, where $s = (s_1, s_2, \ldots, s_p)'$ is a real and constant vector and $\alpha = (\alpha_1, \alpha_2, \ldots, \alpha_p)'$ is a parametric vector. The constant term, $\exp(\alpha_1)$, and a point on the median regression surface, $\exp(\eta)$, $\eta = (x_1, x_2, \ldots, x_p)'\alpha$, are special cases of $\exp(s'\alpha)$.

We shall deal with the full rank lognormal linear model. Given a sample of n observations, the model equations are described in logarithmic form by:

$$y = x\alpha + u$$

where $y = (\ln Y_1, \ln Y_2, \ldots, \ln Y_n)'$ and $x = (x_{ij})$, $x_{ij} = \ln X_{ij}$, is an $n \times p$ matrix with rank $(x) = p$. The error term $u = (u_1, \ldots, u_n)'$ has an n-dimensional normal distribution with mean vector zero and covariance matrix $\sigma^2 I_n$, and

$$E(x_{ij}u_i) = 0$$

Let w be the solution of the normal equations

$$(x'x)w = x'y$$

Then it follows that
(1) The statistic $s'w$ is a best linear unbiased estimator of $\varsigma = s'\alpha$ and has $N(\varsigma, \sigma_z^2)$, where $\sigma_z^2 = \sigma^2 v_z$ and $v_z = s'(x'x)^{-1}s$.
(2) An unbiased estimator of σ^2 is

$$\hat{\sigma}^2 = \frac{1}{n-p}(y - xw)'(y - xw).$$

The variable $(n-p)\hat{\sigma}^2/\sigma^2$ has $\chi^2(n-p)$ and is independent of $s'w$. These situations are similar to that of the estimation in a lognormal distribution $\Lambda(\mu, \sigma^2)$.

Consider a parametric function

$$\xi_{a,b,c} = \sigma^{2c} \exp(a\varsigma + b\sigma^2) \tag{3.3}$$

where a and b are real constants and c is a non-negative integer. The UMVU estimator of (3.3) is

$$\hat{\xi}_{a,b,c} = \frac{\Gamma((n-p)/2)}{\Gamma((n-p)/2 + c)} e^{as'w} \left(\frac{n-p}{2}\hat{\sigma}^2\right)^c \times {}_0F_1\left(\frac{n-p}{2} + c; \frac{(2b - a^2 v_z)}{4}(n-p)\hat{\sigma}^2\right)$$

and its variance is

$$\operatorname{Var}(\hat{\xi}_{a,b,c}) = \xi_{a,b,c}^2 \left\{ \frac{\Gamma((n-p)/2)\Gamma((n-p)/2+2c)}{\Gamma^2((n-p)/2+c)} e^{2(a^2v_z-b)\sigma^2} \right.$$

$$\times {}_2F_2\left(\frac{1}{2}(n-p+2c-1), \frac{n-p}{2}+2c;\right.$$

$$\left.\left.\frac{n-p}{2}+c, n-p+2c-1; 2(2b-a^2v_z)\sigma^2\right) - 1\right\}$$

In particular, if $c = 0$, then the UMVU estimator of $\xi_{a,b} = \exp(a\varsigma + b\sigma^2)$ is

$$\hat{\xi}_{a,b} = e^{as'w}{}_0F_1\left(\frac{n-p}{2}; \frac{(2b-a^2v_z)}{4}(n-p)\hat{\sigma}^2\right)$$

and its variance is

$$\operatorname{Var}(\hat{\xi}_{a,b}) = \xi_{a,b}^2 \left\{ e^{a^2\sigma_z^2}{}_0F_1\left(\frac{n-p}{2}; (b-\frac{a^2}{2}v_z)^2\sigma^4\right) - 1\right\}$$

The proof is similar to that in Sections 2.3 and 2.4.

Duan (1983) has dealt with the efficiency of the "smearing" estimator (a type of bootstrap estimator) for lognormal linear models.

3.3 Two Lognormal Distributions

Let $X^{(1)}$ and $X^{(2)}$ be random variables having $\Lambda(\mu_1, \sigma^2)$ and $\Lambda(\mu_2, \sigma^2)$. To evaluate the effect of cloud-seeding experiments, Crow (1977) considered

$$E\left(X^{(2)}\right) / E\left(X^{(1)}\right) = \exp(\mu_2 - \mu_1)$$

as a measure. Another example of the ratio of means of two lognormal distributions can be seen in Laurent (1963); when the individual price and salary indices with respect to a given base year are lognormally distributed, the global indices used are practically of the expected value type.

We shall take a more general parametric function

$$\gamma_{a,b,c} = \sigma^{2c} \exp\left\{a(\mu_2 - \mu_1) + b\sigma^2\right\} \tag{3.4}$$

where a, b are real constants and c is a non-negative integer.

In the same way, when $X^{(1)}$ and $X^{(2)}$ are random variables having $\Lambda(\mu_1, \sigma_1^2)$ and $\Lambda(\mu_2, \sigma_2^2)$, we shall take a parametric function

$$\gamma_{a,b,c,d,e,f} = \sigma_1^{2c}\sigma_2^{2f}\exp(a\mu_1 + d\mu_2 + b\sigma_1^2 + e\sigma_2^2) \tag{3.5}$$

where a, b, d, e are real constants and c, f are non-negative integers.

We shall deal with the estimation of (3.4) and (3.5).

First, let $X_1^{(1)}, X_2^{(1)}, \ldots, X_m^{(1)}$, $m \geq 2$, be iid random variables having $\Lambda(\mu_1, \sigma^2)$ and let $X_1^{(2)}, X_2^{(2)}, \ldots, X_n^{(2)}$, $n \geq 2$, be iid random variables having $\Lambda(\mu_2, \sigma^2)$. We assume that μ_1, μ_2, and σ^2 are unknown. Put

$$\begin{aligned} \overline{Y}^{(1)} &= \frac{1}{m}\sum_{i=1}^{m} Y_i^{(1)}, \overline{Y}^{(2)} = \frac{1}{n}\sum_{j=1}^{n} Y_j^{(2)} \\ S &= \sum_{i=1}^{m}\left(Y_i^{(1)} - \overline{Y}^{(1)}\right)^2 + \sum_{j=1}^{n}\left(Y_j^{(2)} - \overline{Y}^{(2)}\right)^2 \end{aligned} \tag{3.6}$$

where $Y_i^{(1)} = \ln X_i^{(1)}$ for $i = 1, \ldots, m$ and $Y_j^{(2)} = \ln X_j^{(2)}$ for $j = 1, \ldots, n$. Then $\overline{Y}^{(2)} - \overline{Y}^{(1)}$ has $N(\mu_2 - \mu_1, (m+n)\sigma^2/(mn))$ and S divided by σ^2 has $\chi^2(m+n-2)$. Moreover, the statistics $\overline{Y}^{(2)} - \overline{Y}^{(1)}$ and S are independent. These situations are similar to that of the estimation in a single lognormal distribution.

The UMVU estimator of (3.4) is

$$\begin{aligned} \hat{\gamma}_{a,b,c} &= \frac{\Gamma((m+n-2)/2)}{\Gamma((m+n-2)/2+c)} e^{a(\overline{Y}^{(2)} - \overline{Y}^{(1)})}\left(\frac{S}{2}\right)^c \\ &\times {}_0F_1\left(\frac{m+n-2}{2} + c; \frac{2bmn - a^2(m+n)}{4mn} S\right) \end{aligned}$$

and its variance is

$$\begin{aligned} \mathrm{Var}(\hat{\gamma}_{a,b,c}) = \gamma_{a,b,c}^2 &\left[\frac{\Gamma((m+n-2)/2)\Gamma((m+n-2)/2+2c)}{\Gamma^2((m+n-2)/2+c)}\right. \\ &\times \exp\left\{2\left(a^2\frac{m+n}{mn} - b\right)\sigma^2\right\} {}_2F_2\left(\frac{1}{2}(m+n-3+2c),\right. \\ &\frac{m+n-2}{2} + 2c; \frac{m+n-2}{2} + c, m+n-3+2c; \\ &\left.\left.\frac{2(2bmn - a^2(m+n))}{mn}\sigma^2\right) - 1\right] \end{aligned}$$

In particular, if $c = 0$, then the UMVU estimator of $\gamma_{a,b} = \exp\{a(\mu_2 - \mu_1) + b\sigma^2\}$ is

$$\hat{\gamma}_{a,b} = e^{a(\overline{Y}^{(2)} - \overline{Y}^{(1)})} {}_0F_1\left(\frac{m+n-2}{2}; \frac{2bmn - a^2(m+n)}{4mn} S\right) \tag{3.7}$$

and its variance is

$$\operatorname{Var}(\hat{\gamma}_{a,b}) = \gamma_{a,b}^2$$
$$\times \left\{ e^{\frac{a^2(m+n)}{mn}\sigma^2} {}_0F_1\left(\frac{m+n-2}{2}; \frac{\{2bmn - a^2(m+n)\}^2}{4m^2n^2}\sigma^4\right) - 1 \right\}$$

Second, let $X_1^{(1)}, X_2^{(1)}, \ldots, X_m^{(1)}$, $m \geq 2$, be iid random variables having $\Lambda(\mu_1, \sigma_1^2)$ and let $X_1^{(2)}, X_2^{(2)}, \ldots, X_n^{(2)}$, $n \geq 2$, be iid random variables having $\Lambda(\mu_2, \sigma_2^2)$. We assume that μ_1, μ_2, σ_1^2, and σ_2^2 are unknown. Put

$$\overline{Y}^{(1)} = \frac{1}{m}\sum_{i=1}^{m} Y_i^{(1)}, \overline{Y}^{(2)} = \frac{1}{n}\sum_{j=1}^{n} Y_j^{(2)}$$
$$S_1 = \sum_{i=1}^{m}\left(Y_i^{(1)} - \overline{Y}^{(1)}\right)^2, S_2 = \sum_{j=1}^{n}\left(Y_j^{(2)} - \overline{Y}^{(2)}\right)^2 \tag{3.8}$$

where $Y_i^{(1)} = \ln X_i^{(1)}$ for $i = 1, \ldots, m$ and $Y_j^{(2)} = \ln X_j^{(2)}$ for $j = 1, \ldots, n$. The statistics $\overline{Y}^{(1)}$, $\overline{Y}^{(2)}$, S_1, and S_2 are independent. Since

$$\gamma_{a,b,c,d,e,f} = \sigma_1^{2c}\exp(a\mu_1 + b\sigma_1^2) \times \sigma_2^{2f}\exp(d\mu_2 + e\sigma_2^2)$$

the situations reduce to those of Sections 2.3 and 2.4.

Next we shall treat improved estimation of a parameter

$$\gamma = \exp\left\{a(\mu_1 - \mu_2) + b(\sigma_1^2 - \sigma_2^2)\right\} \tag{3.9}$$

from independent random variables $X_1^{(1)}, \ldots, X_m^{(1)}$ having $\Lambda(\mu_1, \sigma_1^2)$ and $X_1^{(2)}, \ldots, X_n^{(2)}$ having $\Lambda(\mu_2, \sigma_2^2)$, where a and b are real constants. The aim is to derive an estimator of γ having smaller mean square error than the ML and the UMVU estimators. This was considered by Shaban (1981).

First assume that $\sigma_1^2 = \sigma_2^2 = \sigma^2$ and σ^2 is known. Consider the class of estimators of (3.9)

$$\hat{\gamma}_c = e^{a(\overline{Y}^{(1)} - \overline{Y}^{(2)})} f(\sigma^2) \tag{3.10}$$

where $\overline{Y}^{(1)}$ and $\overline{Y}^{(2)}$ are the means of the logarithms of the m $X^{(1)}$-values and n $X^{(2)}$-values, respectively.

The ML estimator

$$\hat{\gamma}_L = \exp\left\{a\left(\overline{Y}^{(1)} - \overline{Y}^{(2)}\right)\right\}$$

and the UMVU estimator

$$\hat{\gamma}_U = \exp\left\{a\left(\overline{Y}^{(1)} - \overline{Y}^{(2)}\right) - \frac{a^2\sigma^2}{2}\left(\frac{m+n}{mn}\right)\right\}$$

are both in the class (3.10).

The mean square error of (3.10) is

$$\begin{aligned} \mathrm{MSE}(\hat{\gamma}_c) = {} & f^2(\sigma^2)\exp\left\{2a(\mu_1 - \mu_2) + 2a^2\sigma^2\left(\frac{m+n}{mn}\right)\right\} \\ & - 2f(\sigma^2)\exp\left\{2a(\mu_1 - \mu_2) + \frac{a^2\sigma^2}{2}\left(\frac{m+n}{mn}\right)\right\} \\ & + \exp\{2a(\mu_1 - \mu_2)\} \end{aligned}$$

which is readily seen to be minimal when

$$f(\sigma^2) = \exp\left\{-\frac{3a^2\sigma^2}{2}\left(\frac{m+n}{mn}\right)\right\}$$

Hence, the minimum mean square error estimator in the class (3.10) is

$$\hat{\gamma}_M = \exp\left\{a\left(\overline{Y}^{(1)} - \overline{Y}^{(2)}\right) + k\sigma^2\right\} \tag{3.11}$$

where

$$k = -\frac{3}{2}a^2\left(\frac{m+n}{mn}\right)$$

We obtain

$$\text{MSE}(\hat{\gamma}_L) > \text{MSE}(\hat{\gamma}_U) > \text{MSE}(\hat{\gamma}_M).$$

Second assume that $\sigma_1^2 = \sigma_2^2 = \sigma^2$ and σ^2 is unknown. Consider the class of estimators of (3.9) given by

$$\hat{\gamma}_C = e^{a(\overline{Y}^{(1)} - \overline{Y}^{(2)})} f(S) \tag{3.12}$$

where S is defined by (3.6). Both the ML and the UMVU estimators of (3.9) are members of this class. The former is

$$\hat{\gamma} = \exp\left\{a\left(\overline{Y}^{(1)} - \overline{Y}^{(2)}\right)\right\}$$

and the latter is, according to (3.7),

$$\hat{\gamma}_U = \exp\left\{a\left(\overline{Y}^{(1)} - \overline{Y}^{(2)}\right)\right\} {}_0F_1\left(\frac{m+n-2}{2}; -\frac{a^2(m+n)}{4mn} S\right)$$

It is not possible to find a minimum mean square error estimator within the class (3.12). However, two methods for obtaining estimators are suggested.

The minimum mean square error estimator of σ^2 within the class of estimators of the form $S/(m+n+c)$ is $S/(m+n)$. Replacing σ^2 in (3.11) by $S/(m+n)$, we have an estimator of (3.9),

$$\hat{\gamma}_1 = \exp\left\{a\left(\overline{Y}^{(1)} - \overline{Y}^{(2)}\right) - \frac{3a^2}{2mn} S\right\} \tag{3.13}$$

When σ^2 is known, the mean square error of the estimator (3.11) with the indicated minimizing value of k is

$$\begin{aligned}\text{MSE}(\hat{\gamma}_M) = E &\left[\exp\left\{a\left(\overline{Y}^{(1)} - \overline{Y}^{(2)}\right) + k\sigma^2\right\}\right. \\ &\left. - \exp(k\sigma^2) E\left\{\exp\left(a\left(\overline{Y}^{(1)} - \overline{Y}^{(2)}\right)\right)\right\}\right]^2 \\ &+ \left[\exp(k\sigma^2) E\left\{\exp\left(a\left(\overline{Y}^{(1)} - \overline{Y}^{(2)}\right)\right)\right\} - \exp\left(a\left(\mu_1 - \mu_2\right)\right)\right]^2\end{aligned} \tag{3.14}$$

From the independence of $\overline{Y}^{(1)}$, $\overline{Y}^{(2)}$, and S, we write the mean square error of $\hat{\gamma}_C$ in the form

$$\begin{aligned}\mathrm{MSE}(\hat{\gamma}_C) = E\Big[&\exp\left(a\left(\overline{Y}^{(1)} - \overline{Y}^{(2)}\right)\right) f(S)\\ &- E\{f(S)\}\, E\left\{\exp\left(a\left(\overline{Y}^{(1)} - \overline{Y}^{(2)}\right)\right)\right\}\Big]^2\\ &+ \Big[E\{f(S)\}\, E\left\{\exp\left(a\left(\overline{Y}^{(1)} - \overline{Y}^{(2)}\right)\right)\right\}\\ &- \exp\left(a\left(\mu_1 - \mu_2\right)\right)\Big]^2 \qquad (3.15)\end{aligned}$$

Equating the respective first terms on the right hand side of (3.14) and (3.15) would lead to the requirement $f(S) = \exp(k\sigma^2)$ which is inadmissible. However, the respective second terms will be equal if we can find $f(S)$ such that

$$E\{f(S)\} = \exp(k\sigma^2)$$

According to (2.3), such an $f(S)$ is

$$f(S) = {}_0F_1\left(\frac{m+n-2}{2}; \frac{k}{2}S\right)$$

This gives the estimator

$$\hat{\gamma}_2 = e^{a(\overline{Y}^{(1)} - \overline{Y}^{(2)})}\, {}_0F_1\left(\frac{m+n-2}{2}; -\frac{3}{4}a^2\left(\frac{m+n}{mn}\right)S\right)$$

From (2.10) and (2.13), the mean square errors of the four estimators are

$$\mathrm{MSE}(\hat{\gamma}_L) = \gamma^2\left[\exp\left\{\frac{2a^2(m+n)\sigma^2}{mn}\right\} - 2\exp\left\{\frac{a^2(m+n)\sigma^2}{2mn}\right\} + 1\right]$$

$$\begin{aligned}\mathrm{MSE}(\hat{\gamma}_U) = \gamma^2\Big[&\exp\left\{\frac{a^2(m+n)\sigma^2}{mn}\right\}\\ &\times {}_0F_1\left(\frac{m+n-2}{2}; \frac{a^4(m+n)^2\sigma^4}{4m^2n^2}\right) - 1\Big]\end{aligned}$$

$$\mathrm{MSE}(\hat{\gamma}_1) = \gamma^2\Big[\exp\left\{\frac{2a^2(m+n)\sigma^2}{mn}\right\}\left(1 + \frac{6a^2\sigma^2}{mn}\right)^{-(m+n-2)/2}$$

$$- 2\exp\left\{\frac{a^2(m+n)\sigma^2}{2mn}\right\}\left(1+\frac{3a^2\sigma^2}{mn}\right)^{-(m+n-2)/2} + 1\Bigg]$$

$$\text{MSE}(\hat{\gamma}_2) = \gamma^2\left[\exp\left\{-\frac{a^2(m+n)\sigma^2}{mn}\right\} {}_0F_1\left(\frac{m+n-2}{2}; \frac{9a^2(m+n)^2\sigma^4}{4m^2n^2}\right)\right.$$
$$\left. - 2\exp\left\{-\frac{a^2(m+n)\sigma^2}{mn}\right\} + 1\right]$$

Numerical computations lead to the fact that except in very few cases $\hat{\gamma}_1$ is optimal in the case in which $n = m$ and $\sigma^2 \leq 3$. The estimator $\hat{\gamma}_1$ has the additional merits of being expressed in terms of the exponential function only.

Third assume that $\sigma_1^2 \neq \sigma_2^2$ and both are unknown. Consider the class of estimators of (3.9) given by

$$\hat{\gamma}_C = e^{a(\overline{Y}^{(1)} - \overline{Y}^{(2)})} f(S_1, S_2) \tag{3.16}$$

where S_1 and S_2 are defined by (3.8). Both the ML and the UMVU estimators of (3.9) are members of this class. The former is

$$\hat{\gamma}_L = \exp\left\{a\left(\overline{Y}^{(1)} - \overline{Y}^{(2)}\right) + \frac{b(nS_1 - mS_2)}{mn}\right\}$$

and the latter is

$$\hat{\gamma}_U = \exp\left\{a\left(\overline{Y}^{(1)} - \overline{Y}^{(2)}\right)\right\}$$
$$\times {}_0F_1\left(\frac{m-1}{2}; \frac{2bm-a^2}{4m}S_1\right) {}_0F_1\left(\frac{n-1}{2}; -\frac{2bn+a^2}{4n}S_2\right)$$

Like the case when $\sigma_1^2 = \sigma_2^2 = \sigma^2$ and σ^2 is unknown, it is not possible to find a minimal mean square error estimator within the class (3.16).

If we find $f(S_1, S_2)$ such that

$$E\{f(S_1, S_2)\} = E\{f(S_1)\}\, E\{f(S_2)\} = \exp(k_1\sigma_1^2)\exp(k_2\sigma_2^2)$$

where

$$k_1 = \frac{2bm - 3a^2}{2m} \qquad \text{and} \qquad k_2 = -\frac{2bn + 3a^2}{2n}$$

we have an estimator

$$\hat{\gamma} = e^{a(\overline{Y}^{(1)} - \overline{Y}^{(2)})} {}_0F_1\left(\frac{m-1}{2}; \frac{k_1}{2}S_1\right) {}_0F_1\left(\frac{n-1}{2}; \frac{k_2}{2}S_2\right)$$

From (2.10) and (2.13), the mean square errors of three estimators are

$$\begin{aligned}
\mathrm{MSE}(\hat{\gamma}_L) = \gamma^2 \Bigg[& \exp\left\{-2b(\sigma_1^2 - \sigma_2^2) + 2a^2(\sigma_1^2/m + \sigma_2^2/n)\right\} \\
& \times \left(1 - \frac{4b}{m}\sigma_1^2\right)^{-(m-1)/2} \left(1 + \frac{4b}{n}\sigma_2^2\right)^{-(n-1)/2} \\
& - 2\exp\left\{-b(\sigma_1^2 - \sigma_2^2) + a^2(\sigma_1^2/m + \sigma_2^2/n)/2\right\} \\
& \times \left(1 - \frac{2b}{m}\sigma_1^2\right)^{-(m-1)/2} \left(1 + \frac{2b}{n}\sigma_2^2\right)^{-(n-1)/2} + 1 \Bigg]
\end{aligned}$$

$$\begin{aligned}
\mathrm{MSE}(\hat{\gamma}_U) = \gamma^2 \Bigg[& \exp\left\{a^2(\sigma_1^2/m + \sigma_2^2/n)\right\} {}_0F_1\left(\frac{m-1}{2}; \frac{(2bm - a^2)^2}{4m^2}\sigma_1^4\right) \\
& \times {}_0F_1\left(\frac{n-1}{2}; \frac{(2bn + a^2)^2}{4n^2}\sigma_2^4\right) - 1 \Bigg]
\end{aligned}$$

$$\begin{aligned}
\mathrm{MSE}(\hat{\gamma}) = \gamma^2 \Bigg[& \exp\left\{-a^2(\sigma_1^2/m + \sigma_2^2/n)\right\} {}_0F_1\left(\frac{m-1}{2}; \frac{(2bm - 3a^2)^2}{4m^2}\sigma_1^4\right) \\
& \times {}_0F_1\left(\frac{n-1}{2}; \frac{(2bn + 3a^2)^2}{4n^2}\sigma_2^4\right) \\
& - 2\exp\left\{-a^2\left(\sigma_1^2/m + \sigma_2^2/n\right)\right\} + 1 \Bigg]
\end{aligned}$$

Numerical study leads to the fact that the suggested estimator $\hat{\gamma}$ has the smallest mean square error within the defined class when $D = |\sigma_1^2 - \sigma_2^2|$ ranges from 0.2 to 2.0 and $n = m$ from 10 to 100 and $(a, b) = (1, 1/2)$, $(1, 0)$, $(1, -1)$.

3.4 Multivariate Lognormal Distributions

Let $(X_1, X_2, \ldots, X_m)'$ be an m-dimensional lognormal distribution with parameters μ and Σ, $\Lambda_m(\mu, \Sigma)$. Then the moment of $(X_1, X_2, \ldots, X_m)'$

about zero is

$$E(X_1^{s_1} X_2^{s_2} \dots X_m^{s_m}) = \exp\left(s'\mu + \tfrac{1}{2}\operatorname{tr}\Sigma s s'\right)$$

where $s = (s_1, s_2, \dots, s_m)'$ is an arbitrary m-dimensional real vector. A generalized parametric function is

$$\theta_{\alpha,B,c} = |\Sigma|^c \exp(\alpha'\mu + \operatorname{tr} B\Sigma) \tag{3.17}$$

where α is an $m \times 1$ real vector, B is an $m \times m$ real symmetric matrix, and c is a real number. This form is a natural generalization of (2.1).

Let $(X_{i1}, X_{i2}, \dots, X_{im})'$, $i = 1, \dots, n$, $n > m$, be iid random vectors having $\Lambda_m(\mu, \Sigma)$, where we assume that μ and Σ are unknown. And let

$$\overline{Y} = (\overline{Y}_{\cdot 1}, \overline{Y}_{\cdot 2}, \dots, \overline{Y}_{\cdot m})'$$

$$T = \{T_{jk}\} = \left\{\sum_{i=1}^{n}(Y_{ij} - \overline{Y}_{\cdot j})(Y_{ik} - \overline{Y}_{\cdot k})\right\}, \qquad j, k = 1, \dots, m$$

where $Y_{ij} = \ln X_{ij}$ and $\overline{Y}_{\cdot j} = n^{-1}\sum_{i=1}^{n} Y_{ij}$. A pair of statistics $(\overline{Y}, T)$ is complete and sufficient for $N_m(\mu, \Sigma)$. Moreover, $\overline{Y}$ and T are mutually independent and distributed as $N_m(\mu, \Sigma/n)$ and the Wishart distribution with parameters Σ and $n - 1$, denoted by $W_m(\Sigma, n - 1)$.

To find the UMVU estimator of (3.17), Iwase, Shimizu, and Suzuki (1982) gave a useful expression. Let T have $W_m(\Sigma, \nu)$, G be an $m \times m$ complex symmetric matrix, ς be a complex number, and b be a real number such that $\nu/2 + b - (m + 1)/2 + 1 > 0$. Then we have

$$E\left[|T|^b {}_0F_1\left(\frac{\varsigma}{2}; \frac{TG}{2}\right)\right] = \frac{|\Sigma^{-1}/2|^{-b}\Gamma_m(\nu/2 + b)}{\Gamma_m(\nu/2)} {}_1F_1\left(\frac{\nu}{2} + b; \frac{\varsigma}{2}; \Sigma G\right) \tag{3.18}$$

where

$$\Gamma_m(\gamma) = \pi^{m(m-1)/4}\Gamma(\gamma)\Gamma\left(\gamma - \frac{1}{2}\right)\cdots\Gamma\left(\gamma - \frac{m-1}{2}\right)$$

As seen in Muirhead (1982), p. 258, the hypergeometric functions of matrix argument are given by

$${}_pF_q(a_1, \dots, a_p; b_1, \dots, b_q; M) = \sum_{k=0}^{\infty}\sum_{\kappa} \frac{(a_1)_\kappa \cdots (a_p)_\kappa}{(b_1)_\kappa \cdots (b_q)_\kappa} \frac{C_\kappa(M)}{k!}$$

where $\sum_\kappa$ denotes summation over all partitions $\kappa = (k_1, \ldots, k_m)$, $k_1 \geq \cdots \geq k_m \geq 0$ of κ, $C_\kappa(M)$ is the zonal polynomial of M corresponding to κ and the generalized hypergeometric coefficient $(a)_\kappa$ is given by

$$(a)_\kappa = \prod_{i=1}^{m} \left(a - \tfrac{1}{2}(i-1)\right)_{k_i}$$

where

$$(a)_k = a(a+1)\cdots(a+k-1) \qquad (a)_0 = 1$$

Assume that $n/2 + c - m/2 > 0$. From (3.18) the UMVU estimator of (3.17) is given by

$$\hat{\theta}_{\alpha,B,c} = \frac{\Gamma_m((n-1)/2)}{2^{mc}\Gamma_m((n-1)/2+c)} e^{\alpha'\overline{Y}} |T|^c {}_0F_1\left(\frac{n-1}{2}+c; \left(B - \frac{\alpha\alpha'}{2n}\right)\frac{T}{2}\right) \tag{3.19}$$

We shall give two examples. First, the jth element of the mode of $\Lambda_m(\mu, \Sigma)$ is

$$\theta_{e_j,-\Phi_j,0} = \exp(e_j'\mu - \operatorname{tr}\Sigma\Phi_j),$$

where

$$e_j = (0,\ldots,0,\underset{(j)}{1},0,\ldots,0)', \Phi_j = \begin{bmatrix} & & & \overset{(j)}{1/2} & & & \\ & 0 & & \vdots & & 0 & \\ & & & 1/2 & & & \\ 1/2 & \cdots & 1/2 & 1 & 1/2 & \cdots & 1/2 \\ & & & 1/2 & & & \\ & 0 & & \vdots & & 0 & \\ & & & 1/2 & & & \end{bmatrix} \quad (j)$$

The UMVU estimator of $\theta_{e_j,-\Phi_j,0}$ is

$$\hat{\theta}_{e_j,-\Phi_j,0} = e^{\overline{Y}_{\cdot j}} {}_0F_1\left(\frac{n-1}{2}; -\frac{1}{4}\left(2\phi_j + \frac{1}{n}e_je_j'\right)T\right)$$

Second, we are concerned with

$$\theta(s_1, \ldots, s_m; r) = \left\{ E\left(\prod_{i=1}^{m} X_i^{s_i}\right)\right\}^r$$
$$= \exp\left(rs'\mu + \tfrac{1}{2}\operatorname{tr}\Sigma rss'\right)$$

where $s = (s_1, \ldots, s_m)'$ is an m-dimensional real vector and r is a real number. For $n > m$, the UMVU estimator of $\theta(s_1, \ldots, s_m; r)$ is

$$\hat{\theta}(s_1, \ldots, s_m; r) = \exp(rs'\overline{Y})\,{}_0F_1\left(\frac{n-1}{2}; \frac{r(n-r)s'Ts}{4n}\right)$$

and its variance is

$$\operatorname{Var}\left\{\hat{\theta}(s_1, \ldots, s_m; r)\right\} = \exp(2rs'\mu + rs'\Sigma s)\left[\exp\left(\frac{r^2 s'\Sigma s}{n}\right)\right.$$
$$\left.\times\ {}_0F_1\left(\frac{n-1}{2}; \left(\frac{r(n-r)s'\Sigma s}{2n}\right)^2\right) - 1\right]$$

Suzuki (1983a) treated the case when $c = 0$ in (3.17). The variance of (3.19) when $c = 0$ is given by

$$\operatorname{Var}(\hat{\theta}_{\alpha,B,0}) = \exp(2\alpha'\mu)\left\{\exp\left(\frac{2}{n}\alpha'\Sigma\alpha\right)\sum_{f=0}^{\infty}\sum_{\phi}\xi_{\phi}\left(\frac{n-1}{2}\right)\right.$$
$$\left.\times\ \frac{C_{\phi}(2G\Sigma)}{f!} - \exp(\operatorname{tr}(2B\Sigma))\right\} \tag{3.20}$$

where

$$G = B - \frac{\alpha\alpha'}{2n} \qquad \xi_{\phi}(c) = \sum_{k+l=f}\sum_{\kappa}\sum_{\lambda}\frac{(c)_{\phi}}{2^f (c)_{\kappa}(c)_{\lambda}}\binom{f}{k} g_{\kappa,\lambda}^{\phi}$$

and $g^{\phi}_{\kappa,\lambda}$ is the coefficient given in Khatri and Pillai (1968). An asymptotic expansion of (3.20) is

$$\begin{aligned}\mathrm{Var}(\hat{\theta}_{\alpha,B,0}) &= \exp(2\alpha'\mu + 2\,\mathrm{tr}\,B\Sigma)\left[\frac{1}{n}\left\{\alpha'\Sigma\alpha + 2tr(B\Sigma)^2\right\}\right.\\ &+ \frac{1}{n^2}\left\{\frac{1}{2}(\alpha'\Sigma\alpha)^2 + 2\alpha'\Sigma\alpha\,\mathrm{tr}(B\Sigma)^2 - 2\alpha'\Sigma B\Sigma\alpha\right.\\ &\left.\left.+ 2(\mathrm{tr}(B\Sigma)^2)^2 + 2tr(B\Sigma)^2\right\}\right] + O(n^{-3})\end{aligned}$$

The ML estimator of $\theta_{\alpha,B,0}$ is

$$\tilde{\theta}_{\alpha,B,0} = \exp\left(\alpha'\overline{Y} + \frac{1}{n}trBT\right)$$

and its (asymptotic) mean square error is

$$\begin{aligned}\mathrm{MSE}(\tilde{\theta}_{\alpha,B,0}) &= \exp(2\alpha'\mu)\left\{\exp\left(\frac{2}{n}\alpha'\Sigma\alpha\right)\left|I_m - \frac{4}{n}B\Sigma\right|^{-(n-1)/2}\right.\\ &- 2\exp\left(\mathrm{tr}\,B\Sigma + \frac{1}{2n}\alpha'\Sigma\alpha\right)\left|I_m - \frac{2}{n}B\Sigma\right|^{-(n-1)/2}\\ &\left.+\ \exp(\mathrm{tr}(2B\Sigma))\right\}\\ &= \exp(2\alpha'\mu + 2\,\mathrm{tr}\,B\Sigma)\left[\frac{1}{n}\left\{\alpha'\Sigma\alpha + 2\,\mathrm{tr}(B\Sigma)^2\right\}\right.\\ &+ \frac{1}{n^2}\left\{\frac{7}{4}(\alpha'\Sigma\alpha)^2 + 7\alpha'\Sigma\alpha\,\mathrm{tr}(B\Sigma)^2\right.\\ &- 3\alpha'\Sigma\alpha\,\mathrm{tr}\,B\Sigma + 7(\mathrm{tr}(B\Sigma)^2)^2\\ &+ 8\,\mathrm{tr}(B\Sigma)^3 - 6\,\mathrm{tr}(B\Sigma)^2\,\mathrm{tr}\,B\Sigma - 2\,\mathrm{tr}(B\Sigma)^2\\ &\left.\left.+\ (\mathrm{tr}\,B\Sigma)^2\right\}\right] + O(n^{-3})\end{aligned}$$

where I_m is an $m \times m$ identity matrix and $|\cdot|$ means the determinant.

Moment estimators of some parametric functions of the bivariate log-normal distribution and their large-sample variances were given by Suzuki

(1983c). UMVU estimation of regression equations for the bivariate lognormal distribution was treated by Mostafa and Mahmoud (1964). It is possible to rewrite their results by use of the results of Sections 2.3 and 2.4. The UMVU estimator for analytic functions of the mean of multivariate normal distribution with known covariance matrix was obtained by Watkins and Kern (1973).

3.5 Multivariate Delta-Lognormal Distributions

This is a generalization of univariate delta-lognormal distribution and is treated by Iwase, Shimizu, and Suzuki (1982). A random vector $(X_1^*, X_2^*, \ldots, X_m^*)'$ is said to have an m-dimensional delta-lognormal distribution with parameters δ, μ, and Σ, denoted by $\Delta_m(\delta, \mu, \Sigma)$, if

$$\Pr\{X_i^* = 0 \text{ for all } i\} = \delta$$

$$\Pr\{X_1^* \leq x_1, \ldots, X_m^* \leq x_m\} = \delta + (1-\delta)\Lambda_m(x \mid \mu, \Sigma) \qquad x_i > 0$$

$$\Pr\{\text{elsewhere}\} = 0$$

where $x = (x_1, \ldots, x_m)'$ and $\Lambda_m(x \mid \mu, \Sigma)$ is the distribution function of $\Lambda_m(\mu, \Sigma)$.

Let $(X_{i1}^*, X_{i2}^*, \ldots, X_{im}^*)'$, $i = 1, \ldots, n$, $n > m$, be iid random vectors having $\Delta_m(\delta, \mu, \Sigma)$, where we assume that δ, μ, and Σ are unknown. Let n_1 be the number of samples whose components are non-zero. Without loss of generality, we assume $X_{ij}^* \neq 0$ for $(X_{i1}^*, \ldots, X_{im}^*)'$, $i = 1, \ldots, n_1$. And let

$$\overline{Y}^* = (\overline{Y}_{\cdot 1}^*, \overline{Y}_{\cdot 2}^*, \ldots, \overline{Y}_{\cdot m}^*)'$$

$$T^* = \left\{ \sum_{i=1}^{n_1} (Y_{ij}^* - \overline{Y}_{\cdot j}^*)(Y_{ik}^* - \overline{Y}_{\cdot k}^*) \right\} \qquad j, k = 1, \ldots, m$$

where

$$Y_{ij}^* = \ln X_{ij}^* \text{ for } X_{ij}^* \neq 0$$

and

$$\overline{Y}_{\cdot j}^* = \frac{1}{n_1} \sum_{i=1}^{n_1} Y_{ij}^* \qquad n_1 > m$$

and if $n_1 \leq m$, then $T^* = 0$.

We shall find the UMVU estimator of a parametric function

$$\eta(s_1,\ldots,s_m;r;c) = (1-\delta)^c \exp\left(rs'\mu + \tfrac{1}{2}rs'\Sigma s\right) \tag{3.21}$$

where r is a real number, $s = (s_1,\ldots,s_m)'$ is an m-dimensional real vector, and c is a positive integer.

The UMVU estimator of (3.21) is

$$\hat{\eta}(s_1,\ldots,s_m;r;c) = \begin{cases} \dfrac{n_{1(c)}}{n_{(c)}} e^{rs'\overline{Y}^*} \\ \quad \times\, {}_0F_1\left(\dfrac{n_1-1}{2}; \dfrac{1}{4n_1} r(n_1-r)s'T^*s\right) & n_1 \geq c \\ 0 & n_1 < c \end{cases}$$

and its variance is

$$\begin{aligned} &\mathrm{Var}\,(\hat{\eta}(s_1,\ldots,s_m;r;c)) \\ &\quad = \exp(2rs'\mu + rs'\Sigma s)\left[\frac{1}{n_{(c)}^2}\sum_{i=c}^{n}\binom{n}{i}(1-\delta)^i\delta^{n-i}i_{(c)}^2\right. \\ &\qquad \left.\times \exp\left(\frac{r^2 s'\Sigma s}{i}\right){}_0F_1\left(\frac{i-1}{2}; \left(\frac{r(i-r)s'\Sigma s}{2i}\right)^2\right) - (1-\delta)^{2c}\right] \end{aligned}$$

3.6 Normal-Lognormal (Semi-Lognormal) Distributions

A random vector $(Y,X)'$ is said to have a normal-lognormal (semi-lognormal) distribution with parameters μ and Σ, denoted by $S_{\mathrm{NL}}(\mu,\Sigma)$, if $(Y, \ln X)'$ has a bivariate normal distribution $N_2(\mu,\Sigma)$ with mean vector $\mu = (\mu_1,\mu_2)'$ and covariance matrix

$$\Sigma = \begin{bmatrix} \sigma_1^2 & \sigma_{12} \\ \sigma_{12} & \sigma_2^2 \end{bmatrix}$$

The cross moment of Y and X about zero is

$$E(YX) = (\mu_1 + \sigma_{12})\exp(\mu_2 + \frac{\sigma_2^2}{2})$$

and the covariance between Y and X is

$$\operatorname{Cov}(Y, X) = \sigma_{12} \exp\left(\mu_2 + \frac{\sigma_2^2}{2}\right)$$

We shall find the UMVU estimators of more general parametric functions

$$\lambda_{\alpha,\beta} = \alpha'\Sigma\beta \exp\left(\beta'\mu + \tfrac{1}{2}\beta'\Sigma\beta\right)$$

and

$$\lambda'_{\alpha,\beta} = (\alpha'\mu + \alpha'\Sigma\beta) \exp\left(\beta'\mu + \tfrac{1}{2}\beta'\Sigma\beta\right)$$

where α and β are two-dimensional real vectors. These forms appeared in Suzuki, Iwase, and Shimizu (1984).

Let $(Y_i, X_i)'$, $i = 1, 2, \ldots, n$, $n \geq 3$, be iid random vectors having $S_{\mathrm{NL}}(\mu, \Sigma)$, where we assume that μ and Σ are unknown. Let

$$\overline{Y} = \frac{1}{n}\sum_{i=1}^{n} Y_i, \quad \overline{\ln X} = \frac{1}{n}\sum_{i=1}^{n} \ln X_i, \quad T = \begin{bmatrix} a_{11} & a_{12} \\ a_{21} & a_{22} \end{bmatrix}$$

where

$$a_{11} = \sum_{i=1}^{n}(Y_i - \overline{Y})^2 \qquad a_{22} = \sum_{i=1}^{n}(\ln X_i - \overline{\ln X})^2$$

$$a_{12} = a_{21} = \sum_{i=1}^{n}(Y_i - \overline{Y})(\ln X_i - \overline{\ln X})$$

The statistics $\overline{Z} = (\overline{Y}, \overline{\ln X})'$ and T are mutually independent and distributed as $N_2(\mu, \Sigma/n)$ and as $W_2(\Sigma, n-1)$.

The UMVU estimators of $\lambda_{\alpha,\beta}$ and $\lambda'_{\alpha,\beta}$ are

$$\hat{\lambda}_{\alpha,\beta} = \frac{1}{n-1}\alpha' T\beta \exp(\beta'\overline{Z})\,{}_0F_1\left(\frac{n+1}{2}; \frac{n-1}{4n}\beta' T\beta\right)$$

$$\hat{\lambda}'_{\alpha,\beta} = \exp(\beta'\overline{Z})\left\{\alpha'\overline{Z}\,{}_0F_1\left(\frac{n-1}{2}; \frac{n-1}{4n}\beta' T\beta\right)\right.$$

$$\left. + \frac{1}{n}\alpha' T\beta\,{}_0F_1\left(\frac{n+1}{2}; \frac{n-1}{4n}\beta' T\beta\right)\right\}$$

In particular, the UMVU estimators of $\lambda = \mathrm{Cov}(Y, X)$ and $\lambda' = E(Y, X)$ are

$$\hat{\lambda} = \frac{1}{n-1} a_{12} \exp(\overline{\ln X})\, {}_0F_1\left(\frac{n+1}{2}; \frac{n-1}{4n} a_{22}\right)$$

$$\hat{\lambda}' = \exp(\overline{\ln X}) \left\{ \overline{Y}\, {}_0F_1\left(\frac{n-1}{2}; \frac{n-1}{4n} a_{22}\right) + \frac{1}{n} a_{12}\, {}_0F_1\left(\frac{n+1}{2}; \frac{n-1}{4n} a_{22}\right) \right\}$$

The variances of $\hat{\lambda}$ and $\hat{\lambda}'$ are

$$\begin{aligned} \mathrm{Var}(\hat{\lambda}) = {} & \frac{1}{n-1} \exp\left(2\mu_2 + \frac{2}{n}\sigma_2^2\right) \left\{ (n\sigma_{12}^2 + \sigma_1^2\sigma_2^2) \exp\left(\frac{n-1}{n}\sigma_2^2\right) \right. \\ & \times {}_0F_1\left(\frac{n+1}{2}; \frac{(n-1)^2}{4n^2}\sigma_2^4\right) + \frac{2(n-1)}{n}\sigma_{12}^2\sigma_2^2 \\ & \left. \times\, {}_1F_1\left(\frac{n+2}{2}; n+1; \frac{2(n-1)}{n}\sigma_2^2\right) \right\} - \sigma_{12}^2 \exp(2\mu_2 + \sigma_2^2) \end{aligned}$$

$$\begin{aligned} \mathrm{Var}(\hat{\lambda}') = {} & \exp\left(2\mu_2 + \frac{2}{n}\sigma_2^2\right) \left[\left\{ \frac{1}{n}\sigma_1^2 + \left(\mu_1 + \frac{2}{n}\sigma_{12}\right)^2 \right\} \right. \\ & \times \exp\left(\frac{n-1}{n}\sigma_2^2\right) {}_0F_1\left(\frac{n-1}{2}; \frac{(n-1)^2}{4n^2}\sigma_2^4\right) + \frac{n-1}{n^2} \\ & \times \left\{ (n\sigma_{12}^2 + \sigma_1^2\sigma_2^2) \exp\left(\frac{n-1}{n}\sigma_2^2\right) \right. \\ & \times {}_0F_1\left(\frac{n+1}{2}; \frac{(n-1)^2}{4n^2}\sigma_2^4\right) + \frac{2(n-1)}{n}\sigma_{12}^2\sigma_2^2 \\ & \left. \times\, {}_1F_1\left(\frac{n+2}{2}; n+1; \frac{2(n-1)}{n}\sigma_2^2\right) \right\} + \frac{2(n-1)}{n}\sigma_{12} \\ & \left. \times \left(\mu_1 + \frac{2}{n}\sigma_{12}\right) {}_1F_1\left(\frac{n}{2}; n-1; \frac{2(n-1)}{n}\sigma_2^2\right) \right] \\ & - (\mu_1 + \sigma_{12})^2 \exp(2\mu_2 + \sigma_2^2) \end{aligned}$$

Asymptotic expansions of the variances of $\hat{\lambda}$ and $\hat{\lambda}'$ in powers to n^{-1} are

$$\mathrm{Var}(\hat{\lambda}) = \frac{1}{n}\exp(2\mu_2+\sigma_2^2)\left\{\sigma_{12}^2\left(\tfrac{1}{2}\sigma_2^4+3\sigma_2^2+1\right)+\sigma_1^2\sigma_2^2\right\}+O(n^{-2})$$

$$\begin{aligned}\mathrm{Var}(\hat{\lambda}') = \frac{1}{n}\exp(2\mu_2+\sigma_2^2)&\left\{\mu_1^2\sigma_2^2\left(1+\tfrac{1}{2}\sigma_2^2\right)+\mu_1\sigma_{12}(\sigma_2^4+4\sigma_2^2+2)\right.\\ &\left.+\sigma_1^2(1+\sigma_2^2)+\sigma_{12}^2\left(\tfrac{1}{2}\sigma_2^4+3\sigma_2^2+3\right)\right\}+O(n^{-2})\end{aligned}$$

Crofts and Owen (1972) considered the following setting:

(1) The marginal distribution of Y is normal with mean μ_1 and variance σ_1^2.

(2) The conditional distribution of X given Y is lognormal with parameters $\mu_2(y)$, σ_2^2, and τ. That is

$$f_{2\cdot 1}(x,y) = \frac{1}{\sqrt{2\pi}\sigma_2(x-\tau)}\exp\left[-\frac{1}{2}\left\{\frac{\ln(x-\tau)-\mu_2}{\sigma_2}\right\}^2\right] \qquad x>\tau$$

(3) $E(X \mid Y=y) = \exp(\gamma y+\beta)+\tau$

From these three assumptions the bivariate density function of Y and X is

$$\begin{aligned}f(y,x) &= \frac{1}{2\pi\sigma_1\sigma_2(x-\tau)}\\ &\quad\times\exp\left[-\frac{1}{2}\left\{\frac{(y-\mu_1)^2}{\sigma_1^2}+\frac{(\ln(x-\tau)-(\gamma y+\beta-\sigma_2^2/2))^2}{\sigma_2^2}\right\}\right]\\ &\qquad -\infty<y<\infty \qquad x>\tau\end{aligned}$$

Thus, the variables Y and $Z=\ln X$ have a bivariate normal distribution with $E(Y)=\mu_1$, $E(Z)=\gamma\mu_1+\beta-\sigma_2^2/2$, and covariance matrix

$$\Sigma = \begin{bmatrix}\sigma_1^2 & \gamma\sigma_1^2\\ \gamma\sigma_1^2 & \gamma^2\sigma_1^2+\sigma_2^2\end{bmatrix}$$

Direct evaluation leads to

$$\mathrm{Cov}(Y,X) = \gamma\sigma_1^2\exp(\gamma\mu_1+\beta+\tfrac{1}{2}\gamma^2\sigma_1^2)$$

The ratio of the correlation coefficient $\rho_{Y,X}$ to $\rho_{Y,Z}$ is

$$\frac{\rho_{Y,X}}{\rho_{Y,Z}} = \sqrt{\frac{\gamma^2\sigma_1^2 + \sigma_2^2}{\exp(\gamma^2\sigma_1^2 + \sigma_2^2) - 1}}$$

This ratio has a property that

$$|\rho_{Y,X}| \leq |\rho_{Y,Z}|$$

with equality only for $\gamma = 0$, in which case both correlations are zero. Moreover, for large values of $\gamma^2\sigma_1^2 + \sigma_2^2$, high correlation between Y and Z does not imply high correlation between Y and X. Further, if the quantities σ_1^2, σ_2^2, and $\gamma^2\sigma_1^2 + \sigma_2^2$ are small, then

$$\frac{\rho_{Y,X}}{\rho_{Y,Z}} \to 1$$

We shall treat estimation of γ and β from independent observations $(y_i, x_i)'$, $i = 1, \ldots, n$, $n > 3$, when τ and σ_2^2 are known. ML estimators of γ and β are

$$\hat{\gamma} = \frac{\sum_{i=1}^n (y_i - \bar{y}) \ln(x_i - \tau)}{\sum_{i=1}^n (y_i - \bar{y})^2} \qquad \hat{\beta} = \frac{1}{n}\sum_{i=1}^n \ln(x_i - \tau) + \frac{\sigma_2^2}{2} - \hat{\gamma}\bar{y}$$

These estimators are unbiased with covariance matrix

$$\Sigma_{\hat{\gamma},\hat{\beta}} = \begin{bmatrix} \dfrac{\sigma_2^2}{(n-3)\sigma_1^2} & -\dfrac{\sigma_2^2\mu_1}{(n-3)\sigma_1^2} \\ -\dfrac{\sigma_2^2\mu_1}{(n-3)\sigma_1^2} & \dfrac{\sigma_2^2}{n} + \dfrac{\sigma_2^2}{(n-3)\sigma_1^2}\left(\dfrac{\sigma_1^2}{n} + \mu_1^2\right) \end{bmatrix}$$

To investigate the possibility of improved estimation of γ and β, construct the unbiased estimators, μ_1 and σ_1^2 also being assumed known,

$$\tilde{\gamma} = \frac{\sum_{i=1}^n (y_i - \mu_1) \ln(x_i - \tau)}{n\sigma_1^2} \qquad \tilde{\beta} = \frac{1}{n}\sum_{i=1}^n \ln(x_i - \tau) + \frac{\sigma_2^2}{2} - \tilde{\gamma}\mu_1$$

Computation of the efficiency of $\tilde{\gamma}$ relative to $\hat{\gamma}$ shows $\tilde{\gamma}$ to be a more efficient estimator when

$$2\gamma^2\sigma_1^2 + (\gamma\mu_1 + \beta - \frac{\sigma_2^2}{2})^2 < \frac{3\sigma_2^2}{n-3}$$

Data were generated for various values for n for which the above inequality held and little advantage was seen for $\tilde{\gamma}$; also, relative to the Cramér-Rao lower bound, $\hat{\gamma}$ is asymptotically efficient while $\tilde{\gamma}$ is always inefficient unless $\gamma = 0$ and $\beta = \sigma_2^2/2$. All of these facts lead to the choice of $\hat{\gamma}$ as a generally better estimator than $\tilde{\gamma}$. Similar arguments show that, in general, $\hat{\beta}$ is better than $\tilde{\beta}$. The more general case in which all six parameters are to be estimated was also considered by Crofts and Owen (1972).

Moment estimators of some parametric functions of the distribution and their large-sample variances were given by Suzuki (1983c).

3.7 Transformations of Univariate Normal Variables

A function f is said to be of recursive type (Neyman and Scott, 1960) if it satisfies the second-order differential equation $f^{(2)} = A + Bf$, where A and B are arbitrary real constants. Clearly, x^2, e^x, $\sin^2 x$, and $\sinh^2 x$ are recursive-type functions.

Let Y be a random variable having a normal distribution with mean μ and variance σ^2, $N(\mu, \sigma^2)$, and let $X = f(Y)$. The mean of X is

$$\theta = E(X) = \begin{cases} f(\mu) + \dfrac{A}{2}\sigma^2 & B = 0 \\ \left\{f(\mu) + \dfrac{A}{B}\right\} e^{B\sigma^2/2} - \dfrac{A}{B} & B \neq 0 \end{cases}$$

and the variance of X is

$$\phi^2 = \mathrm{Var}(X) = \begin{cases} \sigma^2 \left[\left\{f^{(1)}(\mu)\right\}^2 + \dfrac{A^2}{2}\sigma^2\right] & B = 0 \\ \dfrac{1}{2}\left[\left(e^{B\sigma^2} - 1\right)^2 \left\{f(\mu) + \dfrac{A}{B}\right\}^2 \right. & \\ \qquad \left. + \dfrac{1}{B}\left(e^{2B\sigma^2} - 1\right)\left\{f^{(1)}(\mu)\right\}^2\right] & B \neq 0 \end{cases}$$

where $f^{(1)}(\cdot)$ denotes the first order derivative of f.

Let $Y_1, Y_2, \ldots, Y_n$, $n \geq 2$, be iid random variables having $N(\mu, \sigma^2)$, where we assume that μ and σ^2 are unknown. Let $\overline{Y} = n^{-1}\sum_{i=1}^{n} Y_i$ and $S = \sum_{i=1}^{n}(Y_i - \overline{Y})^2$. We shall find the UMVU estimators of θ and ϕ^2 and evaluate their variances. Details are seen in Shimizu (1983a). If we adopt the function $f(x) = e^x (A = 0, B = 1)$, we have the UMVU estimation of the mean and the variance in a lognormal distribution.

The UMVU estimator of θ is

$$\hat{\theta} = \begin{cases} f(\overline{Y}) + \dfrac{A}{2n} S & B = 0 \\ \left\{ f(\overline{Y}) + \dfrac{A}{B} \right\} {}_0F_1\left(\dfrac{n-1}{2}; \dfrac{B}{4}\left(1 - \dfrac{1}{n}\right) S \right) - \dfrac{A}{B} & B \neq 0 \end{cases}$$

which is given by Neyman and Scott (1960) in a different form.

The variance of $\hat{\theta}$ is

$$\operatorname{Var}(\hat{\theta}) = \begin{cases} \dfrac{1}{n}\sigma^2 \left[\left\{ f^{(1)}(\mu) \right\}^2 + \dfrac{A^2}{2}\sigma^2 \right] & B = 0 \\ \dfrac{1}{2}\left[e^{2B\sigma^2/n} \left[\left\{ f(\mu) + \dfrac{A}{B} \right\}^2 + \dfrac{1}{B}\left\{ f^{(1)}(\mu) \right\}^2 \right] \right. \\ \quad \left. + \left\{ f(\mu) + \dfrac{A}{B} \right\}^2 - \dfrac{1}{B}\left\{ f^{(1)}(\mu) \right\}^2 \right] e^{B(1-1/n)\sigma^2} \\ \quad \times {}_0F_1\left(\dfrac{n-1}{2}; \dfrac{B^2}{4}\left(1 - \dfrac{1}{n}\right)^2 \sigma^4 \right) \\ \quad - \left\{ f(\mu) + \dfrac{A}{B} \right\}^2 e^{B\sigma^2} & B \neq 0 \end{cases}$$

The expansion of the variance of $\hat{\theta}$ for $B \neq 0$ in powers to n^{-1} is

$$\operatorname{Var}(\hat{\theta}) = \frac{1}{n}\sigma^2 e^{B\sigma^2} \left[\frac{B^2}{2}\sigma^2 \left\{ f(\mu) + \frac{A}{B} \right\}^2 + \left\{ f^{(1)}(\mu) \right\}^2 \right] + O(n^{-2})$$

The UMVU estimator of ϕ^2 was first obtained by Hoyle (1968), but his expression is complicated. An improved form of the UMVU estimator of ϕ^2 is

$$\hat{\phi}^2 = \begin{cases} \dfrac{1}{n-1}\left[S\left\{f^{(1)}(\overline{Y})\right\}^2 + \dfrac{(n-2)A^2}{2n(n+1)}S^2\right] & B = 0 \\ \dfrac{1}{2}\left[\left\{\Xi\left(4B\left(1-\dfrac{1}{n}\right)\right) - \Xi\left(2B\left(1-\dfrac{2}{n}\right)\right)\right.\right. & \\ \quad \left. - \Xi(2B) + 1\right\}\left\{f(\overline{Y}) + \dfrac{A}{B}\right\}^2 & \\ \quad + \dfrac{1}{B}\left\{\Xi\left(4B\left(1-\dfrac{1}{n}\right)\right) - \Xi\left(2B\left(1-\dfrac{2}{n}\right)\right)\right. & \\ \quad \left.\left. + \Xi(2B) - 1\right\}\left\{f^{(1)}(\overline{Y})\right\}^2\right] & B \neq 0 \end{cases}$$

where

$$\Xi(x) = {}_0F_1\left(\frac{n-1}{2}; \frac{x}{4}S\right)$$

The variance of $\hat{\phi}^2$ is

$$\begin{aligned} \operatorname{Var}(\hat{\phi}^2) &= \frac{2}{n-1}\sigma^4\left[\left\{f^{(1)}(\mu)\right\}^4 + 4A^2\sigma^2\left\{f^{(1)}(\mu)\right\}^2\right. \\ &\quad \left. + \frac{n^3+n^2-3n+6}{n^2(n+1)}A^4\sigma^4\right] \qquad B = 0 \\ \operatorname{Var}(\hat{\phi}^2) &= \frac{1}{8}\left[\left\{\lambda_1(e^{4B\sigma^2/n} + e^{-4B\sigma^2/n}) + 2\lambda_2 + 4\lambda_3\right.\right. \\ &\quad \left. - 2(e^{B\sigma^2}-1)^4\right\}\left\{f(\mu) + \frac{A}{B}\right\}^4 + \frac{2}{B}\left\{\lambda_1\left(3e^{4B\sigma^2/n}\right.\right. \\ &\quad \left.\left. - e^{-4B\sigma^2/n}\right) - 2\lambda_2 - 2(e^{B\sigma^2}-1)^2(e^{2B\sigma^2}-1)\right\} \\ &\quad \times\left\{f(\mu) + \frac{A}{B}\right\}^2\left\{f^{(1)}(\mu)\right\}^2 + \frac{1}{B^2}\left\{\lambda_1(e^{4B\sigma^2/n}\right. \\ &\quad \left. + e^{-4B\sigma^2/n}) + 2\lambda_2 - 4\lambda_3 - 2(e^{2B\sigma^2}-1)^2\right\} \end{aligned}$$

$$\times \left\{ f^{(1)}(\mu) \right\}^4 \Bigg] \qquad B \neq 0$$

where

$$\lambda_1 = e^{4B\sigma^2}\Psi\left(16B^2\left(1-\frac{1}{n}\right)^2\right) + e^{2B\sigma^2}\Psi\left(4B^2\left(1-\frac{2}{n}\right)^2\right)$$

$$- 2e^{3B\sigma^2}\Psi\left(8B^2\left(1-\frac{1}{n}\right)\left(1-\frac{2}{n}\right)\right)$$

$$\lambda_2 = e^{2B\sigma^2}\Psi(4B^2) - 2e^{B\sigma^2} + 1$$

$$\lambda_3 = e^{2B\sigma^2}\Psi\left(4B^2\left(1-\frac{2}{n}\right)\right) - e^{3B\sigma^2}\Psi\left(8B^2\left(1-\frac{1}{n}\right)\right)$$

$$+ e^{2B\sigma^2} - e^{B\sigma^2}$$

$$\Psi(x) = {}_0F_1\left(\frac{n-1}{2}; \frac{x}{4}\sigma^4\right)$$

The expansion of the variance of $\hat{\phi}^2$ for $B \neq 0$ in powers to n^{-1} is

$$\mathrm{Var}(\hat{\phi}^2) = \frac{2}{n}\sigma^2 e^{2B\sigma^2}\left[B^2\sigma^2\left[(e^{B\sigma^2}-1)\left\{f(\mu)+\frac{A}{B}\right\}^2\right.\right.$$

$$\left. + \frac{1}{B}e^{B\sigma^2}\left\{f^{(1)}(\mu)\right\}^2\right]^2 + 2(e^{B\sigma^2}-1)^2$$

$$\left.\times \left\{f(\mu)+\frac{A}{B}\right\}^2 \left\{f^{(1)}(\mu)\right\}^2\right] + O(n^{-2})$$

Notice that the case for $B \neq 0$ reduces to the case for $B = 0$ if B tends to zero. For instance,

$$\lim_{B\to 0}\left[\left\{f(\mu)+\frac{A}{B}\right\}e^{B\sigma^2/2} - \frac{A}{B}\right] = f(\mu) + \frac{A}{2}\sigma^2$$

See also Gray, Watkins, and Schucany (1973), Mehran (1975), Gray, Schucany, and Woodward (1976), and Woodward (1977) as references relating to the subject of this section.

3.8 Transformations of Bivariate Normal Variables

Let f and g be recursive-type functions such that $f^{(2)} = A + Bf$ and $g^{(2)} = \alpha + \beta g$, where A, B, α, and β are arbitrary real constants.

Let $(Y^{(1)}, Y^{(2)})'$ be a random vector having a bivariate normal distribution with mean vector $\mu = (\mu_1, \mu_2)'$ and covariance matrix

$$\Sigma = \begin{bmatrix} \sigma_1^2 & \sigma_{12} \\ \sigma_{12} & \sigma_2^2 \end{bmatrix}$$

$N_2(\mu, \Sigma)$, where $\sigma_{12} = \sigma_1\sigma_2\rho$, $|\rho| < 1$. Let $X^{(1)} = f\left(Y^{(1)}\right)$ and $X^{(2)} = g(Y^{(2)})$. The covariance between $X^{(1)}$ and $X^{(2)}$ is

$$\operatorname{Cov}(X^{(1)}, X^{(2)})$$

$$= \begin{cases} \sigma_{12}\left\{f^{(1)}(\mu_1)g^{(1)}(\mu_2) + \dfrac{A\alpha}{2}\sigma_{12}\right\} & B = \beta = 0 \\[2ex] \sigma_{12}\Big[f^{(1)}(\mu_1)g^{(1)}(\mu_2) & \\ \quad + \dfrac{A\beta}{2}\left\{g(\mu_2) + \dfrac{\alpha}{\beta}\right\}\sigma_{12}\Big]\, e^{\beta\sigma_2^2/2} & B = 0 \quad \beta \neq 0 \\[2ex] \exp\left\{\dfrac{1}{2}(B\sigma_1^2 - 2\sqrt{B\beta}\sigma_{12} + \beta\sigma_2^2)\right\} & \\ \quad \times \Big[\dfrac{1}{2}(e^{\sqrt{B\beta}\sigma_{12}} - 1)^2\left\{f(\mu_1) + \dfrac{A}{B}\right\} & \\ \quad \times \left\{g(\mu_2) + \dfrac{\alpha}{\beta}\right\} + \dfrac{1}{2}(e^{2\sqrt{B\beta}\sigma_{12}} - 1) & \\ \quad \times \dfrac{1}{\sqrt{B\beta}} f^{(1)}(\mu_1)g^{(1)}(\mu_2)\Big] & B \neq 0 \quad \beta \neq 0 \end{cases}$$

where $f^{(1)}(\cdot)$ and $g^{(1)}(\cdot)$ denote the first derivatives of f and g, and the first product moment of $X^{(1)}$ and $X^{(2)}$ is

$$E(X^{(1)}X^{(2)}) = \begin{cases} \left\{f(\mu_1) + \dfrac{A}{2}\sigma_1^2\right\}\left\{g(\mu_2) + \dfrac{\alpha}{2}\sigma_2^2\right\} & \\ \quad + \operatorname{Cov}(X^{(1)}, X^{(2)}) & B = \beta = 0 \\ \left\{f(\mu_1) + \dfrac{A}{2}\sigma_1^2\right\}\left[\left\{g(\mu_2) + \dfrac{\alpha}{\beta}\right\}e^{\beta\sigma_2^2/2} - \dfrac{\alpha}{\beta}\right] & \\ \quad + \operatorname{Cov}(X^{(1)}, X^{(2)}) & B = 0 \quad \beta \neq 0 \\ \left[\left\{f(\mu_1) + \dfrac{A}{B}\right\}e^{B\sigma_1^2/2} - \dfrac{A}{B}\right]\left[\left\{g(\mu_2) + \dfrac{\alpha}{\beta}\right\}\right. & \\ \quad \left. \times\, e^{\beta\sigma_2^2/2} - \dfrac{\alpha}{\beta}\right] + \operatorname{Cov}(X^{(1)}, X^{(2)}) & B \neq 0 \quad \beta \neq 0 \end{cases}$$

Let $(Y_1^{(1)}, Y_1^{(2)})'$, $(Y_2^{(1)}, Y_2^{(2)})'$, ..., $(Y_n^{(1)}, Y_n^{(2)})'$, $n \geq 3$, be iid random vectors having $N_2(\mu, \Sigma)$, where we assume that μ and Σ are unknown. Let

$$\overline{Y}^{(1)} = \frac{1}{n}\sum_{i=1}^{n} Y_i^{(1)} \qquad \overline{Y}^{(2)} = \frac{1}{n}\sum_{i=1}^{n} Y_i^{(2)} \qquad S_1 = \sum_{i=1}^{n}(Y_i^{(1)} - \overline{Y}^{(1)})^2$$

$$S_2 = \sum_{i=1}^{n}(Y_i^{(2)} - \overline{Y}^{(2)})^2 \qquad S_{12} = \sum_{i=1}^{n}(Y_i^{(1)} - \overline{Y}^{(1)})(Y_i^{(2)} - \overline{Y}^{(2)})$$

We shall find the UMVU estimators of $\operatorname{Cov}(X^{(1)}, X^{(2)})$ and $E(X^{(1)}X^{(2)})$. This was done by Shimizu (1983b). If we adopt the functions $f(x) = e^x$ $(A = 0,\ B = 1)$ and $g(y) = e^y$ $(\alpha = 0,\ \beta = 1)$, we have the UMVU estimation corresponding to a bivariate lognormal distribution, and if we adopt the functions $f(x) = x (A = B = 0)$ and $g(y) = e^y (\alpha = 0, \beta = 1)$, we have the UMVU estimation corresponding to a normal-lognormal (semi-lognormal) distribution.

If $B = \beta = 0$, then the UMVU estimators of $\mathrm{Cov}(X^{(1)}, X^{(2)})$ and $E(X^{(1)}X^{(2)})$ are

$$E^{-1}\left[\mathrm{Cov}\left(X^{(1)}, X^{(2)}\right)\right]$$

$$= \frac{1}{n-1} S_{12} f^{(1)}\left(\overline{Y}^{(1)}\right) g^{(1)}\left(\overline{Y}^{(2)}\right) + \frac{A\alpha}{2n(n+1)}\left(S_{12}^2 - \frac{1}{n-1} S_1 S_2\right),$$

$$E^{-1}\left[\left(EX^{(1)}X^{(2)}\right)\right]$$

$$= \left\{f\left(\overline{Y}^{(1)}\right) + \frac{A}{2n} S_1\right\}\left\{g\left(\overline{Y}^{(2)}\right) + \frac{\alpha}{2n} S_2\right\}$$

$$+ \frac{1}{n} S_{12} f^{(1)}\left(\overline{Y}^{(1)}\right) g^{(1)}\left(\overline{Y}^{(2)}\right) + \frac{(n-1)A\alpha}{2n^2(n+1)}\left(S_{12}^2 - \frac{1}{n-1} S_1 S_2\right)$$

If $B = 0$, $\beta \neq 0$, then

$$E^{-1}\left[\mathrm{Cov}\left(X^{(1)}, X^{(2)}\right)\right]$$

$$= \frac{1}{n-1} S_{12} f^{(1)}\left(\overline{Y}^{(1)}\right) g^{(1)}\left(\overline{Y}^{(2)}\right) {}_0F_1\left(\frac{n+1}{2}; \frac{\beta}{4}\left(1 - \frac{1}{n}\right) S_2\right)$$

$$+ \frac{A\beta}{2n(n+1)}\left\{g\left(\overline{Y}^{(2)}\right) + \frac{\alpha}{\beta}\right\}\left(S_{12}^2 - \frac{1}{n-1} S_1 S_2\right)$$

$$\times {}_0F_1\left(\frac{n+3}{2}; \frac{\beta}{4}\left(1 - \frac{1}{n}\right) S_2\right),$$

$$E^{-1}\left[E\left(X^{(1)}X^{(2)}\right)\right]$$

$$= \left[f\left(\overline{Y}^{(1)}\right) {}_0F_1\left(\frac{n-1}{2}; \frac{\beta}{4}\left(1 - \frac{1}{n}\right) S_2\right) + \frac{A}{2n} S_1\right.$$

$$\left.\times {}_0F_1\left(\frac{n+1}{2}; \frac{\beta}{4}\left(1 - \frac{1}{n}\right) S_2\right) + \frac{(n-1)A\beta}{2n^2(n+1)} S_{12}^2\, {}_0F_1\left(\frac{n+3}{2}; \frac{\beta}{4}\left(1 - \frac{1}{n}\right) S_2\right)\right]$$

$$\times \left\{g\left(\overline{Y}^{(2)}\right) + \frac{\alpha}{\beta}\right\} - \frac{\alpha}{\beta}\left\{f\left(\overline{Y}^{(1)}\right) + \frac{A}{2n} S_1\right\}$$

$$+ \frac{1}{n} S_{12} f^{(1)}\left(\overline{Y}^{(1)}\right) g^{(1)}\left(\overline{Y}^{(2)}\right) {}_0F_1\left(\frac{n+1}{2}; \frac{\beta}{4}\left(1 - \frac{1}{n}\right) S_2\right)$$

If $B \neq 0$, $\beta \neq 0$, then, by use of (3.18) with $b = 0$,

$$E^{-1}\left[\mathrm{Cov}\left(X^{(1)}, X^{(2)}\right)\right]$$

$$= \frac{1}{2}\left[\left\{f\left(\overline{Y}^{(1)}\right) + \frac{A}{B}\right\}\left\{g\left(\overline{Y}^{(2)}\right) + \frac{\alpha}{\beta}\right\} + \frac{1}{\sqrt{B\beta}} f^{(1)}\left(\overline{Y}^{(1)}\right) g^{(1)}\left(\overline{Y}^{(2)}\right)\right]$$

$$\times\left\{\Omega\left(\left(1 - \frac{1}{n}\right)(B_1 + B_2)\right) - \Omega\left(\left(1 - \frac{1}{n}\right)B_1 - \frac{1}{n}B_2\right)\right\}$$

$$+ \frac{1}{2}\left[\left\{f\left(\overline{Y}^{(1)}\right) + \frac{A}{B}\right\}\left\{g\left(\overline{Y}^{(2)}\right) + \frac{\alpha}{\beta}\right\} - \frac{1}{\sqrt{B\beta}} f^{(1)}\left(\overline{Y}^{(1)}\right) g^{(1)}\left(\overline{Y}^{(2)}\right)\right]$$

$$\times\left\{\Omega\left(\left(1 - \frac{1}{n}\right)(B_1 - B_2)\right) - \Omega\left(\left(1 - \frac{1}{n}\right)B_1 + \frac{1}{n}B_2\right)\right\}$$

$$E^{-1}\left[E\left(X^{(1)}X^{(2)}\right)\right]$$

$$= \frac{1}{2}\left[\left\{f\left(\overline{Y}^{(1)}\right) + \frac{A}{B}\right\}\left\{g\left(\overline{Y}^{(2)}\right) + \frac{\alpha}{\beta}\right\} + \frac{1}{\sqrt{B\beta}} f^{(1)}\left(\overline{Y}^{(1)}\right) g^{(1)}\left(\overline{Y}^{(2)}\right)\right]$$

$$\times\,\Omega\left(\left(1 - \frac{1}{n}\right)(B_1 + B_2)\right) + \frac{1}{2}\left[\left\{f\left(\overline{Y}^{(1)}\right) + \frac{A}{B}\right\}\left\{g\left(\overline{Y}^{(2)}\right) + \frac{\alpha}{\beta}\right\}\right.$$

$$\left. - \frac{1}{\sqrt{B\beta}} f^{(1)}\left(\overline{Y}^{(1)}\right) g^{(1)}\left(\overline{Y}^{(2)}\right)\right]\Omega\left(\left(1 - \frac{1}{n}\right)(B_1 - B_2)\right) - \frac{\alpha}{\beta}\left\{f\left(\overline{Y}^{(1)}\right) + \frac{A}{B}\right\}$$

$$\times\, {}_0F_1\left(\frac{n-1}{2}; \frac{B}{4}\left(1 - \frac{1}{n}\right)S_1\right) - \frac{A}{B}\left\{g\left(\overline{Y}^{(2)}\right) + \frac{\alpha}{\beta}\right\}$$

$$\times\, {}_0F_1\left(\frac{n-1}{2}; \frac{\beta}{4}\left(1 - \frac{1}{n}\right)S_2\right) + \frac{A\alpha}{B\beta}$$

where

$$B_1 = \begin{bmatrix} B & 0 \\ 0 & \beta \end{bmatrix} \qquad B_2 = \begin{bmatrix} 0 & \sqrt{B\beta} \\ \sqrt{B\beta} & 0 \end{bmatrix}$$

$$\Omega(Z) = {}_0F_1\left(\frac{n-1}{2}; \frac{1}{4}TZ\right) \qquad T = \begin{bmatrix} S_1 & S_{12} \\ S_{12} & S_2 \end{bmatrix}$$

From Section 3.7, we will have the UMVU estimators $\hat{\theta}_1$ and $\hat{\theta}_2$ of $\theta_1 = E(X^{(1)})$ and $\theta_2 = E(X^{(2)})$. We shall give here the covariance between $\hat{\theta}_1$ and $\hat{\theta}_2$. This was obtained by Shimizu (1983c).

If $B = \beta = 0$, then

$$\text{Cov}(\hat{\theta}_1, \hat{\theta}_2) = \frac{1}{n}\sigma_{12}\left\{f^{(1)}(\mu_1)g^{(1)}(\mu_2) + \frac{A\alpha}{2}\sigma_{12}\right\}$$

and if $B = 0$, $\beta \neq 0$, then

$$\text{Cov}(\hat{\theta}_1, \hat{\theta}_2) = \frac{1}{n}\sigma_{12}\left[f^{(1)}(\mu_1)g^{(1)}(\mu_2) + \frac{A\beta}{2}\left\{g(\mu_2) + \frac{\alpha}{\beta}\right\}\sigma_{12}\right]e^{\beta\sigma_2^2/2}$$

To obtain the covariance between $\hat{\theta}_1$ and $\hat{\theta}_2$ in the case $B \neq 0$ and $\beta \neq 0$, we prepare a formula similar to (2.10). Let a and b be arbitrary real numbers. Then we have

$$E\left\{{}_0F_1\left(\frac{n-1}{2}; \frac{a}{2}S_1\right){}_0F_1\left(\frac{n-1}{2}; \frac{b}{2}S_2\right)\right\} = e^{a\sigma_1^2 + b\sigma_2^2}{}_0F_1\left(\frac{n-1}{2}; ab\sigma_{12}^2\right) \tag{3.22}$$

The formula (3.22) is a bivariate extension of (2.10). A sketch of a proof is as follows. Its idea is similar to that of the derivation of (2.10).

The joint frequency function of S_1 and S_2 is, according to Kibble (1941),

$$\begin{aligned} h(x,y) = {} & \frac{1}{4\Gamma((n-1)/2)}\left(\frac{\sqrt{xy}}{2\sigma_1\sigma_2\rho}\right)^{(n-3)/2}\frac{1}{\sigma_1^2\sigma_2^2(1-\rho^2)} \\ & \times e^{-(x/\sigma_1^2 + y/\sigma_2^2)/(2(1-\rho^2))}I_{(n-3)/2}\left(\frac{\rho\sqrt{xy}}{\sigma_1\sigma_2(1-\rho^2)}\right) \end{aligned} \tag{3.23}$$

where

$$I_\nu(z) = \frac{(z/2)^\nu}{\Gamma(\nu+1)}{}_0F_1\left(\nu+1; \frac{1}{4}z^2\right)$$

which is a modified Bessel function of the first kind and the νth order. Let j and k be arbitrary non-negative integers. Then (3.23) leads to

$$\begin{aligned} E(S_1^j S_2^k) = {} & (1-\rho^2)^{(n-1)/2}\left\{2\sigma_1^2(1-\rho^2)\right\}^j\left\{2\sigma_2^2(1-\rho^2)\right\}^k \\ & \times \frac{\Gamma((n-1)/2+j)\Gamma((n-1)/2+k)}{\Gamma^2((n-1)/2)} \\ & \times {}_2F_1\left(\frac{n-1}{2}+j, \frac{n-1}{2}+k; \frac{n-1}{2}; \rho^2\right) \end{aligned}$$

Using Hardy's formula

$$\sum_{l=0}^{\infty} \frac{l!}{\Gamma(l+\nu+1)} L_l^{(\nu)}(x) L_l^{(\nu)}(y) z^l$$
$$= (1-z)^{-1} \exp\left(-z\frac{x+y}{1-z}\right) (xyz)^{-\nu/2} I_\nu \left(2\frac{\sqrt{xyz}}{1-z}\right) \qquad |z| < 1$$

which appears in Kibble (1941), p. 141, as an alternative form, and in Erdélyi et al. (1953), Vol. 2, p. 189, where $L_l^{(\nu)}(x)$ are Laguerre polynomials, we have (3.22).

From (3.22), if $B \neq 0$, $\beta \neq 0$, then

$$\begin{aligned}
\mathrm{Cov}(\hat{\theta}_1, \hat{\theta}_2) = \frac{1}{2} e^{(B\sigma_1^2 + \beta\sigma_2^2)/2} \Bigg[& \left\{ f(\mu_1) + \frac{A}{B} \right\} \left\{ g(\mu_2) + \frac{\alpha}{\beta} \right\} \\
& \times \Bigg\{ \left(e^{\sqrt{B\beta}\sigma_{12}/n} + e^{-\sqrt{B\beta}\sigma_{12}/n} \right) \\
& \times {}_0F_1 \left(\frac{n-1}{2}; \frac{B\beta}{4} \left(1 - \frac{1}{n}\right)^2 \sigma_{12}^2 \right) - 2 \Bigg\} \\
& + \frac{1}{\sqrt{B\beta}} f^{(1)}(\mu_1) g^{(1)}(\mu_2) \left(e^{\sqrt{B\beta}\sigma_{12}/n} - e^{-\sqrt{B\beta}\sigma_{12}/n} \right) \\
& \times {}_0F_1 \left(\frac{n-1}{2}; \frac{B\beta}{4} \left(1 - \frac{1}{n}\right)^2 \sigma_{12}^2 \right) \Bigg]
\end{aligned}$$

and the asymptotic expansion of $\mathrm{Cov}(\hat{\theta}_1, \hat{\theta}_2)$ up to n^{-3} is

$$\begin{aligned}
\mathrm{Cov}(\hat{\theta}_1, \hat{\theta}_2) = \frac{1}{2} e^{(B\sigma_1^2 + \beta\sigma_2^2)/2} \Bigg[& \left\{ f(\mu_1) + \frac{A}{B} \right\} \left\{ g(\mu_2) + \frac{\alpha}{\beta} \right\} \Bigg\{ \frac{1}{n} B\beta\sigma_{12}^2 \\
& + \frac{1}{4n^2} B^2\beta^2\sigma_{12}^4 + \frac{1}{2n^3} B^2\beta^2\sigma_{12}^4 \left(\frac{1}{12} B\beta\sigma_{12}^2 - 1 \right) \Bigg\} \\
& + f^{(1)}(\mu_1) g^{(1)}(\mu_2) \\
& \times \left\{ \frac{2}{n}\sigma_{12} + \frac{1}{n^2} B\beta\sigma_{12}^3 + \frac{1}{n^3} B\beta\sigma_{12}^3 \left(\frac{1}{4} B\beta\sigma_{12}^2 - \frac{2}{3} \right) \right\} \Bigg] \\
& + O(n^{-4})
\end{aligned}$$

4. RELATED TOPICS

4.1 Sequential Estimation

Let $X_1, X_2, \ldots, X_n, \ldots$ be iid random variables having $\Lambda(\mu, \sigma^2)$, where we assume that μ and σ^2 are unknown. Consider the following problem: Let h_n be any estimator of the mean $\mu_1' = \exp(\mu + \sigma^2/2)$ based on $X_1, \ldots, X_n$. Then can we obtain n and h_n such that

$$\Pr(|h_n - \mu_1'| \leq \delta\mu_1') \geq \gamma$$

for all $\delta > 0$ and $0 < \gamma < 1$? Zacks (1966, 1967) considered a sequential approach to the problem. We shall refer here to a paper by Nagao (1980).

Consider the following modified ML estimator as an estimator of μ_1':

$$\hat{\xi}_n = \exp\left(\overline{Y}_n + \tfrac{1}{2}S_n^2\right)$$

where $\overline{Y}_n = n^{-1}\sum_{i=1}^n \ln X_i$ and $S_n^2 = (n-1)^{-1}\sum_{i=1}^n(\ln X_i - \overline{Y}_n)^2$. Zacks (1966) considered the ML estimator instead of $\hat{\xi}_n$. Put $U_n = \sqrt{n}(\overline{Y}_n - \mu)/\sigma$ and $V_n = \sqrt{(n-1)/2}(S_n^2/\sigma^2 - 1)$. The distribution of U_n and the limiting distribution of V_n are $N(0,1)$. From this the limiting distribution of $\hat{\xi}_n$ is $N(\mu_1', \mu_1'^2\sigma^2(1 + \sigma^2/2)/n)$. Since

$$\Pr(|\hat{\xi}_n - \mu_1'| \leq \delta\mu_1') = \Pr\left(\frac{\sqrt{n}|\hat{\xi}_n - \mu_1'|}{\{\mu_1'^2\sigma^2(1 + \sigma^2/2)\}^{1/2}} \leq \frac{\sqrt{n}\delta}{\{\sigma^2(1 + \sigma^2/2)\}^{1/2}}\right)$$

we define the stopping time

$$N = \inf\left\{n \geq n_0 \mid n \geq \chi_\gamma^2(1)\delta^{-2}S_n^2\left(1 + \tfrac{1}{2}S_n^2\right)\right\} \tag{4.1}$$

where $\chi_\gamma^2(1)$ denotes the lower $100\gamma\%$ point of chi-square distribution with one degree of freedom and the fixed constant number n_0 is greater than 3. After we have taken N observations, we estimate μ_1' by $\hat{\xi}_n = \exp(\overline{Y}_n + S_N^2/2)$.

The limiting distribution of the sample mean $\overline{X}_n = n^{-1}\sum_{i=1}^n X_i$ is $N(\mu_1', \mu_2/n)$ by the central limit theorem, where $\mu_2 = \mu_1'^2\{\exp(\sigma^2) - 1\}$. We define the stopping time

$$N^* = \inf\left\{n \geq n_0 \mid n \geq \chi_\gamma^2(1)\delta^{-2}(\exp(S_n^2) - 1)\right\} \tag{4.2}$$

It is shown that the sequential modified ML procedure is superior to the sequential sample mean procedure in an exact sense. That is, for N

and N^* defined by (4.1) and (4.2), we have $N \leq N^*$ almost surely. Zacks (1966) showed that the sequential ML procedure is superior to the sequential sample mean procedure in an asymptotic sense as $\delta \to 0$.

Other properties of sequential estimation were discussed in the papers by Zacks (1966, 1967) and Nagao (1980).

4.2 Lognormal Processes

Box and Jenkins (1976) systematically discussed the forecasting and model specification of the autoregressive integrated moving average process. The method has been used after logarithmic and Box-Cox transformations. Stoica (1980) discussed the one-step prediction of the stationary autoregressive (AR) lognormal process from a Bayesian viewpoint.

Let $y(t) = \exp[x(t)]$, $t = 0, \pm 1, \pm 2, \ldots$, be the stationary AR lognormal process, where

$$A(q^{-1})x(t) + \alpha_{n+1} = e(t)$$
$$A(q^{-1}) = 1 + \alpha_1 q^{-1} + \cdots + \alpha_n q^{-n}$$

assuming that $e(t)$ is a sequence of iid normal random variables with mean zero and known second moment and q^{-1} is the backward shift operator, i.e., $q^{-1}x(t) = x(t-1)$.

Introduce the notations

$$\theta = (\alpha_1, \alpha_2, \ldots, \alpha_n, \alpha_{n+1})'$$
$$f(t) = (-x(t-1), \ldots, -x(t-n), -1)'$$
$$N = \text{sample size}, \ E[e^2(t)] = d^2$$
$$X(N) = (x(N), x(N-1), \ldots, x(0))$$

Then the conditional distribution of θ given $X(N)$ is normal with mean vector

$$\hat{\theta} = \left[\sum_{t=n}^{N} f(t)(f(t))'\right]^{-1} \sum_{t=n}^{N} f(t)x(t)$$

and covariance matrix

$$P = d^2 \left[\sum_{t=n}^{N} f(t)(f(t))'\right]^{-1}$$

The conditional distribution of $x(N+1)$ given $X(N)$ is normal with mean

$$\hat{x} = (f(N+1))'\hat{\theta}$$

and variance

$$s^2 = d^2\left[1 + (f(N+1))'\frac{P}{d^2}f(N+1)\right]$$

Hence, the conditional distribution of $y(N+1)$ given $Y(N)$ is lognormal with parameters $\hat{x}$ and s^2, where $Y(N) = (y(N), y(N-1), \ldots, y(0))$.

Let m_k, m_e, and m_o be, respectively, the kth moment, the median, and the mode of the lognormal distribution $\Lambda(\hat{x}, s^2)$, i.e.,

$$m_k = \exp\left(k\hat{x} + \tfrac{1}{2}k^2s^2\right)$$
$$m_e = \exp(\hat{x})$$
$$m_o = \exp(\hat{x} - s^2)$$

Then m_1 as well as m_e and m_o can be used to predict $y(N+1)$ based on $Y(N)$. Stoica (1980) also gives asymptotic solutions for the case of unknown d and for the prediction of $y(N+k)$.

REFERENCES

Aitchison, J. and Brown, J. A. C. (1957). *The Lognormal Distribution*, Cambridge University Press, Cambridge.

Blight, B. J. N. and Rao, P. V. (1974). The convergence of Bhattacharyya bounds, *Biometrika*, *61*, 137–142.

Box, G. E. P. and Jenkins, G. M. (1976). *Time Series Analysis: Forecasting and Control*, 2nd ed., Holden-Day, San Francisco.

Bradu, D. and Mundlak, Y. (1970). Estimation in lognormal linear models, *J. Amer. Statist. Assoc.*, *65*, 198–211.

Crofts, A. E. Jr. and Owen, D. B. (1972). Large sample maximum likelihood estimation in a normal-lognormal distribution, *S. Afr. Statist. J.*, *6*, 1–10.

Crow, E. L. (1977). Minimum variance unbiased estimators of the ratio of means of two lognormal variates and of two gamma variates, *Commun. Statist.-Theor. Meth.*, *A6*, 967–975.

Duan, N. (1983). Smearing estimate: a nonparametric retransformation method, *J. Amer. Statist. Assoc.*, *78*, 605–610.

Ebbeler, D. H. (1973). A note on large-sample approximations in lognormal linear models, *J. Amer. Statist. Assoc.*, *68*, 231.

Erdélyi, A. et al. (1953). *Higher Transcendental Functions*, McGraw-Hill Book Company, Inc., New York.

Evans, I. G. and Shaban, S. A. (1974). A note on estimation in lognormal models, *J. Amer. Statist. Assoc.*, *69*, 779–781.

Evans, I. G. and Shaban, S. A. (1976). New estimators (of smaller M. S. E.) for parameters of a lognormal distribution, *Biom. Z.*, *18*, 453–466.

Finney, D. J. (1941). On the distribution of a variate whose logarithm is normally distributed, *J. R. Statist. Soc. Suppl.*, *7*, 155-161.

Gleit, A. (1982). Estimation of functions of the parameters of a normal distribution, *Commun. Statist.-Theor. Meth.*, *A11*, 2845-2855.

Gray, H. L., Schucany, W. R., and Woodward, W. A. (1976). Best estimates of functions of the parameters of the Gaussian and gamma distributions, *IEEE Trans. Reliability*, *R-25*, 95–99.

Gray, H. L., Watkins, T. A., and Schucany, W. R. (1973). On the jackknife statistic and its relation to UMVU estimators in the normal case, *Commun. Statist.-Theor. Meth.*, *A2*, 285–308.

Hoyle, M. H. (1968). The estimation of variances after using a Gaussianating transformation, *Ann. Math. Statist.*, *39*, 1125–1143.

Iwase, K., Shimizu, K., and Suzuki, M. (1982). On UMVU estimators for the multivariate lognormal distribution and their variances, *Commun. Statist.-Theor. Meth.*, *A11*, 687–697.

Johnson, N. L. and Kotz, S. (1970). *Distributions in Statistics: Continuous Univariate Distributions-1*, Chap. 14, pp. 112–136, John Wiley & Sons, Inc., New York.

Khatri, C. G. and Pillai, K. C. S. (1968). On the non-central distributions of two test criteria in multivariate analysis of variance, *Ann. Math. Statist.*, *39*, 215–226.

Kibble, W. F. (1941). A two-variate Gamma type distribution, *Sankhyā*, *5*, 137–150.

Laurent, A. G. (1963). The lognormal distribution and the translation method: description and estimation problems, *J. Amer. Statist. Assoc.*, *58*, 231–235.

Likeš, J. (1977). Relations between functions g, G and Φ for estimation in lognormal models, *Biom. J.*, *19*, 361–363.

Likeš, J. (1980). Variance of the MVUE for lognormal variance, *Technometrics*, *22*, 253–258.

Likeš, J. (1983). Efficiency of the sample variance for the lognormal distribution, *Biom. J.*, *25*, 617–620.

Mehran, F. (1973). Variance of the MVUE for the lognormal mean, *J. Amer. Statist. Assoc.*, *68*, 726–727.

Mehran, F. (1975). Relationships between the UMVU estimators of the mean and median of a function of a normal distribution, *Ann. Statistic.*, *3*, 457–460.

Mostafa, M. D. and Mahmoud, M. W. (1964). On the problem of estimation for the bivariate lognormal distribution, *Biometrika*, *51*, 522–527.

Muirhead, R. J. (1982). *Aspects of Multivariate Statistical Theory*, John Wiley & Sons, Inc., New York.

Nagao, H. (1980). On stopping times of sequential estimations of the mean of a log-normal distribution, *Ann. Inst. Statist. Math.*, *32*, 369–375.

Neyman, J. and Scott, E. L. (1960). Correction for bias introduced by a transformation of variables, *Ann. Math. Statist.*, *31*, 643–655.

Oldham, P. D. (1965). On estimating the arithmetic means of lognormally-distributed populations, *Biometrics*, *21*, 235–239.

Pennington, M. (1983). Efficient estimators of abundance, for fish and plankton surveys, *Biometrics*, *39*, 281–286.

Rukhin, A. L. (1986). Improved estimation in lognormal models, *J. Amer. Statist. Assoc.*, *81*, 1046–1049.

Sen, A. R. (1978). Estimation of the population mean when the coefficient of variation is known, *Commun. Statist.-Theor. Meth.*, *A7*, 657–672.

Shaban, S. A. (1981). On the estimation of the ratio of means and other characteristics of two log normal variates, *Biom. J.*, *23*, 357–369.

Shimizu, K. (1982). Estimation of measures related to the Lorenz curve and of measures of the heaviness of tail, *Proc. Inst. Statist. Math.*, *30*, 11–27 (in Japanese).

Shimizu, K. (1983a). Variances of UMVU estimators for means and variances after using a normalizing transformation, *Commun. Statist.-Theor. Meth.*, *A12*, 975–985.

Shimizu, K. (1983b). UMVU estimation for covariances and first product moments of transformed variables, *Commun. Statist.-Theor. Meth.*, *A12*, 1661–1674.

Shimizu, K. (1983c). Covariances between UMVU estimators for means of transformed variables, *Commun. Statist.-Theor. Meth.*, *A12*, 2525–2532.

Shimizu, K. (1983d). Estimation of the quantile of a lognormal distribution, *Rep. Stat. Appl. Res., JUSE*, *30*, 28–33.

Shimizu, K. (1984). Derivation of a useful formula for estimation in a lognormal distribution, *Japanese J. Appl. Statist.*, *13*, 27–29. (in Japanese).

Shimizu, K. and Iwase, K. (1981). Uniformly minimum variance unbiased estimation in lognormal and related distributions, *Commun. Statist.-Theor. Meth.*, *A10*, 1127–1147.

Shimizu, K., Iwase, K., and Ushizawa, K. (1979). Estimation of parameters in a two parameter log-normal distribution, *Japanese J. Appl. Statist.*, *8*, 97–110 (in Japanese).

Stoica, P. (1980). Prediction of autoregressive lognormal processes, *IEEE Trans. Automat. Contr.*, *AC-25*, 292–293.

Suzuki, M. (1983a). Some asymptotic results in the multivariate lognormal estimation theory, *Commum. Statist.-Theor. Meth.*, *A12*, 1761–1774.

Suzuki, M. (1983b). Estimation in a bivariate semi-lognormal distribution, *Behaviormetrika*, *13*, 59–68.

Suzuki, M. (1983c). On estimators given by the method of moments for the bivariate lognormal distribution, *Rep. Stat. Appl. Res., JUSE*, *30*, 1–7.

Suzuki, M., Iwase, K., and Shimizu, K. (1984). Uniformly minimum variance unbiased estimation in a semi-lognormal distribution, *J. Japan Statist. Soc.*, *14*, 63–68.

Thöni, H. (1969). A table for estimating the mean of a lognormal distribution, *J. Amer. Statist. Assoc.*, *64*, 632–636.

Watkins, T. A. and Kern, D. M. (1973). On UMVU estimators related to the multivariate normal distribution, *Commun. Statist.-Theor. Meth., A2*, 321–326.

Woodward, W. A. (1977). Subroutines for computing minimun variance unbiased estimators of functions of the parameters in the normal and gamma distributions, *Commun. Statist.-Simula. Computa., B6*, 63–73.

Zacks, S. (1966). Sequential estimation of the mean of a log-normal distribution having a prescribed proportional closeness, *Ann. Math. Statist.*, *37*, 1688–1696.

Zacks, S. (1967). On the non-existence of a fixed sample estimator of the mean of a log-normal distribution having a prescribed proportional closeness, *Ann. Math. Statist.*, *38*, 949.

3

Hypothesis Tests and Interval Estimates

CHARLES E. LAND Radiation Epidemiology Branch, National Cancer Institute, National Institutes of Health, Bethesda, Maryland

1. INTRODUCTION

As is well known, all quantitative properties of the normal ($N(\mu, \sigma)$) distribution are expressible in terms of parametric functions of μ and σ. The same is true of the lognormal ($\Lambda(\mu, \sigma)$) distribution and, indeed, of any distribution that can be derived from the normal distribution by a completely specified transformation of variables. The difference between working with the $\Lambda(\mu, \sigma)$ as opposed to the $N(\mu, \sigma)$ distribution is that interest tends to be focused on a different set of parametric functions $\phi(\mu, \sigma)$.

For example, if the monetary value X of a product is lognormally distributed, and if one's income derives from the sale of these products, one is likely to be interested in the expected value of X,

$$E(X) = \exp\left(\mu + \tfrac{1}{2}\sigma^2\right)$$

but not as much in its median,

$$M(X) = \exp(\mu)$$

or the expected value, μ, of its logarithm. If, on the other hand, X itself were normally distributed, there would be little reason to be concerned with the exponential of the sum of its mean and half its variance.

The approach of this chapter is to consider optimal hypothesis tests and confidence interval procedures for various functions of μ and σ, grouped according to how easily results for one function can be extended to another. For each grouping, optimal tests and confidence interval procedures are given first for that function which corresponds most closely to the scale of μ, σ, or σ^2, and then extended to other functions. Optimality here is expressed in terms of statistical power for hypothesis tests or critical regions, and in terms of coverage probabilities for confidence interval procedures.

The approach taken is only mildly heuristic, since a detailed and rigorous development would require something on the order of Lehmann's (1959) text, *Testing Statistical Hypotheses*, to which the reader is referred for many details. The aim is to show the reader why a thing is true, without going through every detail. In general, simple cases are given rather more space than might be thought strictly necessary, in order to simplify the exposition of more complex cases which can be presented as extensions.

The initial theoretical development is in terms of a slight generalization of the single-sample model, in which the complete, sufficient statistic for (μ, σ) is (Y, S), where Y is normally distributed with mean μ and standard deviation σ/γ, where γ is known, and S is independently distributed as $\sigma/\nu^{1/2}$ times the square root of a chi-square variate with ν degrees of freedom, for known ν. Extensions to more complex sampling models are given later in the chapter.

Unless otherwise stated, uppercase letters are used to denote random variables (e.g., Y, S) and the corresponding lowercase letters (y, s) denote sample values.

2. DEFINITIONS

2.1 Hypothesis Testing

Let H and K denote disjoint regions in the parameter space

$$\Omega = \{(\mu, \sigma) : -\infty < \mu < \infty,\ \sigma > 0\}$$

and let C and C' denote subsets of the sample space $\mathcal{S} = \{(y, s) : -\infty < y < \infty, s > 0\}$. In what follows, $P\{C \mid \mu, \sigma\} = P\{(Y, S) \in C\}$ for a particular $(\mu, \sigma) \in \Omega$.

(1) *Level*: C is a critical region of level α for testing the hypothesis H $((\mu, \sigma) \in H)$ if $P\{C \mid \mu, \sigma\} \leq \alpha$ whenever $(\mu, \sigma) \in H$.

(2) *Unbiasedness*: C is unbiased for testing H vs. K if $P\{C \mid \mu, \sigma\} \leq P\{C \mid \mu', \sigma'\}$ whenever $(\mu, \sigma) \in H$, $(\mu', \sigma') \in K$.

(3) *Invariance*: Let G be a group of one-to-one transformations g mapping $\mathcal{S}$ onto itself. For each $g \in G$, define $\mathbf{g}$ mapping Ω onto itself, such that $\mathbf{g}(\mu, \sigma)$ characterizes the distribution of $g(Y, S)$ whenever (μ, σ) characterizes that of (Y, S). We say that the problem of testing H vs. K is invariant under G if $\mathbf{g}(H) = H$ and $\mathbf{g}(K) = K$, for all $g \in G$. In that case, we also say the critical region C is invariant under G provided that $g(C) = C$ for all $g \in G$.

(4) *Uniformly most powerful (UMP) tests*: C is UMP among some class $\mathcal{C}$ of critical regions for testing H vs. K if $C \in \mathcal{C}$ and if, for any $C' \in \mathcal{C}$,

$$P(C \mid \mu, \sigma) \geq P(C' \mid \mu, \sigma)$$

for all $(\mu, \sigma) \in K$. Thus, C might be UMP among all level α critical regions for testing H vs. K, or just among those that are unbiased, or invariant under group G, or both.

2.2 Confidence Interval Estimation

Let $\theta = \theta(\mu, \sigma)$ denote a real-valued function on Ω or, alternatively, a point in $\theta(\Omega)$. For each specific value θ_0 of θ, let $H_1(\theta_0) = \{(\mu, \sigma) : \theta \leq \theta_0\}$ and $K_1(\theta_0) = \{(\mu, \sigma) : \theta > \theta_0\}$. Similarly, let $H_2(\theta_0)$ and $K_2(\theta_0)$ correspond to $\theta \geq \theta_0$ and $\theta < \theta_0$, and $H_3(\theta_0)$ and $K_3(\theta_0)$ to $\theta = \theta_0$ and $\theta \neq \theta_0$, respectively. Let $\mathcal{U}_1$, $\mathcal{U}_2$, and $\mathcal{U}_3$ be rules defining, for each $(y, s) \in \mathcal{S}$, a set in $\theta(\Omega)$ denoted by $U_i(y, s)$, for $i = 1, 2, 3$.

(1) *Level*: We say that $\mathcal{U}_i$ is a confidence procedure of level $1 - \alpha$ for θ, corresponding to $\{H_i(\theta_0), K_i(\theta_0) : \theta_0 \in \theta(\Omega)\}$, if for all $\theta_0 \in \theta(\Omega)$ and all $(\mu, \sigma) \in H_i(\theta_0)$,

$$P\{\theta_0 \in U_i(Y, S) \mid \mu, \sigma\} \geq 1 - \alpha$$

Note that if U_i is a level $1 - \alpha$ confidence procedure, the sets

$$C_i(\theta_0) = \{(y, s) : \theta_0 \in U_i(y, s)\}$$

for all $\theta_0 \in \theta(\Omega)$, are level α critical regions for testing $H_i(\theta_0)$ vs. $K_i(\theta_0)$, and conversely.

(2) *Unbiasedness*: The confidence procedure $\mathcal{U}_i$ is unbiased with respect to $\{H_i(\theta_0),\ K_i(\theta_0) : \theta_0 \in (\Omega)\}$, if, for all $\theta_0 \in \theta(\Omega)$,

$$P\{\theta_0 \in U_i(Y, S) \mid \mu, \sigma\} \leq P\{\theta_0 \in U_i(Y, S) \mid \mu', \sigma'\}$$

for $(\mu, \sigma) \in K_i(\theta_0)$, $(\mu', \sigma') \in H_i(\theta_0)$. Unbiasedness of $\mathcal{U}_i$ is equivalent to unbiasedness of $C_i(\theta_0)$ for all $\theta_0 \in \theta(\Omega)$.

(3) *Invariance*: If G is a group of transformations g mapping $\mathcal{S}$ onto itself, and if for each $g \in G$ we define $\mathbf{g}$ mapping Ω onto itself such that $\mathbf{g}(\mu, \sigma)$ characterizes the distribution of $g(Y, S)$ whenever (μ, σ) characterizes that of (Y, S), $\mathcal{U}_i$ is invariant under G if, for all $(y, s) \in \mathcal{S}$ and all $g \in G$,

$$\{\mathbf{g}(\mu, \sigma) : \theta(\mu, \sigma) \in U_i(y, s)\} = \{(\mu, \sigma) : \theta(\mu, \sigma) \in U_i(g(y, s))\}$$

If the problem of testing $H_i(\theta_0)$ vs. $K_i(\theta_0)$ is invariant under the group $G(\theta_0)$, and if G is the group generated by $\{G(\theta_0) : \theta_0 \in \theta(\Omega)\}$, then $\mathcal{U}_i$ is invariant under G if and only if $C_i(\theta_0)$ is invariant under $G(\theta_0)$, for each $\theta_0 \in \theta(\Omega)$.

(4) *Uniformly most accurate (UMA) confidence procedures*: $\mathcal{U}_i$ is UMA (with respect to $\{H_i(\theta_0), K_i(\theta_0) : \theta_0 \in \theta(\Omega)\}$) among some class $\boldsymbol{\mathcal{U}}_i$ of confidence procedures for θ if, for any $\mathcal{U}_i' \in \boldsymbol{\mathcal{U}}_i$,

$$P\{\theta_0 \in U_i(Y, S) \mid \mu, \sigma\} \leq P\{\theta_0 \in U_i'(Y, S) \mid \mu, \sigma\}$$

for all $(\mu, \sigma) \in K_i(\theta_0)$ and all $\theta_0 \in \theta(\Omega)$. For example, $\mathcal{U}_i$ might be UMA among all confidence procedures of level $1-\alpha$, those that are unbiased, those that are invariant under some group G, or those that are both unbiased and invariant under G.

Note that $\mathcal{U}_i$ is UMA among level $1-\alpha$ confidence procedures having certain properties if and only if for each $\theta_0 \in \theta(\Omega)$ the critical region $C_i(\theta_0)$ corresponding to $\mathcal{U}_i$ is UMP among all level α critical regions having the corresponding properties for critical regions.

(5) *Transformability of confidence procedures*: If $\mathcal{U}_i$ is a confidence procedure of level $1-\alpha$ for a parametric function $\theta(\mu, \sigma)$, and if $\psi(\theta)$ is a function defined on $\theta(\Omega)$ for which $\psi(H_i(\theta_0)) \cap \psi(K_i(\theta_0))$ is empty, for all $\theta_0 \in \theta(\Omega)$ and for $i = 1, 2, 3$, then

$$\psi(\mathcal{U}_i) = \{\psi(U_i(y, s)) : (y, s) \in \mathcal{S}\}$$

is a level $1-\alpha$ confidence procedure for $\psi(\theta)$ in terms of $\{\psi(H_i(\theta_0)), \psi(K_i(\theta_0)) : \theta_0 \in \theta(\Omega)\}$. If $\mathcal{U}_i$ is UMA unbiased or UMA invariant, then so is $\psi(\mathcal{U}_i)$. Moreover, the critical regions $C_i(\theta_0)$ also serve, and retain their probabilistic properties of level, unbiasedness, invariance, and power whether the hypotheses are stated in terms of θ or $\psi(\theta)$.

(6) *Confidence interval procedures*: Virtuallly all confidence procedures used in practice for real-valued parametric functions define confidence sets

that are intervals. This reflects the fact that the usual hypotheses $H_i(\theta_0)$ and $K_i(\theta_0)$ are intervals or complements of intervals, and the general inconvenience of confidence sets that are not intervals. In this chapter, the confidence procedures under discussion are of the form

$$\begin{aligned} U_1(y,s) &= [u_1(y,s),\infty) \cap \theta(\Omega) \\ U_2(y,s) &= (-\infty, u_2(y,s)] \cap \theta(\Omega) \\ U_3(y,s) &= [u_{31}(y,s), u_{32}(y,s)] \cap \theta(\Omega) \end{aligned}$$

where u_1, u_2, u_{31}, and u_{32}, with $u_{31} < u_{32}$, are functions mapping S into $\theta(\Omega)$. It should not be assumed, however, that all confidence procedures define intervals. Although it is often obvious, it can on occasion be quite difficult to determine the form of a particular confidence procedure defined in terms of a family of critical regions.

3. FUNCTIONS OF μ ALONE

3.1 The Case $\theta(\mu,\sigma) = \mu$

3.1.1 For known σ, the statistic Y is sufficient for μ and, because its distribution has monotone likelihood ratio in μ, it follows from the Neyman-Pearson Lemma (Lehmann, 1959, sec. 3.2) that the conventional level α likelihood ratio tests are UMP for testing $H_i(\theta_0)$ vs. $K_i(\theta_0)$, for $i = 1, 2$. That is, the UMP level α critical regions are

$$\begin{aligned} C_1(\theta_0) &= \left\{ y : \frac{\gamma}{\sigma}(y-\theta_0) > \mathbf{z}(1-\alpha) \right\} \\ C_2(\theta_0) &= \left\{ y : \frac{\gamma}{\sigma}(y-\theta_0) < \mathbf{z}(\alpha) \right\} \end{aligned}$$

respectively, where $\mathbf{z}(\alpha)$ is the αth quantile of the standard normal ($N(0,1)$) distribution.

There is no UMP level α test of $H_3(\theta_0)$ vs. $K_3(\theta_0)$. There is, however, a UMP unbiased test of level α, whose critical region is

$$C_3(\theta_0) = \left\{ y : \frac{\gamma}{\sigma}(y-\theta_0) \notin \left[\mathbf{z}\left(\frac{1}{2}\alpha\right), \mathbf{z}\left(1-\frac{1}{2}\alpha\right) \right] \right\}$$

For any critical region C for testing $H_3(\theta_0)$ vs. $K_3(\theta_0)$, unbiasedness requires that the power function

$$\beta(\theta) = P\{Y \in C \mid \mu = \theta)\}$$

have its minimum at $\theta = \theta_0$ (and therefore that $\beta'(\theta_0) = 0$). Differentiation under the integral sign of

$$\beta(\theta) \propto \int_C \exp\left\{-\left(\frac{1}{2}\right)\left(\frac{\gamma}{\sigma}\right)^2 (y-\theta)^2\right\} dy$$

gives

$$\beta'(\theta) \propto \int_C (y-\theta)\exp\left\{-\frac{1}{2}\left(\frac{\gamma}{\sigma}\right)^2 (y-\theta)^2\right\} dy$$

from which $\beta'(\theta_0) = 0$ and $\beta(\theta_0) = \alpha$ together imply

$$\int_C y\phi(y)\,dy = \alpha\theta_0 \int_{-\infty}^{\infty} \phi(y)\,dy$$

where

$$\phi(y) \propto \exp\left\{-\frac{1}{2}\left(\frac{\gamma}{\sigma}\right)^2 (y-\theta_0)^2\right\} \qquad -\infty < y < \infty$$

Let C be an arbitrary critical region satisfying the two requirements $\beta(\theta_0) = \alpha$ and $\beta'(\theta_0) = 0$. It is convenient to treat C as the union of subregions C_1, to the left of the point $y = \theta_0$, and C_2, to the right of that point, with levels α_1 and α_2, respectively (note $\alpha_1 + \alpha_2 = \alpha$). If neither C_1 nor C_2 is an interval of infinite length, power can be increased uniformly by shifting C_1, or parts of it, to the left and C_2, or parts of it, to the right without changing α_1 and α_2 or violating unbiasedness. Eventually either C_1 or C_2 will become an interval of infinite length. If only one of them (C_1, say) is an interval of infinite length, then $\alpha_1 < \alpha_2$ because $\phi(y)$ is symmetric about $y = \theta_0$, and power can be increased uniformly by extending C_1 to the right (and increasing α_1), and moving C_2 to the right (until $\alpha_2 = \alpha - \alpha_1$). Eventually, (in this case, when α_1 reaches the value $\frac{1}{2}\alpha$), both C_1 and C_2 will be intervals of infinite length. Furthermore, C will be identical to $C_3(\theta_0)$, which is the only unbiased level α critical region whose left and right subregions are intervals of infinite length.

The confidence interval procedures $\mathcal{U}_i$ corresponding to $\mathcal{C}_i = \{C_i(\theta_0) : -\infty < \theta_0 < \infty\}$, $i = 1, 2, 3$, are as follows:

$$U_1(y) = \left[y - \frac{\sigma}{\gamma}\mathbf{z}(1-\alpha), \infty\right]$$

$$U_2(y) = \left[-\infty, y - \left(\frac{\sigma}{\gamma}\right)\mathbf{z}(\alpha)\right]$$

$$U_3(y) = \left[y - \frac{\sigma}{\gamma}\mathbf{z}\left(1 - \frac{1}{2}\alpha\right), y - \frac{\sigma}{\gamma}\mathbf{z}\left(\frac{1}{2}\alpha\right)\right]$$

3.1.2 For unknown σ, the optimal critical regions and confidence interval procedures are defined analogously to those for the case of known σ, with s replacing σ and the quantiles $\mathbf{t}_\nu(\alpha)$ of Student's $\mathbf{t}$ distribution with ν degrees of freedom replacing the normal quantiles $\mathbf{z}(\alpha)$:

$$C_1(\theta_0) = \left\{(y, s) : \frac{\gamma(y - \theta_0)}{s} > \mathbf{t}_\nu(1-\alpha)\right\} \tag{3.1}$$

$$C_2(\theta_0) = \left\{(y, s) : \frac{\gamma(y - \theta_0)}{s} < \mathbf{t}_\nu(\alpha)\right\} \tag{3.2}$$

$$C_3(\theta_0) = \left\{(y, s) : \frac{\gamma(y - \theta_0)}{s} \notin \left[\mathbf{t}_\nu\left(\frac{1}{2}\alpha\right), \mathbf{t}_\nu\left(1 - \frac{1}{2}\alpha\right)\right]\right\} \tag{3.3}$$

The critical regions $C_i(\theta_0)$ are UMP unbiased level α for testing $H_i(\theta_0)$ vs. $K_i(\theta_0)$ for $i = 1, 2, 3$. Unlike the case of known σ, there are no UMP level α critical regions for $i = 1, 2$ since the method for known σ defines a critical region that is better than all others for any particular value of σ, but which can be improved upon for any other value of σ. Optimality of the regions in (3.1)–(3.3) above follows from completeness and sufficiency of (Y, S) for (μ, σ), and the fact that $Y - \theta_0$ and $V^2 = \nu S^2 + \gamma^2(Y - \theta_0)^2$ are the sufficient statistics in the exponential form of the joint density of $(Y - \theta_0, S)$, with natural parameters $\xi = (\mu - \theta_0)/\sigma^2$ and $\varsigma = -\frac{1}{2}/\sigma^2$:

$$f(y, s) \propto s^{\nu-1} \exp\left\{-\frac{1}{2\sigma^2}[\gamma^2(y-\mu)^2 + \nu s^2]\right\}$$

$$= s^{\nu-1} \exp\left\{-\frac{1}{2\sigma^2}[\gamma^2((y-\theta_0)^2 - 2(y-\theta_0)(\mu-\theta_0) + (\mu-\theta_0)^2) + \nu s^2]\right\}$$

$$\propto s^{\nu-1} \exp \left\{ -\frac{1}{2\sigma^2}(\gamma^2(y-\theta_0)^2 + \nu s^2) + \gamma^2[(\mu-\theta_0)/\sigma^2](y-\theta_0) \right\}$$

The transformation of variables

$$T = \frac{\gamma(Y-\theta_0)}{S}$$
$$V = (\nu S^2 + \gamma^2(Y-\theta_0)^2)^{1/2}$$

preserves completeness and sufficiency, and (T, V) has the joint density

$$f_{\nu,\varsigma,\xi}(t,v) \propto v^{\nu}(\nu+t^2)^{-(\nu+1)/2} \exp \frac{\varsigma v^2 + \gamma\xi v t}{(\nu+t^2)^{1/2}}$$

for $v > 0$, $-\infty < t < \infty$. The hypotheses $H_i(\theta_0)$ and $K_i(\theta_0)$ ($H_1(\theta_0) : \mu \leq \theta_0$, $K_1(\theta_0) : \mu > \theta_0$, etc.) can be specified in terms of ξ, as $H_1(\theta_0) : \xi \leq 0$, $K_1(\theta_0) : \xi > 0$, etc. Furthermore, the conditional density of T given $V = v$,

$$f_{\nu,\xi}(t \mid v) \propto (\nu+t^2)^{-(\nu+1)/2} \exp \frac{\gamma\xi v t}{(\nu+t^2)^{1/2}} \qquad -\infty < t < \infty \tag{3.4}$$

does not depend upon ς and has monotone likelihood in ξ. Thus the critical regions given by (3.1)–(3.3) above are UMP ($i = 1, 2$) and UMP unbiased ($i = 3$) level α, conditionally given $V = v$, by the reasoning of section 3.1.1. Unconditionally, they are UMP unbiased of level α (Lehmann, 1959, sec. 4.4).

When $\xi = 0$ the conditional density $f_{\nu,\xi}(t \mid v)$ does not depend upon v, and therefore the critical values $\mathbf{t}_\nu(\alpha)$ are obtained by solving the following equation:

$$\int_{-\infty}^{\mathbf{t}_\nu(\alpha)} (\nu+t^2)^{-(\nu+1)/2}\, dt = \alpha \int_{-\infty}^{\infty} (\nu+t^2)^{-(\nu+1)/2}\, dt \tag{3.5}$$

As in the case where σ is known, the null density function of the test statistic is symmetric about zero, yielding (3.3).

The above critical regions are also obtained when invariance, rather than unbiasedness, is used to restrict the universe of possible critical regions. The problems of testing $H_i(\theta_0)$ vs. $K_i(\theta_0)$ are invariant under the group G_1 of scale changes, $g_a(y,s) = (ay, as)$, $a > 0$, for $i = 1$, 2, and the group G_2 of transformations $g_b(y,s) = (by, |b|s)$, $b \neq 0$, for $i = 3$ (the corresponding transformations on the parameter space are $\mathbf{g}_a(\mu,\sigma) = (a\mu, a\sigma)$

and $\mathbf{g}_b(\mu, \sigma) = (b\mu, |b|\sigma)$, respectively). Specifically, $C_1(\theta_0)$ and $C_2(\theta_0)$ are UMP among level α critical regions that are invariant under G_1, and $C_3(\theta_0)$ is UMP among level α critical regions that are invariant under G_2.

Briefly, the statistics $T = \gamma(Y - \theta_0)/S$ and $|T|$ are maximal invariant under G_1 and G_2, respectively. That is, considered as transformations on the sample space after translation of y by θ_0, $T(y, s) = T(y', s')$ if and only if $(y', s') = g_a(y, s)$ for some $a > 0$, and $|T(y, s)| = |T(y', s')|$ if and only if $(y', s') = g_b(y, s)$ for some $b \neq 0$. Thus, all invariant critical regions must be defined in terms of T and, in the case $i = 3$, must be symmetric about zero. T has (unconditionally) a noncentral $\mathbf{t}$ distribution with ν degrees of freedom and noncentrality $\delta = \gamma(\mu - \theta_0)/\sigma$, and with probability density function

$$f_{\nu,\delta}(t) \propto \int_{-\infty}^{\infty} \exp\left\{-\frac{1}{2}\left(t\left(\frac{w}{\nu}\right)^{1/2} - \delta\right)^2\right\} w^{(\nu-1)/2} \exp\left(-\frac{1}{2}w\right)\, dw \tag{3.6}$$

for $-\infty < t < \infty$. For $\delta = 0$ $f_{\nu,\delta}(t)$ reduces to the density for Student's $\mathbf{t}$,

$$f_{\nu,0}(t) \propto (\nu + t^2)^{-(\nu+1)/2} \qquad -\infty < t < \infty \tag{3.7}$$

The UMA unbiased (UMA invariant) level $1 - \alpha$ confidence interval procedures $\mathcal{U}_i$ are defined by

$$U_1(y, s) = \left[y - \frac{s}{\gamma}\mathbf{t}_\nu(1 - \alpha), \infty\right)$$

$$U_2(y, s) = \left(-\infty, y - \frac{s}{\gamma}\mathbf{t}_\nu(\alpha)\right]$$

$$U_3(y, s) = \left[y - \frac{s}{\gamma}\mathbf{t}_\nu\left(1 - \frac{1}{2}\alpha\right), y - \frac{s}{\gamma}\mathbf{t}_\nu\left(\frac{1}{2}\alpha\right)\right]$$

3.2 Other Functions of μ

Hypotheses about the lognormal median, $\exp(\mu)$, can be specified in terms of its logarithm, μ. For example, the hypothesis $H_1(\theta_0) : \exp(\mu) \leq \theta_0$ can also be restated as $H_1(\theta_0) : \mu \leq \log(\theta_0)$. The exponentials of the endpoints of optimal confidence intervals for μ define optimal confidence limits for $\exp(\mu)$.

4. FUNCTIONS OF σ ALONE

4.1 The Case $\theta(\mu, \sigma) = \sigma$

4.1.1 If μ is known, then

$$V = (\nu S^2 + \gamma^2(y-\mu)^2)^{1/2}$$

is sufficient for σ, and V^2/σ^2 is distributed as chi square with $\nu+1$ df. The probability density of V^2/σ^2 is

$$f_{\nu+1}(u) \propto u^{\nu/2-1} \exp(-\tfrac{1}{2}u) \qquad u > 0$$

The distribution of V^2/θ_0^2 belongs to a 1-parameter exponential family with natural parameter $\varsigma = -\frac{1}{2}(\theta_0/\sigma)^2$; its density is

$$f_{\nu+1,\varsigma}(u) \propto u^{\nu/2-1} \exp \varsigma u \qquad u > 0$$

V^2/θ_0^2 has monotone likelihood ratio in ς, and the usual hypotheses $H_1(\theta_0)$: $\sigma \le \theta_0$, $K_1(\theta_0) : \sigma > \theta_0$, etc., can be restated in terms of ς as $H_1(\theta_0)$: $\varsigma \le -\frac{1}{2}$, $K_1(\theta_0) : \varsigma > \frac{1}{2}$, etc. Thus, by reasoning similar to that of section 3.1.1, UMP level α critical regions for testing $H_i(\theta_0)$ vs. $K_i(\theta_0)$, $i = 1, 2$, are

$$C_1(\theta_0) = \left\{(y, s) : \frac{v}{\theta_0} > \chi_{\nu+1}(1-\alpha)\right\}$$

$$C_2(\theta_0) = \left\{(y, s) : \frac{v}{\theta_0} < \chi_{\nu+1}(\alpha)\right\}$$

where for $0 < \alpha < 1$, $\chi_{\nu+1}(\alpha)$ is defined by the equation

$$\int_0^{(\chi_{\nu+1}(\alpha))^2} f_{\nu+1}(u)\,du = \alpha \int_0^{\infty} f_{\nu+1}(u)\,du \tag{4.1}$$

As in the case $\theta(\mu, \sigma) = \mu$, there is no UMP level α test of $H_3(\theta_0)$: $\theta(\mu, \sigma) = \theta_0$ vs. $K_3(\theta_0) : \theta(\mu, \sigma) \ne \theta_0$. There is, however, a UMP unbiased level α test, with critical region

$$C_3(\theta_0) = \left\{(y, s) : \frac{v}{\theta_0} \notin [\chi_{\nu+1}(\alpha_1), \chi_{\nu+1}(1-\alpha_2)]\right\}$$

where

$$\int_{(\chi_{\nu+1}(\alpha_1))^2}^{(\chi_{\nu+1}(1-\alpha_2))^2} f_{\nu+1}(u)\,du = (1-\alpha)\int_0^{\infty} f_{\nu+1}(u)\,du \tag{4.2}$$

and

$$\int_{(\chi_{\nu+1}(\alpha_1))^2}^{(\chi_{\nu+1}(1-\alpha_2))^2} u f_{\nu+1}(u)\,du = (1-\alpha)\int_0^{\infty} u f_{\nu+1}(u)\,du \tag{4.3}$$

The values $\chi_{\nu+1}(\alpha_1)$ and $\chi_{\nu+1}(1-\alpha_2)$ (or, alternatively, the values α_1 and α_2) are determined by the requirements that the integral of the null density of V^2/θ_0^2 over $C_3(\theta_0)$ must be equal to α (4.2) and that the derivative with respect to ς of the power function must be zero when $\varsigma = -\frac{1}{2}$ (4.3).

Note that, unlike the case $\theta(\mu, \sigma) = \mu$, the null distribution is not symmetric, and that therefore α_1 and α_2 are not equal to $\frac{1}{2}\alpha$. As a practical matter, however, unless ν is very small $\frac{1}{2}\alpha$ is a good approximation to α_1 and α_2.

Confidence interval procedures corresponding to $\mathcal{C}_i = \{C_i(\theta_0) : 0 < \theta_0 < \infty\}$ are

$$U_1(y,s) = \left[\frac{v}{\chi_{\nu+1}(1-\alpha)}, \infty\right)$$

$$U_2(y,s) = \left(0, \frac{v}{\chi_{\nu+1}(\alpha)}\right]$$

$$U_3(y,s) = \left[\frac{v}{\chi_{\nu+1}(1-\alpha_2)}, \frac{v}{\chi_{\nu+1}(\alpha_1)}\right]$$

where $v = (\nu s^2 + \gamma^2(y-\mu)^2)^{1/2}$.

4.1.2 For the case of unknown μ, we can proceed in a manner analogous to section 3.1.2, beginning with the exponential form of the joint density of (Y, S):

$$f(y,s) \propto s^{\nu-1}\exp\left\{\varsigma(\nu s^2 + \gamma^2 y^2) + \gamma^2 \xi y\right\}, \sigma > 0 \qquad -\infty < y < \infty$$

where $\varsigma = -\frac{1}{2}/\sigma^2$ and $\xi = \mu/\sigma^2$. Following the approach of section 3.1.2, we obtain UMP unbiased level α critical regions for testing $H_1(\theta_0) : \varsigma \leq -\frac{1}{2}\theta_0^2$

vs. $K_1(\theta_0) : \varsigma > -\frac{1}{2}\theta_0^2$, etc., by the likelihood ratio method applied to the conditional distribution of $V^2 = \nu S^2 + \gamma^2 Y^2$ given $Y = y$. Now conditionally on $Y = y$, V^2 is equal to νS^2 plus a constant, and so the likelihood ratio method can be applied just as well to the conditional distribution of νS^2 given $Y = y$. But S^2 and Y are independently distributed, and therefore the UMP unbiased level α tests are simple likelihood ratio tests based on S.

The statistic $\nu S^2/\theta_0^2$ is distributed as σ^2/θ_0^2 times chi square with ν df, differing in only a single degree of freedom from the distribution of $(\nu S^2 + \gamma^2(Y - \mu)^2/\theta_0^2$ in the case of known μ. Thus the UMP unbiased level α critical regions are analogous to those in the previous section:

$$C_1(\theta_0) = \left\{(y, s) : \frac{\nu^{1/2}s}{\theta_0} > \chi_\nu(1 - \alpha)\right\}$$

$$C_2(\theta_0) = \left\{(y, s) : \frac{\nu^{1/2}s}{\theta_0} < \chi_\nu(\alpha)\right\}$$

$$C_3(\theta_0) = \left\{(y, s) : \frac{\nu^{1/2}s}{\theta_0} \notin [\chi_\nu(\alpha_1), \chi_\nu(1 - \alpha_2)]\right\}$$

where $\chi_\nu(\alpha)$, $0 < \alpha < 1$, and α_1 and α_2 are defined according to (4.1)–(4.3), with ν rather than $\nu + 1$ df. UMA unbiased level $1 - \alpha$ confidence procedures correspond to

$$U_1(y, s) = \left[\frac{\nu^{1/2}s}{\chi_\nu(1 - \alpha)}, \infty\right)$$

$$U_2(y, s) = \left(0, \frac{\nu^{1/2}s}{\chi_\nu(\alpha)}\right]$$

$$U_3(y, s) = \left[\frac{\nu^{1/2}s}{\chi_\nu(1 - \alpha_2)}, \frac{\nu^{1/2}s}{\chi_\nu(\alpha_1)}\right]$$

(The critical region $C_1(\theta_0)$ is UMP among all level α tests of $H_1(\theta_0)$ vs. $K_1(\theta_0)$, and the confidence interval procedure $\mathcal{U}_1$ is UMA among all level $1 - \alpha$ confidence procedures for σ, whereas unbiasedness is required to obtain $C_i(\theta_0)$ and $\mathcal{U}_i$ for $i = 2, 3$ (Lehmann, 1959, section 5.2).)

The above critical regions and confidence interval procedures can also be obtained as optimal solutions invariant under the group of transformations $g_a(y, s) = (y + a, s)$, $-\infty < a < \infty$. The hypotheses $H_i(\theta_0)$, $K_i(\theta_0)$, $i = 1, 2, 3$, are invariant under this group, and S is maximal invariant.

4.2 Other Functions of σ

The geometric standard deviation (GSD), $\exp(\sigma)$, corresponds to σ in exactly the same way as the median, $\exp(\mu)$, corresponds to μ. Therefore critical regions and confidence interval procedures for the GSD are defined in terms of those for σ in the same way as critical regions and confidence interval procedures for the median are defined in terms of those for μ. Other functions of interest that depend on σ alone, and which can be treated analogously, are the shape factors b_1 and b_2, the coefficient of variation, the Theil coefficient and modified Theil coefficients, and Gini's coefficient of concentration (Chapter 1, formulae 4.8–4.13, 4.15).

5. PARAMETRIC FUNCTIONS DEPENDING ON LINEAR FUNCTIONS OF μ AND σ

5.1 The Case $(\theta(\mu,\sigma) = \mu + \kappa\sigma$

For fixed $\kappa \neq 0$, hypotheses $H_1(\theta_0) : \mu + \kappa\sigma \leq \theta_0$, $K_1(\theta_0) : \mu + \kappa\sigma > \theta_0$, etc., are invariant under the group of transformations $g_a(y-\theta_0, s) = (a(y-\theta_0), as)$, $a > 0$. The statistic

$$T = \frac{\gamma(Y - \theta_0)}{S}$$

is maximal invariant under this group, and is distributed as noncentral $\mathbf{t}$ with ν df and noncentrality parameter $\delta = \gamma(\mu-\theta_0)/\sigma$, with density $f_{\nu,\delta}(t)$ given by (3.6). The hypotheses $H_i(\theta_0)$, $K_i(\theta_0)$, $i = 1, 2, 3$, can be stated in terms of δ, as $H_1(\theta_0) : \delta \leq \eta$, $K_1(\theta_0) : \delta > \eta$, etc., where $\eta = -\gamma\kappa$. Since T has monotone likelihood ratio in δ, the UMP level α invariant critical regions for testing $H_i(\theta_0)$ vs. $K_i(\theta_0)$, $i = 1, 2$, are

$$C_1(\theta_0) = \left\{(y,s) : \frac{\gamma(y-\theta_0)}{s} > \mathbf{t}_{\nu,\eta}(1-\alpha)\right\}$$

$$C_2(\theta_0) = \left\{(y,s) : \frac{\gamma(y-\theta_0)}{s} < \mathbf{t}_{\nu,\eta}(\alpha)\right\}$$

respectively, where the values $\mathbf{t}_{\nu,\eta}(\alpha)$, $0 < \alpha < 1$, are found by solving

$$\int_{-\infty}^{\mathbf{t}_{\nu,\eta}(\alpha)} f_{\nu,\eta}(t)\,dt = \int_{-\infty}^{\infty} f_{\nu,\eta}(t)\,dt \qquad (5.1)$$

Solutions $\mathbf{t}_{\nu,\eta}(\alpha)$ have been extensively tabulated for commonly used test levels α and values of η corresponding to certain quantiles of the $N(0,1)$ distribution, for certain integer values ν (Johnson and Welch, 1940; Resnikoff and Lieberman, 1957; van Eeden, 1961; Locks, Alexander, and Byars, 1963; Owen and Amos, 1963; Scheuer and Spurgeon, 1963; Owen, 1965; Pearson and Hartley, 1972), while asymptotic approximations have been devised for large values of ν (see Johnson and Kotz, 1970, chapter 31). More useful are computer algorithms for calculating $\mathbf{t}_{\nu,\delta}(\alpha)$ for arbitrary ν, δ, and α (Owen and Amos, 1963; Land et al., 1987). A helpful symmetry relationship is $\mathbf{t}_{\nu,\eta}(\alpha) = -\mathbf{t}_{\nu,-\eta}(1-\alpha)$.

Unlike the case $\kappa = 0$, an optimal invariant test cannot be obtained for $H_3(\theta_0)$ vs. $K_3(\theta_0)$ without invoking unbiasedness to further restrict the universe of tests. Unbiasedness implies that the derivative of the power function $\beta(\delta)$ be zero at $\delta = \eta$. The UMP level α invariant, unbiased critical region is

$$C_3(\theta_0) = \left\{(y,s) : \gamma(y-\theta_0)/s \notin [\mathbf{t}_{\nu,\eta}(\alpha_1), \mathbf{t}_{\nu,\eta}(1-\alpha_2)]\right\};$$

the defining equations for α_1 and α_2 are

$$\alpha_1 + \alpha_2 = \alpha \tag{5.2}$$

$$\nu^{-1/2} \int_{\mathbf{t}_{\nu,\eta}(\alpha_1)}^{\mathbf{t}_{\nu,\eta}(1-\alpha_2)} t f_{\nu+1,\eta}\left(\left(\frac{\nu+1}{\nu}\right)^{1/2} t\right) dt = \eta(1-\alpha) \int_{-\infty}^{\infty} f_{\nu,\eta}(t)\, dt \tag{5.3}$$

(The power function $\beta(\delta)$ is the integral over $C_3(\theta_0)$ of

$$f_{\nu,\delta}(t) \propto \int_0^{\infty} \exp\left\{-\frac{1}{2}\left[t\left(\frac{w}{\nu}\right)^{1/2} - \delta\right]^2\right\} w^{(\nu-1)/2} \exp\left\{-\frac{1}{2}w\right\} dw;$$

the derivative $\beta'(\delta)$ is the integral of

$$df_{\nu,\delta}(t)/d\delta \propto \int_0^{\infty} \left(t\left(\frac{w}{\nu}\right)^{1/2} - \delta\right) \exp\left\{-\frac{1}{2}\left[t\left(\frac{w}{\nu}\right)^{1/2} - \delta\right]^2\right\} w^{(\nu-1)/2}$$
$$\times \exp\left\{-\tfrac{1}{2}w\right\} dw$$

$$= \nu^{-1/2} t \int_0^\infty \exp\left\{-\frac{1}{2}\left[t\left(\frac{\nu+1}{\nu}\right)^{1/2}\left(\frac{w}{\nu+1}\right)^{1/2} - \delta\right]^2\right\}$$

$$\times\, w^{\nu/2} \exp\left\{-\frac{1}{2}w\right\}\, dw$$

$$- \delta \int_0^\infty \exp\left\{-\frac{1}{2}\left[t\left(\frac{w}{\nu}\right)^{1/2} - \delta\right]^2\right\} w^{(\nu-1)/2} \exp\left\{-\tfrac{1}{2}w\right\}\, dw$$

$$\propto \nu^{-1/2} t f_{\nu+1,\delta}\left(t\left(\frac{\nu+1}{\nu}\right)^{1/2}\right) - \delta f_{\nu,\delta}(t)$$

from which setting $\beta'(\eta) = 0$ gives (5.3).)

For small ν and large η, α_1 and α_2 may not be well approximated by $\frac{1}{2}\alpha$, although equal tail probabilities may be preferred nonetheless for reasons that are probably best described as esthetic.

UMA level $1-\alpha$ invariant and level $1-\alpha$ unbiased invariant confidence interval procedures are as follows:

$$U_1(y,s) = \left[y - \frac{s}{\gamma}\mathrm{t}_{\nu,\eta}(1-\alpha), \infty\right]$$

$$U_2(y,s) = \left[-\infty, y - \frac{s}{\gamma}\mathrm{t}_{\nu,\eta}(\alpha)\right]$$

$$U_3(y,s) = \left[y - \frac{s}{\gamma}\mathrm{t}_{\nu,\eta}(1-\alpha_2), y - \frac{s}{\gamma}\mathrm{t}_{\nu,\eta}(\alpha_1)\right]$$

5.2 Other Functions of Linear Functions of μ and σ

Lognormal quantiles are of the form $\theta(\mu,\sigma) = \exp\{\mu + \kappa\sigma\}$, where κ is a standard normal quantile, $\kappa = \mathbf{z}(p) = \Phi^{-1}(p)$, $0 < p < 1$. Thus UMA invariant and UMA unbiased invariant confidence intervals for $\theta(\mu,\sigma)$ are obtained by taking the exponentials of the appropriate confidence intervals for $\mu + \kappa\sigma$, and hypothesis tests of the form $H_1(\theta_0) : \exp(\mu + \kappa\sigma) \le \theta_0$, etc., are tested in terms of $\mu + \kappa\sigma$ (e.g., $H_1(\theta_0) : \mu + \kappa\sigma \le \log(\theta_0)$).

Optimal tests and confidence interval procedures for the lognormal distribution function,

$$\theta(\mu,\sigma;x) = \Phi\left(\frac{(\log(x) - \mu)}{\sigma}\right)$$

can be derived from the procedures for linear functions of μ and σ. For example, the hypothesis $H_1(\theta_0) : \Phi((\log(x) - \mu)/\sigma \leq \theta_0$ can be restated in terms of a linear function of μ and σ:

$$H_1'(\theta_0) : \mu + \Phi^{-1}(\theta_0)\sigma \geq \log(x)$$

and other inequalities have analogous correspondences. Note that $H_1(\theta_0)$ corresponds to $H_1'(\theta_0) : \mu + \kappa\sigma \geq \log(x)$, where $\kappa = \Phi^{-1}(\theta_0)$, while $H_2(\theta_0)$ corresponds to $H_2'(\theta_0) : \mu + \kappa\sigma \leq \log(x)$. Thus critical regions for testing $H_i(\theta_0)$ vs. $K_i(\theta_0)$, for $i = 1, 2, 3$, are

$$C_1(\theta_0) = \{(y, s) : t_x < \mathbf{t}_{\nu,\eta}(\alpha)\}$$

$$C_2(\theta_0) = \{(y, s) : t_x > \mathbf{t}_{\nu,\eta}(1 - \alpha)\}$$

$$C_3(\theta_0) = \{(y, s) : t_x \notin [\mathbf{t}_{\nu,\eta}(\alpha_1), \mathbf{t}_{\nu,\eta}(1 - \alpha_2)]\}$$

where $t_x = \gamma(y - \log(x))/s$, $\eta = -\gamma\Phi^{-1}(\theta_0)$, and $\mathbf{t}_{\nu,\eta}(\alpha)$, α_1, α_2 are defined by equations (5.1)–(5.3).

The confidence procedures $\mathcal{U}_i$ defined by the families $\mathcal{C}_i$ of critical regions $C_i(\theta_0)$, $0 < \theta_0 < 1$, define intervals. For fixed θ_0, the integral of the noncentral $\mathbf{t}$ density, $f_{\nu,\delta}(t)$, over the interval $(-\infty, \mathbf{t}_{\nu,\eta}(\theta_0)(\alpha))$ is a decreasing function of δ; therefore, $\mathbf{t}_{\nu,\delta}(\alpha)$ is an increasing function of δ. Since $\eta(\theta_0) = -\gamma\Phi^{-1}(\theta_0)$ is a decreasing function of θ_0, it follows that $\mathbf{t}_{\nu,\eta}(\theta_0)(\alpha)$ is a decreasing function of θ_0. Thus,

$$U_1(y, s) = \left\{\theta_0 : \frac{\gamma(y - \log(x))}{s} > \mathbf{t}_{\nu,\eta(\theta_0)}(\alpha)\right\}$$
$$= [u_1(y, s), 1]$$

where $u_1(y, s)$ is found by solving

$$\mathbf{t}_{\nu,\eta(\theta_0)}(\alpha) = t_x$$

for θ_0, a problem for which computer algorithms are available (Owen, 1963; Land et al., 1987). Similarly,

$$U_2(y, s) = (0, u_2(y, s))$$

and

$$U_3(y, s) = [u_{31}(y, s), u_{32}(y, s)]$$

where u_2, u_{31}, and u_{32} are found by solving

$$\mathrm{t}_{\nu,\eta(\theta_0)}(\beta) = t_x$$

for θ_0, where $\beta = 1 - \alpha$, α_1, and $1 - \alpha_2$, respectively.

6. FUNCTIONS OF $\mu + \lambda\sigma^2$, $\lambda \neq 0$

6.1 The Case $\theta(\mu, \sigma) = \mu + \lambda\sigma^2$, $\lambda \neq 0$

This case is a generalization of the case $\lambda = 0$ (section 3.1.2). The hypotheses $H_1(\theta_0) : \mu + \lambda\sigma^2 \leq \theta_0$, $K_1(\theta_0) : \mu + \lambda\sigma^2 > \theta_0$, etc., can be restated in terms of $\xi = (\mu - \theta_0)/\sigma^2$, as $H_1(\theta_0) : \xi \leq -\lambda$, $K_1(\theta_0) : \xi > -\lambda$, etc. The natural parameterization of the exponential form of the joint distribution of $(Y - \theta_0, S)$ is (ξ, ς), where $\varsigma = -\frac{1}{2}/\sigma^2$, and, by the same reasoning as in 3.1.2, the UMP unbiased level α critical regions for testing $H_i(\theta_0)$ vs. $K_i(\theta_0)$, $i = 1, 2, 3$, are based on the conditional distribution of

$$T = \frac{\gamma(Y - \theta_0)}{S}$$

given $V = v$, where

$$V = (\nu S^2 + \gamma^2(Y - \theta_0)^2)^{1/2}$$

In the present case, however, the boundary value of ξ, between $H_i(\theta_0)$ and $K_i(\theta_0)$, is $-\lambda \neq 0$ rather than 0 as in 3.1.2. For $\xi \neq 0$ the conditional density of T given $V = v$, shown in (3.4), does not depend upon ς, but it does depend upon v, and it has monotone likelihood ratio in ξ.

The critical regions for the conditional likelihood ratio tests of $H_i(\theta_0)$ vs. $K_i(\theta_0)$, $i = 1, 2, 3$, are determined by

$$f_{\nu,-\lambda}(t \mid v) \propto (\nu + t^2)^{-1/2(\nu+1)} \exp\left\{\frac{-\gamma\lambda v t}{(\nu + t^2)^{1/2}}\right\} \qquad -\infty < t < \infty$$

which depends upon v, λ, and γ through their product. Since V^2 is a random variable whose mean, for small $\mu - \theta_0$, is roughly proportional to $\nu + 1$, and since in the single-sample case $\gamma^2 = \nu + 1$, it is convenient to tabulate the quantiles of this distribution in terms of

$$z = \frac{-\gamma v \lambda}{(\nu + 1)}$$

a value that does not vary systematically with ν. Denoting these quantiles by $\mathbf{t}_{\nu,z|V}(\alpha)$, $0 < \alpha < 1$, to emphasize that they pertain to a conditional distribution, the UMP unbiased critical regions of level α are

$$C_1(\theta_0) = \left\{ (y, s) : \frac{\gamma(y - \theta_0)}{s} > \mathbf{t}_{\nu,z|V}(1 - \alpha) \right\}$$

$$C_2(\theta_0) = \left\{ (y, s) : \frac{\gamma(y - \theta_0)}{s} < \mathbf{t}_{\nu,z|V}(\alpha) \right\}$$

$$C_3(\theta_0) = \left\{ (y, s) : \frac{\gamma(y - \theta_0)}{s} \notin [\mathbf{t}_{\nu,z|V}(\alpha_1), \mathbf{t}_{\nu,z|V}(1 - \alpha_2)] \right\}$$

For $0 < \alpha < 1$, $\mathbf{t}_{\nu,z|V}(\alpha)$ is obtained by solving the equation

$$\int_{-\infty}^{\mathbf{t}_{\nu,z|V}(\alpha)} f_{\nu,z|V}(t)\, dt = \alpha \int_{-\infty}^{\infty} f_{\nu,z|V}(t)\, dt$$

where

$$f_{\nu,z|V}(t) \propto (\nu + t^2)^{-(\nu+1)/2} \exp\left\{(\nu + 1)zt/(\nu + t^2)^{1/2}\right\} \qquad -\infty < t < \infty$$

The tail probabilities in the definition of $C_3(\theta_0)$, α_1 and α_2, depend upon θ_0. They are defined by the two equations $\alpha_1 + \alpha_2 = \alpha$ and

$$\int_{\mathbf{t}_{\nu,z|V}(\alpha_1)}^{\mathbf{t}_{\nu,z|V}(1-\alpha_2)} t(\nu + t^2)^{-1/2} f_{\nu,z|V}(t)\, dt = (1 - \alpha) \int_{-\infty}^{\infty} t(\nu + t^2)^{-1/2} f_{\nu,z|V}(t)\, dt$$

It is not immediately clear that the UMA unbiased level $1 - \alpha$ confidence procedures for $\mu + \lambda\sigma^2$, based on $\mathcal{C}_i = \{C_i(\theta_0) : -\infty < \theta_0 < \infty\}$, define confidence sets that are intervals. The result holds uniformly for $\nu \geq 2$, for $\mathcal{U}_1$, $\mathcal{U}_2$, and $\mathcal{U}_3$, but it appears that confidence sets other than intervals may be obtained in the case $\nu = 1$ (Land, 1971a; Land, Johnson, and Joshi, 1973). For $\nu \geq 2$, therefore, confidence limits u_1, u_2, u_{31} and u_{32} are obtained by solving

$$\gamma(y - \theta_0)/s = \mathbf{t}_{\nu,z(\theta_0)|V}(\beta)$$

for θ_0, where

$$z(\theta_0) = -\gamma\lambda(\nu s^2 + \gamma^2(y - \theta_0)^2)^{1/2}$$

and $\beta = \alpha$, $1 - \alpha$, $1 - \alpha_2(\theta_0)$, and $\alpha_1(\theta_0)$, respectively.

Tables of the critical values $\mathrm{t}_{\nu,z|V}(\alpha)$, $\mathrm{t}_{\nu,z|V}(\alpha_1)$, and $\mathrm{t}_{\nu,z|V}(1 - \alpha_2)$ are available (Land, 1971a, 1971b, 1974b). Confidence limits for $\mu + \lambda\sigma^2$, based on the critical regions $C_i(\theta_0)$, $-\infty < \theta_0 < \infty$, $i = 1, 2, 3$, have been tabulated in more detail, however, and are easier to use (Land, 1973, 1975). That is, if $u_1(y, s)$ and $u_2(y, s)$ and $u_{31}(y, s)$ and $u_{32}(y, s)$ are, respectively, lower and upper one-sided and lower and upper two-sided confidence limits corresponding to the UMA unbiased confidence procedures of level $1 - \alpha$ for $\mu + \lambda\sigma^2$, the critical regions can be described as follows:

$$C_1(\theta_0) = \{(y, s) : \theta_0 < u_1(y, s)\}$$
$$C_2(\theta_0) = \{(y, s) : \theta_0 > u_2(y, s)\}$$
$$C_3(\theta_0) = \{(y, s) : \theta_0 \notin [u_{31}(y, s), u_{32}(y, s)]\}$$

The tabulated confidence limits are in standardized form, corresponding to the special case $\lambda = \frac{1}{2}$, $y = -\frac{1}{2}s^2$, and $\gamma = (\nu + 1)^{1/2}$, and are tabulated by ν and

$$w = \frac{2|\lambda|\gamma s}{(\nu + 1)^{1/2}}$$

Once standardized confidence limits $\mathrm{u}(\alpha)$, $0 < \alpha < 1$, have been obtained (by interpolation if necessary), the desired limit for a particular α is calculated as

$$u(y, s) = y + \lambda s^2 + \left(\frac{\lambda}{|\lambda|}\right)\left(\frac{\nu + 1}{\nu}\right)^{1/2}\left(\frac{s}{\gamma}\right)\mathrm{u}(\alpha')$$

where $\alpha' = \alpha$ if $\lambda > 0$ and $\alpha' = 1 - \alpha$ if $\lambda < 0$. A less laborious approach is to calculate the confidence limit directly, using a computer program designed for that purpose (Land et al., 1987).

For small s and large sample size (say, $s = .5$ and $\nu = 100$), approximate confidence limits for $\mu + \lambda\sigma^2$ can be obtained by treating the minimum variance unbiased estimate, $y + \lambda s^2$, as an observation on a random variable that is approximately normally distributed with standard deviation equal to $s(1/\gamma^2 + \frac{1}{2}s^2/(\nu + 2))^{1/2}$, a method suggested by D. R. Cox (Land, 1972).

6.2 Other Parametric Functions of $\mu + \lambda\sigma^2$, $\lambda \neq 0$

The $\Lambda(\mu, \sigma)$ mode, $\exp(\mu - \sigma^2)$, the mean, $\exp(\mu + \frac{1}{2}\sigma^2)$, and higher moments about zero,

$$\mu_k = \exp\left(k\mu + \tfrac{1}{2}k^2\sigma^2\right)$$

are all exponentials of linear functions of μ and σ^2, and critical regions and confidence limits for these parametric functions are easily obtained from the solutions for the corresponding functions $\mu + \lambda\sigma^2$.

7. MISCELLANEOUS FUNCTIONS OF μ AND σ

7.1 Some Functions of Interest

Exact and optimal hypothesis tests and confidence interval procedures have been presented in sections 3–6 for parametric functions depending upon μ alone and σ alone, on linear functions of μ and σ, and on linear functions of μ and σ^2. The methods of sections 5 and 6 also define optimal procedures for functions of $(\mu+x)/\sigma$ and $(\mu+x)/\sigma^2$, for arbitrary x. All of these procedures depend upon the ability to restate hypotheses in terms of parameters that occur naturally in the probability density functions of (Y, S) or various statistics, like $T = \gamma Y/S$, based on (Y, S).

There are parametric functions of interest that are not covered by sections 3–6. Examples include the lognormal central moments,

$$\begin{aligned}
\mu_2 &= \exp(2\mu)\omega(\omega - 1) \\
\mu_3 &= \exp(3\mu)\omega^{3/2}(\omega - 1)^2(\omega + 2) \\
\mu_4 &= \exp(4\mu)\omega^2(\omega - 1)^2(\omega^4 + 2\omega^3 + 3\omega^2 - 3)
\end{aligned}$$

etc., where $\omega = \exp(\sigma^2)$. Another is Gini's coefficient of mean difference (see Chapter 1, (4.14)),

$$G = 2\exp\left(\mu + \frac{1}{2}\sigma^2\right)\left\{2\Phi\left(\frac{\sigma}{\sqrt{2}}\right) - 1\right\}$$

7.2 Approximate Methods

7.2.1 Approximation of $\theta(\mu, \sigma)$ by a Linear Function of μ and σ^2. An approximate method of confidence interval estimation was developed for the family of parametric functions expressible as $E(f(X))$, where X is a

$N(\mu, \sigma)$ random variable and f is a nonlinear, twice-differentiable, real-valued function such that the expected values of $f(X)$ and $f'(X)$ both exist, for $-\infty < \mu < \infty$ and $\sigma > 0$. In such cases, $E(f(X))$ depends nontrivially on both μ and σ^2; moreover, it depends on them as a function of $\mu + \lambda\sigma^2$, for some $\lambda \neq 0$, if and only if $f(x) = \exp(x^k)$, for some $k \neq 0$, and it cannot be a function of $\mu + \kappa\sigma$, for any κ (Land, 1971a). Examples of such parametric functions are

$$\begin{aligned} E(X^2) &= \mu^2 + \sigma^2 \\ E(X^3) &= \mu^3 + 3\mu\sigma^2 \\ E(\sin^2(X)) &= \tfrac{1}{2}(1 - \cos(2\mu)\exp(-2\sigma^2)) \\ E(\sinh^2(X) &= \tfrac{1}{2}(\cosh(2\mu)\exp(2\sigma^2) - 1) \end{aligned}$$

corresponding to the square root, cube root, arcsine, and hyperbolic arcsine data transformations, respectively, when used to achieve normality.

The approach, as described in (Land, 1975), is to approximate the parametric function $\theta(\mu, \sigma) = E(f(X))$ by the linear function $\mu + \lambda\sigma^2$, where λ is the ratio of partial derivatives, $\lambda = \theta_{\sigma^2}/\theta_{\mu}$, evaluated at $(\hat{\mu}, \hat{\sigma}^2) = (y, s^2)$. If $u(y, s)$ is a confidence limit for $\mu + \lambda\sigma^2$, the corresponding approximate confidence limit for $\theta(\mu, \sigma)$ is $\theta(\tilde{\mu}, \tilde{\sigma})$, where $(\tilde{\mu}, \tilde{\sigma}^2)$ is that point on the line $\mu + \lambda\sigma^2 = u(y, s)$ for which the likelihood function is maximized. The approach is motivated by the following considerations. First, any smooth function of (μ, σ) can be approximated locally by a linear function of μ and σ^2. Second, confidence intervals for $\mu + \lambda\sigma^2$ and $\theta(\mu, \sigma)$ correspond to regions, bounded by one or two contours of the appropriate function, in the half-plane of points (μ, σ^2), $-\infty < \mu < \infty$, $\sigma^2 > 0$; in terms of coverage probabilities, the degree to which a particular region bounded by contours of $\theta(\mu, \sigma)$ corresponds to another region bounded by contours of $\mu + \lambda\sigma^2$ is important only within a region of non-negligible likelihood. As the likelihood surface becomes steeper (e.g., with increasing sample size), the region of non-negligible likelihood becomes smaller, and the correspondence within that region between exact confidence regions defined in terms of $\mu + \lambda\sigma^2$ and approximate confidence regions defined in terms of $\theta(\mu, \sigma)$ becomes closer.

The linearization approximation has been evaluated by Monte Carlo simulation for the four parametric functions identified above as the expected values of random variables that can be transformed to normality by the square root, cube root, arcsine, and arcsinh transformations. For these functions, the method performed well for a range of values of ν,

μ, and σ^2, and substantially better than methods based on transformations ofconfidence limits for μ or on normal theory methods applied to the MVUE of $\theta(\mu, \sigma)$ and the MVUE of the variance of that estimate (Land, 1974a). The linearization method is illustrated below for the lognormal variance.

Example: For

$$\theta(\mu, \sigma) = \exp(2\mu + \sigma^2)(\exp(\sigma^2) - 1)$$

we have

$$\lambda = \theta_{\sigma^2}/\theta_{\mu} = \frac{\exp(\sigma^2) - \frac{1}{2}}{\exp(\sigma^2) - 1}$$

Let $(y, s) = (2, 1)$ be the sample mean and standard deviation of the logarithms of a $\Lambda(\mu, \sigma)$ random sample of size 16. Then $\nu = 15$ and $\gamma = (\nu + 1)^{1/2} = 4$; further, $\lambda = 1.291$ and $\theta(y, s) = 255.0$. The tangent line to the contour $\theta(\mu, \sigma) = 255.0$ at the point $(\hat{\mu}, \hat{\sigma}^2) = (2, 1)$ is

$$\mu + 1.291\sigma^2 = 3.291$$

The UMAU level .95 upper confidence limit for $\mu + 1.291\sigma^2$ is 4.736 (by direct calculation [Land et al., 1987]). For arbitrary u, λ, the likelihood function on the line $\mu + \lambda\sigma^2 = u$ is maximized at the point $(\tilde{\mu}, \tilde{\sigma}^2) = (u - \lambda\tilde{\sigma}^2, \tilde{\sigma}^2)$, where

$$\tilde{\sigma}^2 = \frac{\{-|k| + [k^2 + (y - u)^2 + \nu s^2/\gamma^2]^{1/2}\}}{|\lambda|}$$

and

$$k = \frac{\nu + 1}{2\lambda\gamma^2}$$

In the present example, $k = 1/(2\lambda) = .3873$, $u = 4.736$, and $\tilde{\sigma}^2 = 1.968$. The confidence limit $u = 4.736$ for $\mu + 1.291\sigma^2$ corresponds to the approximate confidence limit

$$\begin{aligned}\theta(4.736 - (1.291)(1.968), (1.968)^{1/2}) \\ &= \theta(2.195, (1.968)^{1/2}) \\ &= \exp((2)(2.195) + 1.968)(\exp(1.968) - 1) \\ &= 3553.7\end{aligned}$$

for μ_2.

The essential aspects of the method are illustrated for the above example in Figure 1. The figure shows an ovoid likelihood contour in the half-plane $-\infty < \mu < \infty$, $\sigma^2 > 0$, and the point $(\hat{\mu}, \hat{\sigma}^2) = (2.0, 1.0)$ inside the contour. The contour $\theta(\mu, \sigma) = \theta(2.0, 1.0) = 255.0$ is shown, and also the line $\mu + 1.291\sigma^2 = 3.291$, which is tangent to it at $(\hat{\mu}, \hat{\sigma}^2)$. The level .95 upper confidence limit for $\mu + 1.291\sigma^2$, 4.736, is represented by its (linear) contour, and the point of maximum likelihood on that line, $(\tilde{\mu}, \tilde{\sigma}^2) = (2.195, 1.968)$, at which the depicted likelihood contour is tangent, is shown. Finally, the approximate confidence limit for $\theta(\mu, \sigma)$, 3553.7, is represented as the contour of $\theta(\mu, \sigma)$ passing through $(\tilde{\mu}, \tilde{\sigma}^2)$. The example is one in which the statistical properties of the approximate method should be fairly good, since the degree of approximation of the family of contours of $\theta(\mu, \sigma)$ by the family of lines is rather close compared to other parametric functions that have been evaluated by Monte Carlo simulation (Land, 1974a).

The figure suggests an iterative refinement of the method just described. Note that, in the figure, the line corresponding to the exact confidence limit for $\mu + \lambda\sigma^2$ is not tangent to the contour corresponding to the approximate limit for $\theta(\mu, \sigma)$. If, however, we adjust λ to correspond to the tangent line at $(\tilde{\mu}, \tilde{\sigma}^2)$, and compute the confidence limit for the new linear function $\mu + \lambda\sigma^2$, we obtain a new approximate limit for $\theta(\mu, \sigma)$. The procedure can be continued as long as desired, but convergence should be rapid given reasonable regularity of the contours of $\theta(\mu, \sigma)$, as in the present example. The purpose of the refined method is to obtain a better correspondence between the boundaries of the exact confidence set for $\mu + \lambda\sigma^2$ and the associated approximate confidence set for $\theta(\mu, \sigma)$, in a region of non-negligible likelihood.

Continuing with the example, the tangent line at $(\tilde{\mu}, \tilde{\sigma}^2)$ corresponds to $\lambda = 1.081$, compared to 1.291 for the tangent at $(\hat{\mu}, \hat{\sigma}^2)$. The exact level .95 confidence limit for $\mu + 1.081\sigma^2$ is 4.315. The likelihood function on the line $\mu + 1.081\sigma^2 = 4.315$ is maximized at (2.226,1.933), and the

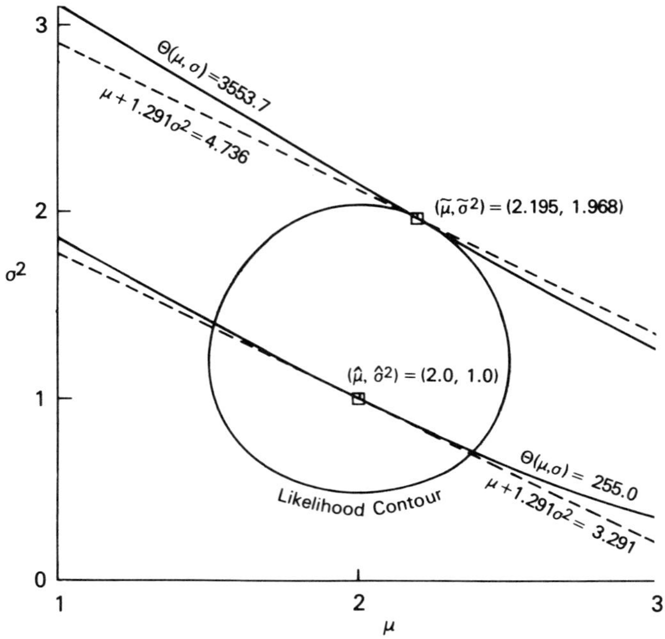

Figure 1 Illustrations of an approximate confidence interval procedure for $\theta(\mu, \sigma) = \exp(2\mu + \sigma^2)(\exp(\sigma^2) - 1)$, given sample estimates $(\hat{\mu}, \hat{\sigma}^2) = (2.0, 1.0)$ from a random sample of size 16. Contours of $\theta(\mu, \sigma)$ are approximated by contours of the linear function $\mu + \lambda\sigma^2$, where $\lambda = 1.291$ is chosen so that the two contours passing through $(\hat{\mu}, \hat{\sigma}^2)$ are tangent at that point. Given level .95 upper confidence limit 4.736 for $\mu + \lambda\sigma^2$, the corresponding approximate confidence limit for $\theta(\mu, \sigma)$, 3553.7, is chosen such that the contours $\theta(\mu, \sigma) = 3553.7$ and $\mu + 1.291\sigma^2 = 4.736$ intersect at the point of maximum likelihood, $(\tilde{\mu}, \tilde{\sigma}^2)$, on the latter contour.

value of θ at that point is 3499.7, which may be compared to the previous limit of 3553.7. (The contours corresponding to the two limits are almost indistinguishable in the scale of Figure 1, as the logarithms of the limits, 8.160 and 8.176, respectively, are rather close.) A further iteration produces only a negligible difference, as the next change in λ is from 1.081 to 1.085; this time, the change in the approximate limit is only from 3499.7 to 3500.8 or, in the logarithmic scale, from 8.1604 to 8.1607. It would appear from this example that the iterative procedure is a worthwhile refinement of the method.

7.2.2 Approximation of $\theta(\mu, \sigma)$ by a linear function of μ and σ. There is no reason why the approach of section 7.2.1 could not also be applied to approximation of $\theta(\mu, \sigma)$ by a linear function of μ and σ. One would expect the choice to be based on the availability of tables, computational algorithms, and computer programs, and on the shape of the contours of $\theta(\mu, \sigma)$ in the (μ, σ^2) half-plane (especially, that part of the half-plane corresponding to the desired confidence limits). In the example illustrated in Figure 1, the curvature of the contours of the function $\mu + \kappa\sigma$, where κ is chosen to give contours tangent to those of $\theta(\mu, \sigma)$ at certain points, is such that approximation by linear functions of μ and σ might be expected to be marginally superior to approximation by linear functions of μ and σ^2 in the lower left of the figure, i.e., for lower confidence bounds, and marginally inferior in the upper right of the figure, i.e., for upper confidence bounds.

8. LINEAR SAMPLING MODELS

In this section we consider sampling models in which the logarithms of the random variables have jointly a spherical normal distribution, or one which can be transformed to such a distribution by a known linear transformation. That is, it is assumed that sampling is from n random variables whose logarithms are jointly multivariate normal, with mean vector $\mathbf{X}\boldsymbol{\beta}$ and covariance matrix $\sigma^2\Sigma$, where $\mathbf{X}$ and Σ are known matrices of size $n \times k$ and $n \times n$, respectively, $k < n$, and where the positive number σ and the $k \times 1$ vector $\boldsymbol{\beta}$ are unknown parameters.

If we let $\mu = \mathbf{a}'\boldsymbol{\beta} = \Sigma a_i \beta_i$ for some vector $\mathbf{a}$ of real numbers a_i, $i = 1, \ldots, k$, and if we let Y and S^2 be the MVUEs of μ and σ^2, respectively, it is clear that sections 3–7 apply to hypothesis tests and confidence interval procedures for parametric functions $\theta(\mu, \sigma)$, since Y and S are independently distributed as a $N(\mu, \sigma/\gamma)$ variate and as $\sigma/\nu^{1/2}$ times the square root of a chi-square variate with ν df, where γ and ν are known. For example, the methods of sections 3–7 are appropriate for the distribution of a lognormal variate corresponding to a particular set of values of the independent variables in a regression model for the mean of its logarithm, or for the distribution of a geometric contrast $\prod Z_{ij} a_{ij}$ in lognormal variates Z_{ij} whose logarithms correspond to an analysis of variance model.

REFERENCES

Eeden, Constance van (1961). Some approximations to the percentage points of the noncentral t-distribution, *Revue de l'Inst. Internat. de Statist.*, *29*, 4–31.

Johnson, N. L. and Kotz, S. (1970). *Continuous Univariate Distributions, Vol 2*, John Wiley & Sons, Inc., New York, Chap. 31, 201–219.

Johnson, N. L. and Welch, B. L. (1940). Applications of the noncentral t-distribution, *Biometrika, 31*, 362–389.

Land, C. E. (1971a). Confidence intervals for linear functions of the normal mean and variance, *Ann. Math. Statist., 42*, 1187–1205.

Land, C. E. (1971b). Critical values for hypothesis test about linear functions of the normal mean and variance, *Unpublished Mathematical Tables File, Math. of Computation, 25*, 44.

Land, C. E. (1972). An evaluation of approximate confidence interval methods for lognormal means, *Technometrics, 14*, 145–158.

Land, C. E. (1973). Standard confidence limits for linear functions of the normal mean and variance, *J. Amer. Statist. Assoc., 68*, 960–963.

Land, C. E. (1974a). Confidence interval estimation for means after data transformations to normality, *J. Amer. Statist. Assoc., 69*, 795–802.

Land, C. E. (1974b). Tables of critical values for hypothesis tests about linear functions of the normal mean and variance. *Unpublished Mathematical Tables File, Math. of Computation, 28, 127.*

Land, C. E. (1975). Tables of confidence limits for linear functions of the normal mean and variance, in Harter, H. L., and Owen, D. B., eds., *Selected Tables in Mathematical Statistics, vol III*, Providence, R.I., American Mathematical Society, 385–419.

Land, C. E., Johnson, B. R., and Joshi, V. M. (1973). A note on two-sided confidence intervals for linear functions of the normal mean and variance, *Ann. Statist., 1*, 940–943.

Land, C. E., Greenberg, L. M., Hall, C., and Drzyzgula, C. C. (1987). Interactive computer algorithms for computing confidence limits for linear functions of the mean and variance, or the mean and standard deviation, of a normal distribution, unpublished manuscript (to be submitted for publication).

Lehmann, E. L. (1959). *Testing Statistical Hypotheses*, New York, John Wiley & Sons, Inc.

Locks, M. O., Alexander, M. J., and Byars, B. J. (1963). New Tables of the Noncentral t-Distribution, Report ARL63-19, Wright-Patterson Air Force Base.

Owen, D. B. (1965). A survey of properties and applications of the noncentral t-distribution, *Technometrics, 10*, 445–478.

Owen, D. B. and Amos, D. E. (1963). Programs for Computing Percentage Points of the Noncentral t-Distributions, Sandia Corporation Monograph SCR-551.

Pearson, E. S. and Hartley, H. O. (1972). *Biometrika Tables for Statisticians, vol. II*, 58–65, 242–248, Cambridge, Cambridge University Press.

Resnikoff, G. J. and Lieberman, G. J. (1957). *Tables of the Non-Central t-Distribution*, Stanford, Stanford University Press.

Scheuer, E. M. and Spurgeon, R. A. (1963). Some percentage points of the noncentral t-distribution, *J. Amer. Statist. Assoc., 58*, 176–182.

4

Three-Parameter Estimation

A. CLIFFORD COHEN Department of Statistics, University of Georgia, Athens, Georgia

1. THE THREE-PARAMETER DISTRIBUTION

The random variable X is said to have a three-parameter lognormal distribution if the random variable $Y = \ln(X - \gamma)$, where $X > \gamma$, is distributed normally (μ, σ^2), $\sigma > 0$. Accordingly, the probability density function (p.d.f.) of X becomes

$$f(x; \gamma, \mu, \sigma) = \frac{1}{\sigma\sqrt{2\pi}(x-\gamma)} \exp \frac{-1}{2\sigma^2}[\ln(x-\gamma) - \mu]^2$$

$$\gamma < x < \infty \qquad \sigma > 0 \quad (1.1)$$

zero otherwise

In this notation, γ is a threshold parameter (i.e. the lower limit), μ is a scale parameter, and σ is a shape parameter. It is sometimes desirable to employ an alternate scale parameter $\beta = \exp(\mu)$, in which case the p.d.f. is written as

$$f(x;\gamma,\beta,\sigma) = \frac{1}{\sigma\sqrt{2\pi}(x-\gamma)} \exp \frac{-1}{2\sigma^2}\left[\ln\left(\frac{x-\gamma}{\beta}\right)\right]^2$$

$$\gamma < x < \infty \qquad \sigma > 0 \quad (1.2)$$

zero otherwise

The cumulative distribution function of X may conveniently be expressed as

$$F(x;\gamma,\mu,\sigma) = \Phi\left[\frac{\ln(x-\gamma)-\mu}{\sigma}\right] \quad (1.3)$$

where $\Phi(\cdot)$ is the cumulative standard normal distribution function.

1.1 Some Characteristics (Measures of Location, Dispersion, and Shape)

If we follow the lead of Yuan (1933), and introduce an alternate shape parameter $\omega = \exp(\sigma^2)$, the expected value (mean), median, mode, variance, third standard moment (skewness), fourth standard moment (kurtosis) and coefficient of variation C.V. $= \sqrt{V(X)}/[E(X) - \gamma]$ may be expressed in terms of γ, β, and ω as

$$\begin{aligned}
E(X) &= \gamma + \beta\omega^{1/2} \\
Me(X) &= \gamma + \beta \\
Mo(X) &= \gamma + \beta\omega^{-1} \\
V(X) &= \beta^2[\omega(\omega-1)] \\
\alpha_3(X) &= (\omega+2)(\omega-1)^{1/2} \\
\alpha_4(X) &= \omega^4 + 2\omega^3 + 3\omega^2 - 3 \\
\text{C.V.}(X) &= (\omega-1)^{1/2}
\end{aligned} \quad (1.4)$$

where $E(\cdot)$ is the expectation symbol, $V(\cdot)$ is the variance, α_3 is the third standard moment and α_4 is the fourth standard moment. Note that α_3, α_4 and C.V. are functions of ω only.

1.2 The Standard Distribution

If we make the standardizing transformation

$$Z = \frac{X - E(X)}{\sqrt{V(X)}}$$

the p.d.f. of the standard lognormal distribution with mean zero, unit variance and shape parameter ω becomes

$$g(z;0,1,\omega) = \frac{\sqrt{(\omega-1)}}{\sqrt{2\pi \ln \omega}[1 + z\sqrt{(\omega-1)}]} \exp \frac{-1}{2(\ln \omega)} \tag{1.5}$$

$$\times [\ln \left\{ \sqrt{\omega} + z\sqrt{\omega(\omega-1)} \right\}]^2 \qquad -(\omega-1)^{-1/2} < z, \omega > 1, \text{ zero otherwise}$$

The p.d.f. of (1.5) can also be expressed in terms of the standard normal p.d.f. $\phi(\cdot)$ as

$$g(z;0,1,\omega) = \frac{1}{\sqrt{\ln \omega}} \left(\frac{\sqrt{\omega-1}}{1 + z\sqrt{\omega-1}} \right) \phi \left[\frac{\ln \left\{ \sqrt{\omega} + z\sqrt{\omega(\omega-1)} \right\}}{\sqrt{\ln \omega}} \right] \tag{1.6}$$

Its cumulative distribution function becomes

$$G(z;0,1,\omega) = \Phi \left[\frac{\ln[\sqrt{\omega} + z\sqrt{\omega(\omega-1)}]}{\sqrt{\ln \omega}} \right] \tag{1.7}$$

In order to facilitate comparisons with other skewed distributions, it is sometimes desirable to employ α_3 rather than ω or σ as the primary shape parameter. Accordingly, we need to express ω as a function of α_3. The fifth equation of (1.4) reduces to the cubic equation

$$\omega^3 + 3\omega^2 - (4 + \alpha_3^2) = 0 \tag{1.8}$$

This equation has a single positive real root ($\omega > 1$) which may be written as

$$\omega = \sqrt[3]{1 + \frac{\alpha_3}{2}\left\{\alpha_3 + \sqrt{\alpha_3^2 + 4}\right\}} + \sqrt[3]{1 + \frac{\alpha_3}{2}\left\{\alpha_3 - \sqrt{\alpha_3^2 + 4}\right\}} - 1 \tag{1.9}$$

Details concerning the solution of equation (1.8) may be found in Abramowitz and Stegun (1964), page 17.

A relation between α_3 and σ that leads to a simpler cubic equation can be based on the coefficient of variation, which we now designate by the symbol v. It follows that

$$\omega = 1 + v^2 \qquad \sigma = \sqrt{\ln(1+v^2)} \qquad \text{and} \qquad \alpha_3 = v(v^2+3) \tag{1.10}$$

Accordingly v is the positive root of the cubic equation

$$v^3 + 3v - \alpha_3 = 0 \tag{1.11}$$

which we write as

$$v = \sqrt[3]{\frac{1}{2}\left[\alpha_3 + \sqrt{4+\alpha_3^2}\right]} + \sqrt[3]{\frac{1}{2}\left[\alpha_3 - \sqrt{4+\alpha_3^2}\right]} \tag{1.12}$$

The required value of σ then follows from the second equation of (1.10).

2. THE HAZARD FUNCTION

In reliability and time-to-failure studies, the instantaneous failure rate [i.e., the hazard function] is often an item of concern to investigators. For this reason, a brief introduction to this function is included here. The hazard function is defined as

$$h(x) = \frac{f(x)}{1-F(x)} \qquad \gamma < x \tag{2.1}$$

where $f(x)$ and $F(x)$ are the p.d.f. and the distribution function of X. Since $f(\gamma) = 0$ and $F(\gamma) = 0$, it follows that $h(\gamma) = 0$.

The hazard function of the lognormal distribution can thus be written as

$$h(x;\gamma,\mu,\sigma) = \frac{\frac{1}{\sigma(x-\gamma)}\phi\left[\frac{\ln(x-\gamma)-\mu}{\sigma}\right]}{1-\Phi\left[\frac{\ln(x-\gamma)-\mu}{\sigma}\right]} \tag{2.2}$$

In standard units where $Z = (X - E(X))/\sqrt{V(X)}$, the hazard function can be written as

$$h(z) = \frac{g(z)}{1-G(z)} \qquad \frac{-1}{\sqrt{\omega-1}} < z \tag{2.3}$$

Note that $h(-1/\sqrt{\omega - 1}) = 0$.

A more complete account of the lognormal hazard function has been given by Nelson (1982).

Entries of the standardized lognormal hazard function for selected values of α_3 have been included in Table 1. These values are plotted as a collection of graphs in Figure 1. We observe that the lognormal hazard function is an increasing function until it reaches a maximum from which it then decreases. The maximum is attained sooner (i.e. for smaller values of z) as the skewness α_3 becomes larger. This characteristic is in contrast with the constant hazard function of the exponential distribution. For the Weibull distribution, the hazard function is an increasing function for $\alpha_3 < 2$ and a decreasing function for $\alpha_3 > 2$. When $\alpha_3 = 2$, the Weibull becomes the exponential distribution with its constant hazard function.

2.1 Skewness and Kurtosis

In practical applications, analysts must often choose from among the Weibull, gamma, lognormal, inverse Gaussian, or other skewed distributions when selecting a model for life test data that are to be analyzed. It is therefore important to understand variations between distributions of these types for which means, variances and degrees of skewness are identical. As a common measure of skewness, α_3 or $\beta_1 = \alpha_3^2$ can be employed. As a measure of kurtosis, $\beta_2 = \alpha_4$ is available, where β_1 and β_2 are Pearson's Betas. The lognormal distribution is generally perceived to be a "heavy-tailed" distribution that is unlikely to be used unless skewness is large ($\alpha_3 \geq 1$). The β_1 vs. β_2 curve of this distribution lies in Pearson's Type VI region near the Type V line which forms the boundary between the Type IV and the Type VI regions.

3. COMPLICATION IN THE ESTIMATION PROCESS

Introduction of the threshold parameter, γ, creates complications when we seek to estimate lognormal parameters from sample data. Heyde (1963) demonstrated that this distribution is not uniquely determined by its moments and this raised questions about moment estimators. Hill (1963) has shown that global maximum likelihood estimators lead to inadmissible estimates. For these and other reasons, various investigators have turned their attention to alternate estimators that might be free from the regularity problems of the moment and maximum likelihood estimators.

Cohen (1951) and Harter and Moore (1966) proposed local maximum likelihood estimators, LMLE. Hill (1963) proposed Bayesian estimators. Box and Cox (1964) used a scoring technique to graphically estimate

Table 1 The Lognormal Hazard Function

z	α_3								
	0	0.5	1.0	1.5	2.0	2.5	3.0	4.0	4.5
-3.00	.00444	.00019	.00000						
-2.80	.00794	.00079	.00000						
-2.60	.01365	.00268	.00000						
-2.40	.02258	.00751	.00005						
-2.20	.03597	.01791	.00116						
-2.00	.05525	.03719	.00851	.00000					
-1.80	.08189	.06860	.03306	.00130					
-1.60	.11735	.11449	.08528	.02708	.00001				
-1.40	.16288	.17561	.16744	.11823	.03017	.00000			
-1.20	.21944	.25109	.27302	.26742	.20861	.08143	.00003		
-1.00	.28760	.33870	.39146	.43600	.45783	.43457	.33160	.00000	
-.80	.36756	.43550	.51284	.59435	.67442	.74692	.80380	.81072	.69935
-.60	.45915	.53837	.62995	.72960	.83367	.93981	1.04659	1.25863	1.36208
-.40	.56188	.64440	.73851	.83934	.94299	1.04712	1.15064	1.35462	1.45522
-.20	.67507	.75112	.83646	.92579	1.01510	1.10212	1.18591	1.34316	1.41696
.00	.79788	.85658	.92323	.99259	1.06081	1.12588	1.18704	1.29771	1.34776
.20	.92942	.95932	.99913	1.04339	1.08803	1.13087	1.17099	1.24264	1.27449
.40	1.06876	1.05832	1.06490	1.08138	1.10232	1.12452	1.14639	1.18677	1.20501
.60	1.21503	1.15289	1.12149	1.10921	1.10753	1.11132	1.11777	1.13340	1.14141
.80	1.36740	1.24263	1.16991	1.12898	1.10635	1.09403	1.08761	1.08363	1.08389
1.00	1.52514	1.32736	1.21113	1.14239	1.10066	1.07441	1.05727	1.03768	1.03199
1.20	1.68755	1.40703	1.24603	1.15073	1.09181	1.05354	1.02750	.99542	.98512
1.40	1.85406	1.48171	1.27542	1.15507	1.08075	1.03216	.99870	.95657	.94266
1.60	2.02413	1.55152	1.30001	1.15622	1.06818	1.01072	.97107	.92082	.90406
1.80	2.19731	1.61666	1.32043	1.15483	1.05459	.98952	.94470	.88785	.86884
2.00	2.37322	1.67733	1.33723	1.15142	1.04036	.96875	.91960	.85738	.83657
2.20	2.55150	1.73377	1.35087	1.14641	1.02575	.94853	.89575	.82914	.80691
2.40	2.73186	1.78621	1.36176	1.14013	1.01097	.92892	.87310	.80291	.77954
2.60	2.91406	1.83490	1.37027	1.13284	.99615	.90996	.85159	.77849	.75420
2.80	3.09787	1.88004	1.37670	1.12476	.98142	.89166	.83116	.75568	.73067
3.00	3.28310	1.92189	1.38131	1.11608	.96684	.87403	.81174	.73433	.70876
3.20	3.46959	1.96063	1.38434	1.10692	.95248	.85705	.79328	.71432	.68830
3.40	3.65720	1.99649	1.38599	1.09741	.93837	.84071	.77570	.69550	.66914
3.60	3.84581	2.02964	1.38643	1.08764	.92455	.82498	.75895	.67778	.65117
3.80	4.03531	2.06027	1.38581	1.07769	.91103	.80984	.74298	.66106	.63427
4.00	4.22561	2.08855	1.38428	1.06762	.89782	.79527	.72774	.64526	.61834
4.20	4.41662	2.11464	1.38194	1.05748	.88493	.78124	.71318	.63029	.60330
4.40	4.60827	2.13868	1.37890	1.04733	.87237	.76772	.69925	.61610	.58908
4.60	4.80051	2.16080	1.37524	1.03719	.86013	.75470	.68592	.60261	.57560
4.80	4.99327	2.18115	1.37105	1.02709	.84821	.74215	.67315	.58979	.56281
5.00	5.18650	2.19982	1.36639	1.01705	.83660	.73004	.66090	.57757	.55065
5.20	5.38017	2.21695	1.36133	1.00710	.82530	.71836	.64915	.56592	.53908
5.40	5.57424	2.23262	1.35592	.99724	.81430	.70708	.63785	.55479	.52805
5.60	5.76867	2.24694	1.35020	.98750	.80359	.69618	.62699	.54415	.51752
5.80	5.96342	2.25998	1.34422	.97788	.79317	.68565	.61655	.53397	.50746
6.00	6.15848	2.27185	1.33801	.96838	.78302	.67547	.60649	.52421	.49784
L.L.	$-\infty$	-6.05456	-3.10380	-2.14491	-1.67765	-1.40316	-1.22290	-1.00000	-.92589

power-shift transformation parameters which were LMLE. Tiku (1968) suggested approximate linearization of the maximum likelihood equation based on the approximate formula $F(x)[1 - \Phi(x)]^{-1} \doteqdot \alpha + \beta x$ for Mill's ratio. Weighted least square estimators were proposed by Munro and Wixley (1970) while Calitz (1973) used simulation procedures to compare likelihood, percentile, and moment estimators. Penalty and barrier functions were introduced by Wingo (1975, 1976) to increase convergence in calculating LMLE. Giesbrecht and Kempthorne (1976) obtained maximum likelihood estimators from a discrete model by considering a grouping interval. Cohen and Whitten (1980) reexamined results of Cohen (1951) and proposed several modifications of maximum likelihood and of moment estimators. Kane (1982) considered a class of weighted order statistic estimators which he compared with LMLE, percentile, and Cohen modified estimators. More recently Rukhin (1984) has resorted to Bayesian procedures in an effort to improve on estimators given by Finney (1941) and later studied by Bradu and Mundlak (1970) and by Evans and Shaban (1974). Aitchison and Brown (1957) contains an excellent summary of estimation in the lognormal distribution and contains many additional references. Johnson and Kotz (1970) also contains an excellent summary plus additional references. Stedinger (1980) and Hoshi et al. (1984) have been concerned with estimation for the three-parameter lognormal distribution in connection with fitting hydrologic data. Lawrence (1979, 1980, 1984) has been concerned with applications involving marketing and sociological studies. In what follows here, we limit consideration to moment, maximum likelihood, and modified moment estimators of Cohen et al. (1985).

4. MOMENT ESTIMATORS

On equating the first three moments of the lognormal distribution as given in equation (1.4) to corresponding sample moments, we have

$$\begin{aligned} \gamma + \beta\omega^{1/2} &= \bar{x} \\ \beta^2[\omega(\omega - 1)] &= s^2 \\ (\omega + 2)(\omega - 1)^{1/2} &= a_3 \end{aligned} \tag{4.1}$$

where

$$\bar{x} = \frac{\Sigma_1^n x_i}{n} \quad s^2 = \frac{\Sigma_1^n (x_i - \bar{x})^2}{n - 1} \quad a_3 = \frac{\Sigma_1^n (x_i - \bar{x})^3}{n} \div \left(\frac{\Sigma_1^n (x_i - \bar{x})^2}{n} \right)^{3/2}$$

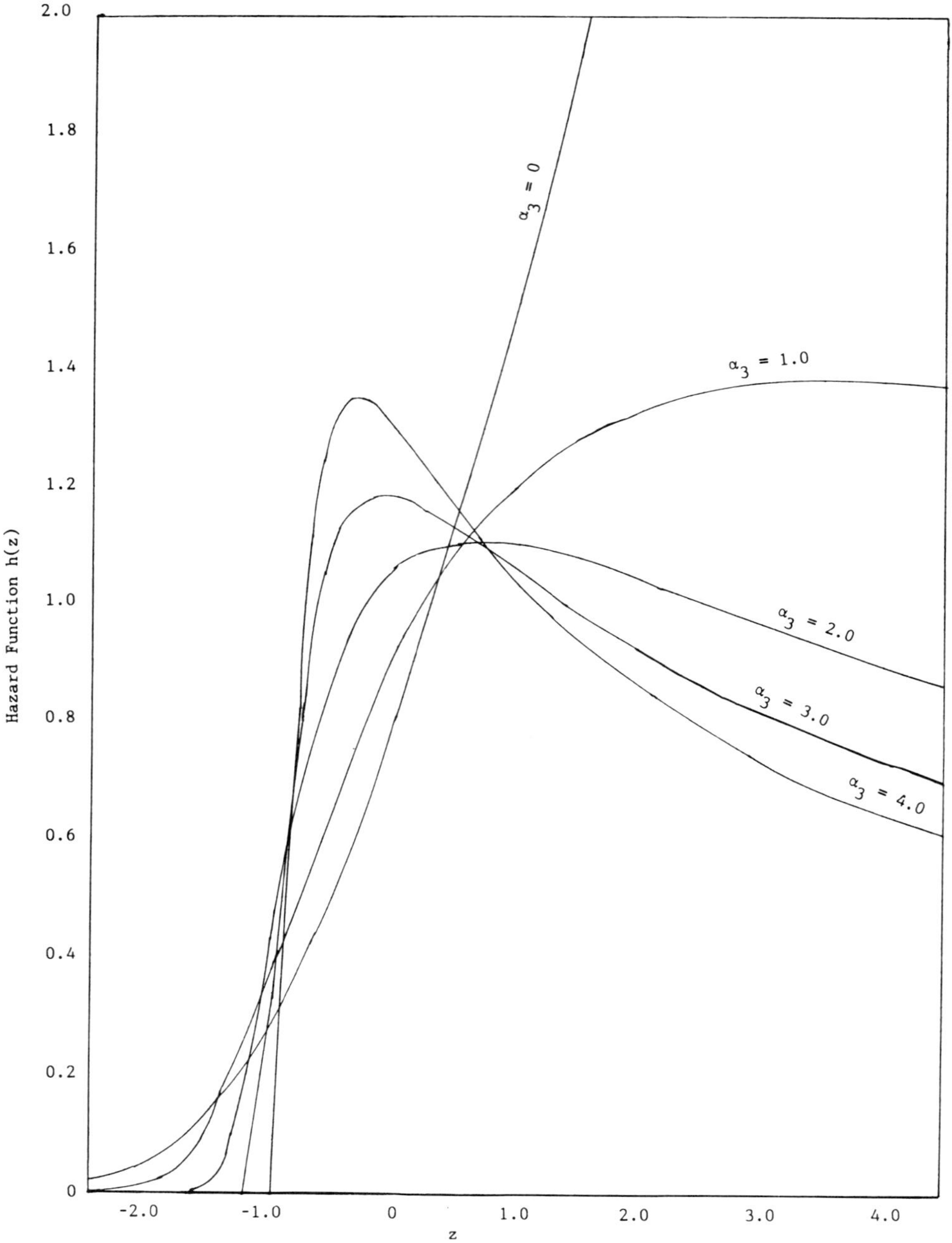

Figure 1 Lognormal hazard functions.

The third equation of (4.1) can be solved for the estimate ω^* and of course $\sigma^* = \sqrt{\ln \omega^*}$. Estimates β^* and γ^* then follow from the first two equations of (4.1) as

$$\beta^* = \frac{s}{\sqrt{\omega^*(\omega^* - 1)}}$$
$$\gamma^* = \bar{x} - \beta^* \sqrt{\omega^*} \tag{4.2}$$

When required, we calculate $\mu^* = \ln \beta^*$.

Various iterative techniques might be used in solving the third equation of (4.1) for ω^*. However, we might employ (1.9) or (1.12) with α_3 replaced by a_3 to obtain an explicit solution. We might even resort to a "trial and error" approach coupled with linear interpolation in calculating ω^*. As an aid in providing a first approximation for use in iterative calculations, Table 2 contains entries for ω, α_3, α_4 and the lower limit as functions of the shape parameter σ.

The principal objection to moment estimators centers about the large sampling errors due to use of the second and third moments, plus the fact that as shown by Heyde (1963) the lognormal distribution is not uniquely determined by its moments. We are therefore led to consider maximum likelihood estimation.

Table 2 Values of ω, α_3, α_4, and the Lower Limit as Functions of σ for the Lognormal Distribution

σ	ω	α_3	α_4	L.L
0.0	1.00000	0.	3.00000	$-\infty$
0.1	1.01005	0.30176	3.16232	−9.97501
0.2	1.04081	0.61429	3.67837	−4.95008
0.3	1.09417	0.94953	4.64491	−3.25862
0.4	1.17351	1.32191	6.26001	−2.40069
0.5	1.28403	1.75019	8.89845	−1.87638
0.6	1.43333	2.26008	13.27335	−1.51912
0.7	1.63232	2.88836	20.79117	−1.25757
0.8	1.89648	3.68929	34.36765	−1.05616
0.9	2.24791	4.74533	60.41076	−.89518
1.0	2.71828	6.18488	113.93639	−.76287
1.5	9.48774	33.46805	10078.25285	−.34324
2.0	54.59815	414.35934	9220559.97731	−.13659

5. MAXIMUM LIKELIHOOD ESTIMATION

We seek estimates that will maximize the likelihood function of a random sample consisting of observations $\{x_i\}$, $i = 1, \ldots, n$. Without any loss of generality it will be assumed that the sample is ordered so that x_1 is the smallest sample observation. The likelihood function may be written as

$$L(x_1, \ldots, x_n; \gamma, \mu, \sigma) = \prod_{i=1}^{n} f(x_i; \gamma, \mu, \sigma)$$

It is immediately obvious that $L(\ \ )$ approaches infinity as $\gamma \to x_1$. It would thus appear that we should take $\hat{\gamma} = x_1$ as our estimate. Hill (1963), however, demonstrated the existence of paths along which the likelihood function of any ordered sample $x_1, \ldots, x_n$ tends to ∞ as (γ, μ, σ^2) approach $(x_1, -\infty, \infty)$. This global maximum thereby leads to the inadmissible estimates $\hat{\mu} = -\infty$ and $\hat{\sigma}^2 = \infty$ regardless of the sample. We therefore seek other estimators that at least produce reasonable estimates.

The writer's attention has recently been called to the lognormal article by Antle (1985) in the *Encyclopedia of Statistical Sciences*, *5*, p. 135, in which reference is made to a proof by Monlezum, Antle and Klimko (1975) to the effect that maximum likelihood estimates of γ and μ always exist when the shape parameter σ is known.

As an alternative to global MLE, we might equate partial derivatives of the log-likelihood function to zero as was done by Cohen (1951), Cohen and Whitten (1980) and Harter and Moore (1966) to obtain local maximum likelihood estimators (LMLE) which in most cases would be considered reasonable in comparison with corresponding moment estimators. Harter and Moore and later Calitz (1973) noted that these LMLE appear to possess most of the desirable properties ordinarily associated with maximum likelihood estimation.

On differentiating the log-likelihood function and equating to zero, we obtain the LMLE estimating equations:

$$\begin{aligned}
\frac{\partial \ln L}{\partial \mu} &= \frac{1}{\sigma^2} \sum_{1}^{n} [\ln(x_i - \gamma) - \mu] = 0 \\
\frac{\partial \ln L}{\partial \sigma} &= \frac{n}{\sigma} + \frac{1}{\sigma^3} \sum_{1}^{n} [\ln(x_i - \gamma) - \mu]^2 = 0 \qquad (5.1) \\
\frac{\partial \ln L}{\partial \gamma} &= \frac{1}{\sigma^2} \sum_{1}^{n} \frac{[\ln(x_i - \gamma) - \mu]}{(x_1 - \gamma)} + \sum_{1}^{n} (x_i - \gamma)^{-1} = 0
\end{aligned}$$

When σ^2 and μ are eliminated from these equations as was done by Cohen (1951), the resulting equation in γ becomes

$$\lambda(\hat{\gamma}) = \left[\sum_1^n (x_i - \hat{\gamma})^{-1}\right]\left[\sum_1^n (x_i - \hat{\gamma}) - \sum_1^n \ln^2(x_i - \hat{\gamma}) + \frac{1}{n}\left\{\sum_1^n \ln(x_i - \hat{\gamma})\right\}^2\right] - n\sum_1^n \frac{\ln(x_i - \hat{\gamma})}{(x_i - \hat{\gamma})} = 0 \tag{5.2}$$

Equation (5.2) may be solved iteratively for $\hat{\gamma}$. It then follows from the first two equations of (5.1) that

$$\hat{\mu} = \frac{1}{n}\sum_1^n \ln(x_i - \hat{\gamma})$$
$$\hat{\sigma}^2 = \frac{1}{n}\sum_1^n \ln^2(x_i - \hat{\gamma}) - \left[\frac{1}{n}\sum_1^n \ln(x_i - \hat{\gamma}\right]^2 \tag{5.3}$$

In solving (5.2) for $\hat{\gamma}$, we accept only admissible roots for which $\gamma < x_1$. Usually, only a single admissible root will be found. In the event that multiple admissible roots occur, we choose as our estimate the root which results in closest agreement between $\bar{x}$ and $\hat{E}(X)$.

Standard iterative procedures such as the Newton-Raphson method are satisfactory for solving equation (5.2), but in many instances it is more convenient to use the "trial and error" technique with linear interpolation. We begin with a first approximation $\gamma_1 < x_1$ and evaluate $\lambda(\gamma_1)$. If this value is zero then no further calculations are required. Otherwise we continue until we find a pair of values γ_i and γ_j in a sufficiently narrow interval such that $\lambda(\gamma_i) \gtrless 0 \gtrless \lambda(\gamma_j)$, and interpolate for the final estimate, $\hat{\gamma}$.

Wilson and Worcester (1945) and Lambert (1964) attempted to solve the three equations of (5.1) simultaneously without simplifying to the form given in (5.2) but they encountered convergence problems. Calitz (1973) examined the convergence problem further and concluded that the simplification of Cohen (1951) using equation (5.2) led to fewer convergence problems and was therefore to be recommended.

5.1 Asymptotic Variances and Covariances

In the absence of regularity restrictions, asymptotic variances and covariances of maximum likelihood estimators of distribution parameters can be obtained by inverting the Fisher information matrix, in which elements are negatives of expected values of second partial derivatives of the likelihood function with respect to the parameters. Unfortunately, the lognormal distribution is subject to regularity problems as previously mentioned, and this raises questions about the validity of asymptotic variances and covariances thus obtained as they might apply to the local maximum likelihood estimators. Nevertheless, they have been found to be in reasonably close agreement with estimate variances obtained in simulation studies by Cohen and Whitten (1980), by Harter and Moore (1966) and by Cohen et al. (1985). They are therefore offered here as possible useful approximations to the applicable variances and covariances. Based on the information matrix for $\hat{\gamma}$, $\hat{\beta}$, $\hat{\sigma}$ given by Cohen (1951) or that for $\hat{\gamma}$, $\hat{\mu}$, $\hat{\sigma}^2$ given by Hill (1963), we have

$$
\begin{aligned}
V(\hat{\gamma}) &\doteq \frac{\sigma^2}{n}\left(\frac{\beta^2}{\omega}\right)H & \mathrm{Cov}(\hat{\gamma},\hat{\beta}) &\doteq \frac{-\sigma^3}{n}\left(\frac{\beta^2}{\sqrt{\omega}}\right)H \\
V(\hat{\beta}) &\doteq \frac{\sigma^2}{n}\beta^2[1+H] & \mathrm{Cov}(\hat{\gamma},\hat{\sigma}) &\doteq \frac{\sigma^3}{n}\left(\frac{\beta^2}{\sqrt{\omega}}\right)H \\
V(\hat{\sigma}) &\doteq \frac{\sigma^2}{2n}[1+2\sigma^2 H] & \mathrm{Cov}(\hat{\beta},\hat{\sigma}) &\doteq \frac{-\sigma^3}{n}\beta^2 H \\
V(\hat{\mu}) &\doteq [1+H] & V(\hat{\sigma}^2) &\doteq \frac{2\sigma^4}{n}[1+2\sigma^2 H]
\end{aligned}
\tag{5.4}
$$

where

$$H = [\omega(1+\sigma^2) - (1+2\sigma^2)]^{-1}$$

For large samples, the central limit theorem can, of course, be employed to approximate the variance of the maximum likelihood estimate $\hat{m}$ of the distribution mean $[m = E(X)]$. The moment estimate is the sample mean, and as is well-known,

$$V(\bar{x}) = \frac{V(X)}{n}$$

In large samples, it seems reasonable to expect the maximum likelihood estimate, $\hat{m}$, to be closely approximated by $\bar{x}$ and thus $V(\hat{m})$ to be closely approximated by $V(\bar{x})$. The variance $V(X)$ is given in (1.4).

The variances and covariances of (5.4) can be expressed in a simpler format as

$$
\begin{aligned}
V(\hat{\gamma}) &= \frac{\sigma^2}{n}\beta^2\phi_{11} & V(\hat{\sigma}^2) &= \frac{2\sigma^4}{n}\phi_{33} \\
V(\hat{\beta}) &= \frac{\sigma^2}{n}\beta^2\phi_{22} & \mathrm{Cov}(\hat{\gamma},\hat{\beta}) &= \frac{\sigma^2}{n}\beta^2\phi_{12} \\
V(\hat{\sigma}) &= \frac{\sigma^2}{2n}\phi_{33} & \mathrm{Cov}(\hat{\gamma},\hat{\sigma}) &= \frac{\sigma^2}{n}\beta^2\phi_{13} \\
V(\hat{\mu}) &= \frac{\sigma^2}{n}\phi_{22} & \mathrm{Cov}(\hat{\beta},\hat{\sigma}) &= \frac{\sigma^2}{n}\beta^2\phi_{23}
\end{aligned}
\tag{5.5}
$$

where

$$
\begin{aligned}
\phi_{11} &= H/\omega & \phi_{22} &= 1 + H & \phi_{22} &= 1 + 2\sigma^2 H \\
\phi_{12} &= -(\sigma/\sqrt{\omega})H & \phi_{13} &= -\phi_{12} & \phi_{23} &= -\sigma H
\end{aligned}
\tag{5.6}
$$

The ϕ_{ij} of (5.6) are functions of σ alone. Values of these factors which will facilitate the calculations of variances and covariances in practical application are included as Table 3 with α_3 as the argument. This choice of the argument is possible since α_3 and ω are functions of σ alone.

6. MODIFIED MOMENT ESTIMATORS

In these estimators the third moment, which is subject to somewhat large sampling errors, is replaced by a function of the first order statistic. This replacement was suggested by the realization that the first order statistic contains more information about the threshold parameter than any of the other sample observations and often more than all the other observations combined. Cohen and Whitten (1980) first proposed the modified moment estimators under consideration here. Cohen et al. (1985) further investigated these estimators and presented tabular and graphical aids which greatly simplify calculations in practical applications. The presentation which follows is to a great extent based on the 1985 paper.

For an ordered random sample of size n, the estimating equations are

$$E(X) = \bar{x} \qquad V(X) = s^2 \qquad \text{and} \qquad E[\ln(X_1 - \gamma)] = \ln(x_1 - \gamma) \tag{6.1}$$

where $\bar{x}$ and s^2 are the sample mean and variance (unbiased). X_1 is the first order statistic (a random variable) in a random sample of size n and x_1 is the corresponding sample value.

Table 3 Variance-Covariance Factors for Maximum Likelihood Estimates of Lognormal Parameters

α_3	ω	ϕ_{11}	ϕ_{22}	ϕ_{33}	ϕ_{12}	ϕ_{23}
0.50	1.02728	885.23263	910.38125	49.95010	-147.19413	-149.18831
0.55	1.03289	607.65422	628.63763	41.61692	-111.08790	-112.89975
0.60	1.03898	431.38171	449.19715	35.27801	-85.98520	-87.64504
0.65	1.04555	315.03444	330.38447	30.34403	-67.98669	-69.51786
0.70	1.05258	235.69195	249.08577	26.42824	-54.74136	-56.16220
0.75	1.06007	180.05004	191.86570	23.26838	-44.77401	-46.09919
0.80	1.06799	140.07179	150.59584	20.68148	-37.12694	-38.36839
0.85	1.07634	110.73199	120.18559	18.53674	-31.15990	-32.32744
0.90	1.08510	88.79275	97.34930	16.73867	-26.43369	-27.53552
0.95	1.09426	72.11193	79.90929	15.21622	-22.64022	-23.68324
1.00	1.10380	59.23866	66.38784	13.91565	-19.55896	-20.54904
1.05	1.11372	49.16952	55.76087	12.79571	-17.02923	-17.97141
1.10	1.12398	41.19733	47.30516	11.82430	-14.93205	-15.83068
1.15	1.13460	34.81515	40.50113	10.97615	-13.17803	-14.03690
1.20	1.14554	29.65392	34.96965	10.23114	-11.69914	-12.52156
1.25	1.15679	25.44114	30.43014	9.57311	-10.44293	-11.23182
1.30	1.16835	21.97300	26.67221	8.98892	-9.36858	-10.12653
1.35	1.18020	19.09523	23.53622	8.46783	-8.44393	-9.17324
1.40	1.19233	16.68980	20.89971	8.00100	-7.64347	-8.34620
1.45	1.20472	14.66545	18.66773	7.58107	-6.94674	-7.62472
1.50	1.21736	12.95095	16.76599	7.20190	-6.33721	-6.99210
1.55	1.23025	11.49026	15.13585	6.85830	-5.80144	-6.43475
1.60	1.24336	10.23888	13.73062	6.54590	-5.32841	-5.94150
1.65	1.25669	9.16122	12.51284	6.26099	-4.90903	-5.50313
1.70	1.27023	8.22860	11.45225	6.00038	-4.53575	-5.11201
1.75	1.28397	7.41778	10.52422	5.76133	-4.20229	-4.76172
1.80	1.29790	6.70977	9.70861	5.54149	-3.90336	-4.44691
1.85	1.31200	6.08899	8.98879	5.33881	-3.63449	-4.16304
1.90	1.32628	5.54257	8.35100	5.15152	-3.39191	-3.90627
1.95	1.34071	5.05983	7.78379	4.97805	-3.17240	-3.67330
2.00	1.35530	4.63135	7.27756	4.81705	-2.97322	-3.46134
2.25	1.43024	3.08881	5.41775	4.16174	-2.20975	-2.64270
2.50	1.50791	2.16869	4.27019	3.68628	-1.70671	-2.09578
2.75	1.58758	1.58683	3.51921	3.32880	-1.35930	-1.71271
3.00	1.66869	1.20085	3.00385	3.05208	-1.11001	-1.43389
3.25	1.75079	0.93447	2.63607	2.83263	-.92535	-1.22440
3.50	1.83355	0.74441	2.36492	2.65498	-.78485	-1.06276
3.75	1.91669	0.60491	2.15943	2.50865	-.67550	-.93519
4.00	2.00000	0.50000	2.00000	2.38629	-.58871	-.83255
4.25	2.08331	0.41942	1.87378	2.28263	-.51863	-.74858
4.50	2.16650	0.35637	1.77207	2.19380	-.46121	-.67886
5.00	2.33211	0.26578	1.61983	2.04972	-.37349	-.57037
6.00	2.65871	0.16333	1.43425	1.84925	-.26335	-.42941
7.00	2.97764	0.11031	1.32847	1.71682	-.19884	-.34311
8.00	3.28840	0.07956	1.26164	1.62290	-.15742	-.28546

$\phi_{13} = -\phi_{12}$

The third equation of (6.1) reduces to

$$\gamma + \beta \exp[\sqrt{\ln \omega} E(Z_{1,n})] = x_1$$

where $E(Z_{1,n})$ is the expected value of the first order statistic in a random sample of size n from a standard normal distribution $(0, 1)$. Values of $E(Z_{1,n})$ can be obtained from tables compiled by Harter (1961, 1969). Selected values from this source are reproduced here as Table 4. Linear interpolation in this table will usually provide sufficient accuracy for most practical applications.

Appropriate substitutions from equations (1.4) into (6.1) enable us to write the modified estimating equations as

$$\begin{aligned} \gamma + \beta\omega^{1/2} &= \bar{x} \\ \beta^2\omega(\omega - 1) &= s^2 \\ \gamma + \beta \exp[\sqrt{\ln \omega} E Z_{1,n})] &= x_1 \end{aligned} \tag{6.2}$$

Following a few simple algebraic manipulations, the three equations of

Table 4 Expected Values of the First Order Statistic from the Standard Normal Distribution (0,1)

n	$E(Z_{1,n})$	n	$E(Z_{1,n})$	n	$E(Z_{1,n})$
5	−1.16296	34	−2.09471	90	−2.46970
10	−1.53875	36	−2.11812	95	−2.48920
12	−1.62923	38	−2.14009	100	−2.50759
14	−1.70338	40	−2.16078	125	−2.58634
16	−1.76599	45	−2.20772	150	−2.64925
18	−1.82003	50	−2.24907	175	−2.70148
20	−1.86748	55	−2.28598	200	−2.74604
22	−1.90969	60	−2.31928	225	−2.78485
24	−1.94767	65	−2.34958	250	−2.81918
26	−1.98216	70	−2.37736	280	−2.85572
28	−2.01371	75	−2.40299	300	−2.87777
30	−2.04276	80	−2.42677	350	−2.92651
32	−2.06967	85	−2.44894	400	−2.96818

Extracted from Harter's (1961) tables.

(6.2) are reduced to

$$\begin{aligned} \frac{s^2}{(\bar{x}-x_1)^2} &= \frac{\hat{\omega}(\hat{\omega}-1)}{[\sqrt{\hat{\omega}}-\exp\left\{\sqrt{\ln\hat{\omega}}E(Z_{1,n})\right\}]^2} \\ \hat{\gamma} &= \bar{x}-s(\hat{\omega}-1)^{-1/2} = \bar{x}-s(\exp\hat{\sigma}^2-1)^{-1/2} \\ \hat{\beta} &= s[\hat{\omega}(\hat{\omega}-1)]^{-1/2} = s[\exp\hat{\sigma}^2\left\{\exp\hat{\sigma}^2-1\right\}]^{-1/2} \end{aligned} \tag{6.3}$$

Equivalent equations for $\hat{\beta}$ and $\hat{\gamma}$, which in some applications might be more convenient to use, are

$$\begin{aligned} \hat{\beta} &= \frac{\bar{x}-x_1}{[\sqrt{\hat{\omega}}-\exp\left\{\sqrt{\ln\hat{\omega}}E(Z_{1,n})\right\}]} \\ \hat{\gamma} &= \bar{x}-\hat{\beta}\sqrt{\hat{\omega}} \end{aligned} \tag{6.4}$$

With $\bar{x}$, s^2, and x_1 available from sample data, we calculate $s^2/(\bar{x}-x_1)^2$ and solve the first equation of (6.3) for $\hat{\omega}$. We subsequently calculate $\hat{\beta}$ and $\hat{\gamma}$ from the second and third equations of (6.3) or from equations (6.4). Estimates $\hat{\mu}$ and $\hat{\sigma}$ follow as

$$\hat{\mu} = \ln\hat{\beta} \qquad \text{and} \qquad \hat{\sigma} = \sqrt{\ln\hat{\omega}} \tag{6.5}$$

In the special case where ω and thus σ are known, we calculate $\hat{\beta}$ and $\hat{\gamma}$ from (6.4) with ω substituted for $\hat{\omega}$.

As in the case of moment and local maximum likelihood estimators, we need to solve a non-linear estimating equation for a single unknown. In this instance the first equation of (6.3) involves only the single unknown, $\hat{\omega}$, or if you prefer, $\hat{\sigma}$. Again the Newton-Raphson technique is satisfactory for this purpose. There are, of course, various other iterative schemes that might serve as well. A simple "trial and error" scheme with linear interpolation will often suffice. We need only find two values ω_i and ω_j, in a sufficiently narrow interval such that $J(n,\omega_i) \lesseqgtr s^2/(\bar{x}-x_1)^2 \gtreqless J(n,\omega_j)$, and then interpolate for the final estimate $\hat{\omega}$, where $J(n,\omega)$ designates the right side of the equation. The only problem likely to be encountered is that of finding a good (close) first approximation. To meet this requirement, Cohen et al. (1985) presented a table and a chart of the function on the right side of the first equation of (6.3) for values of σ and n that are most likely to be encountered in practice. These aids to computation are reproduced here as Table 5 and Figure 2. Note that $J(n,\omega) \equiv J(n,\sigma)$, since $\sigma = \sqrt{\ln\omega}$.

In using these aids, we enter the table or the chart with the sample values $s^2/(\bar{x}-x_1)^2$ and n. Inverse interpolation in Table 5 or a direct

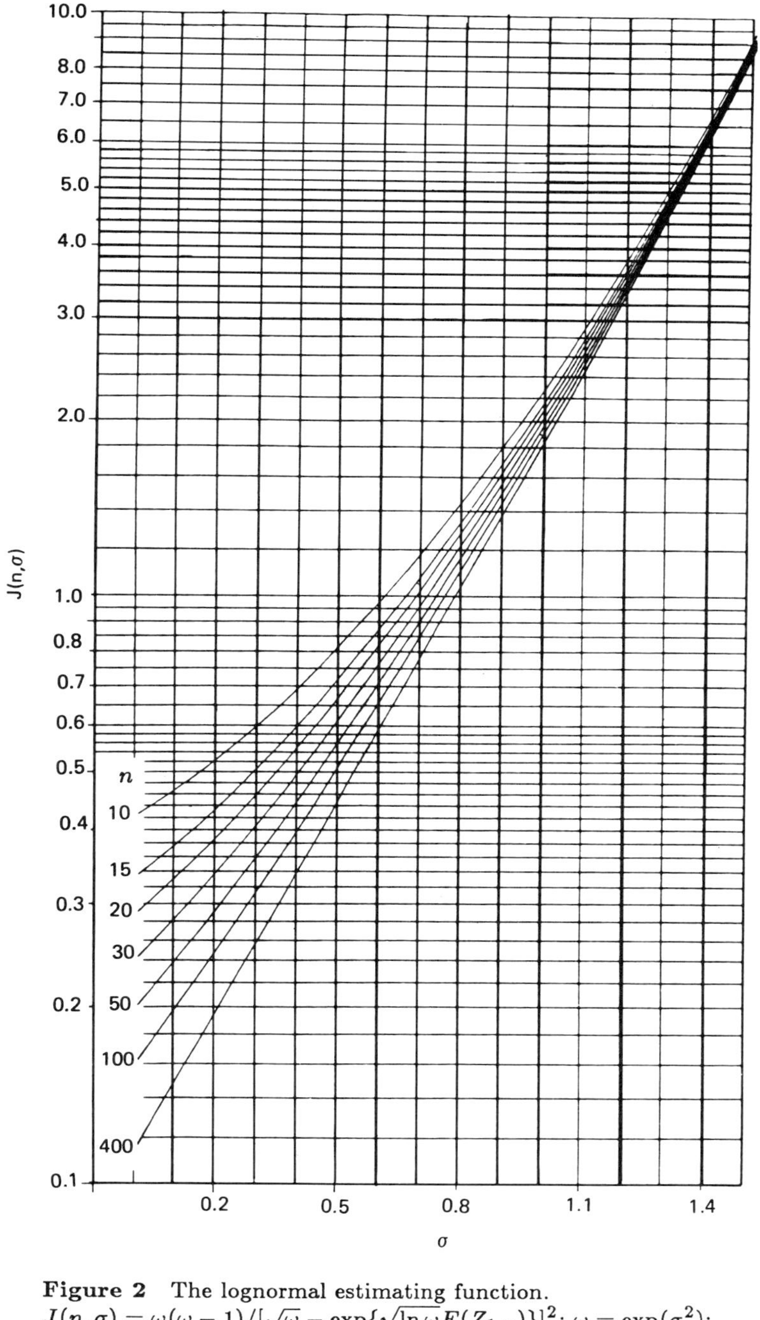

Figure 2 The lognormal estimating function.
$J(n,\sigma) = \omega(\omega - 1)/[\sqrt{\omega} - \exp\{\sqrt{\ln \omega} E(Z_{1,n})\}]^2$; $\omega = \exp(\sigma^2)$;
$\sigma = \sqrt{\ln \omega}$; $J(n, \hat{\sigma}) = s^2/(\bar{x} - x_1)^2$

Table 5 The Lognormal Estimating Function
$J(n,\sigma) = \omega(\omega - 1)/[\sqrt{\omega} - \exp\{\sqrt{\ln\omega}E(Z_{1,n})\}]^2$; $\sigma = \sqrt{\ln\omega}$

$\sigma \backslash n$	10	15	20	25	30	35	40	50	60
.01	.42615	.33575	.29061	.26272	.24341	.22906	.21786	.20131	.18946
.02	.43007	.33975	.29457	.26664	.24727	.23288	.22164	.20501	.19311
.03	.43411	.34386	.29864	.27065	.25123	.23679	.22551	.20881	.19685
.04	.43826	.34807	.30281	.27476	.25529	.24080	.22947	.21270	.20068
.05	.44253	.35239	.30708	.27897	.25945	.24491	.23354	.21669	.20461
.10	.46574	.37572	.33011	.30168	.28186	.26706	.25546	.23823	.22584
.15	.49230	.40216	.35617	.32736	.30722	.29214	.28030	.26267	.24995
.20	.52257	.43212	.38565	.35643	.33593	.32055	.30845	.29039	.27735
.25	.55700	.46603	.41901	.38933	.36845	.35275	.34038	.32189	.30850
.30	.59609	.50444	.45678	.42660	.40532	.38928	.37663	.35769	.34395
.35	.64046	.54793	.49958	.46885	.44715	.43077	.41783	.39843	.38434
.40	.69079	.59723	.54811	.51682	.49467	.47793	.46470	.44483	.43038
.45	.74792	.65317	.60322	.57132	.54871	.53161	.51807	.49774	.48293
.50	.81279	.71670	.66587	.63334	.61026	.59278	.57894	.55813	.54298
.55	.88655	.78897	.73719	.70402	.68045	.66259	.64844	.62717	.61166
.60	.97050	.87129	.81852	.78467	.76061	.74237	.72792	.70618	.69034
.65	1.06620	.96522	.91142	.87688	.85231	.83369	.81894	.79674	.78057
.70	1.17548	1.07261	1.01773	.98248	.95741	.93840	.92335	.90070	.88422
.75	1.30051	1.19562	1.13962	1.10364	1.07806	1.05868	1.04333	1.02025	1.00346
.80	1.44387	1.33683	1.27966	1.24295	1.21686	1.19710	1.18145	1.15796	1.14089
.85	1.60860	1.49928	1.44091	1.40345	1.37685	1.35672	1.34079	1.31690	1.29955
.90	1.79834	1.68661	1.62700	1.58879	1.56168	1.54119	1.52499	1.50071	1.48311
.94	1.97105	1.85730	1.79667	1.75785	1.73034	1.70955	1.69314	1.66857	1.65078
.95	2.01746	1.90319	1.84230	1.80333	1.77571	1.75486	1.73840	1.71375	1.69592
.96	2.06524	1.95045	1.88931	1.85018	1.82247	1.80154	1.78502	1.76031	1.74243
.97	2.11446	1.99914	1.93774	1.89845	1.87064	1.84964	1.83307	1.80829	1.79036
.98	2.16115	2.04929	1.98763	1.94819	1.92027	1.89921	1.88259	1.85774	1.83977
.99	2.21736	2.10097	2.03904	1.99945	1.97143	1.95029	1.93362	1.90870	1.89069
1.00	2.27116	2.15422	2.09202	2.05228	2.02416	2.00295	1.98623	1.96124	1.94319
1.05	2.56573	2.44598	2.38244	2.34191	2.31329	2.29174	2.27477	2.24945	2.23119
1.10	2.90875	2.78605	2.72112	2.67980	2.65068	2.62879	2.61158	2.58595	2.56751
1.15	3.30941	3.18361	3.11725	3.07513	3.04551	3.02328	3.00584	2.97992	2.96131
1.20	3.77887	3.64980	3.58196	3.53903	3.50891	3.48635	3.46868	3.44248	3.42372
1.25	4.33073	4.19823	4.12886	4.08509	4.05447	4.03159	4.01369	3.98722	3.96833
1.30	4.98165	4.84552	4.77456	4.72995	4.69882	4.67561	4.65750	4.63077	4.61174
1.35	5.75206	5.61211	5.53950	5.49401	5.46236	5.43883	5.42050	5.39352	5.37437
1.40	6.66717	6.52316	6.44882	6.40244	6.37026	6.34639	6.32785	6.30062	6.28136
1.45	7.75811	7.60980	7.53366	7.48634	7.45362	7.42941	7.41064	7.38317	7.36380
1.50	9.06354	8.91066	8.83263	8.78433	8.75104	8.72649	8.70750	8.67978	8.66029
1.55	10.63161	10.47387	10.39383	10.34451	10.31064	10.28573	10.26650	10.23852	10.21892
1.60	12.52251	12.35959	12.27743	12.22703	12.19254	12.16725	12.14778	12.11953	12.09981
1.65	14.81178	14.64330	14.55889	14.50735	14.47221	14.44652	14.42680	14.39827	14.37842
1.70	17.59453	17.42010	17.33329	17.28054	17.24471	17.21860	17.19860	17.16976	17.14977
1.75	20.99102	20.81020	20.72082	20.66678	20.63021	20.60365	20.58335	20.55419	20.53405
2.00	54.27287	54.05161	53.94619	53.88409	53.84292	53.81349	53.79132	53.75998	53.73874

σ\n	70	80	90	100	150	200	300	400
.01	.18044	.17327	.16738	.16243	.14576	.13582	.12385	.11056
.02	.18404	.17682	.17090	.16592	.14913	.13911	.12704	.10770
.03	.18773	.18047	.17451	.16950	.15259	.14249	.13031	.10492
.04	.19151	.18421	.17822	.17317	.15614	.14596	.13368	.10222
.05	.19539	.18805	.18202	.17694	.15979	.14953	.13715	.09959
.10	.21637	.20881	.20259	.19734	.17959	.16892	.15601	.08753
.15	.24021	.23243	.22601	.22060	.20223	.19115	.17771	.07707
.20	.26733	.25932	.25270	.24712	.22812	.21664	.20267	.06800
.25	.29821	.28996	.28314	.27739	.25776	.24588	.23138	.06014
.30	.33338	.32489	.31788	.31195	.29170	.27942	.26441	.05332
.35	.37348	.36476	.35755	.35144	.33058	.31791	.30240	.04740
.40	.41924	.41029	.40287	.39660	.37515	.36209	.34612	.04227
.45	.47151	.46233	.45472	.44828	.42625	.41284	.39642	.03781
.50	.53127	.52187	.51408	.50748	.48489	.47115	.45432	.03394
.55	.59969	.59007	.58210	.57534	.55224	.53818	.52097	.03059
.60	.67810	.66827	.66013	.65323	.62964	.61529	.59776	.02767
.65	.76809	.75805	.74975	.74272	.71868	.70408	.68626	.02513
.70	.87150	.86128	.85282	.84566	.82122	.80640	.78834	.02293
.75	.99052	.98012	.97153	.96425	.93946	.92445	.90620	.02102
.80	1.12773	1.11718	1.10845	1.10108	1.07598	1.06082	1.04243	.01937
.85	1.28620	1.27550	1.26667	1.25920	1.23384	1.21857	1.20009	.01794
.90	1.46958	1.45875	1.44982	1.44228	1.41671	1.40135	1.38283	.01670
.94	1.63712	1.62620	1.61719	1.60960	1.58389	1.56850	1.54998	.01584
.95	1.68223	1.67128	1.66226	1.65465	1.62892	1.61352	1.59501	.01564
.96	1.72871	1.71774	1.70870	1.70109	1.67533	1.65993	1.64142	.01544
.97	1.77661	1.76562	1.75657	1.74895	1.72316	1.70776	1.68926	.01526
.98	1.82599	1.81498	1.80591	1.79828	1.77247	1.75706	1.73857	.01507
.99	1.87688	1.86585	1.85677	1.84913	1.82330	1.80789	1.78942	.01490
1.00	1.92935	1.91830	1.90921	1.90155	1.87571	1.86030	1.84184	.01473
1.05	2.21723	2.20610	2.19694	2.18925	2.16334	2.14795	2.12958	.01396
1.10	2.55343	2.54223	2.53303	2.52531	2.49937	2.48403	2.46580	.01331
1.15	2.94713	2.93587	2.92664	2.91889	2.89298	2.87771	2.85964	.01278
1.20	3.40946	3.39815	3.38889	3.38114	3.35528	3.34011	3.32224	.01235
1.25	3.95399	3.94264	3.93337	3.92562	3.89984	3.88480	3.86715	.01202
1.30	4.59734	4.58596	4.57669	4.56894	4.54328	4.52837	4.51097	.01179
1.35	5.35991	5.34852	5.33924	5.33151	5.30598	5.29123	5.27408	.01165
1.40	6.26685	6.25544	6.24617	6.23845	6.21308	6.19849	6.18162	.01159
1.45	7.34924	7.33782	7.32856	7.32087	7.29566	7.28124	7.26465	.01163
1.50	8.64569	8.63426	8.62501	8.61734	8.59230	8.57806	8.56176	.01176
1.55	10.20428	10.19284	10.18361	10.17596	10.15109	10.13703	10.12101	.01199
1.60	12.08512	12.07368	12.06445	12.05683	12.03214	12.01825	12.00252	.01233
1.65	14.36367	14.35221	14.34300	14.33540	14.31087	14.29716	14.28170	.01279
1.70	17.13497	17.12349	17.11428	17.10670	17.08233	17.06878	17.05360	.01338
1.75	20.51918	20.50767	20.49846	20.49089	20.46667	20.45328	20.43836	.01413
2.00	53.72329	53.71150	53.70216	53.69456	53.67075	53.65796	53.64410	.02124

reading from Figure 2 yields a first approximation σ_1 that is sufficiently accurate to become the final estimate in many applications. When greater accuracy is required, this approximation is close enough to the required solution to produce rapid convergence of the iterative process to the final estimate. With $\hat{\sigma}$ or $\hat{\omega}$ thus calculated, we calculate $\hat{\beta}$ and $\hat{\gamma}$ from equations (6.4) or from the last two equations of (6.3).

In many applications, modified moment estimators are the preferred estimators. They are unbiased with respect to the population mean and variance. They are easy to calculate and their variances are minimal or at least near minimal. They do not suffer from regularity problems. However, instances will arise in which an investigator might prefer maximum likelihood estimators or rather local MLE. In those cases the MME will at least provide excellent first approximations from which to begin iterations toward the LMLE.

6.1 Sampling Errors

Simulation results reported by Cohen and Whitten (1980) and by Cohen et al. (1985) disclose that asymptotic estimate variances and covariances calculated using (5.4) are reasonably close to corresponding simulated variances and covariances of the MME. We may therefore use (5.4) or (5.5) to calculate approximate variances of the MME when these are required.

7. ILLUSTRATIVE EXAMPLES

As illustrative examples, we choose the same two sets of data that were used by Cohen et al. (1985). Although these examples are from real life situations, their sole purpose here is to illustrate details concerning the calculation of estimates.

Example 1 The maximum flood level in millions of cubic feet per second for the Susquehanna River at Harrisburg, Pennsylvania, over 20 four-year intervals from 1890 to 1969 as given by Dumonceaux and Antle (1973) are tabulated below.

0.654	0.613	0.315	0.449	0.297
0.402	0.379	0.423	0.379	0.3235
0.269	0.740	0.418	0.412	0.494
0.416	0.338	0.392	0.484	0.265

In summary, $n = 20$, $\bar{x} = 0.423125$, $s^2 = 0.0156948$, $s = 0.12528$, $a_3 = 1.0673243$, $x_1 = 0.265$, and $s^2/(\bar{x} - x_1)^2 = 0.6277$. Dumonceaux and

Antle considered this sample to be from an extreme value distribution. For our purpose we assume it to be from a three-parameter lognormal distribution. Accordingly we enter the graphs of Figure 2 with $n = 20$ and $s^2/(\bar{x} - x_1)^2 = 0.628$ (rounded off) to read $\sigma_1 = 0.47$. This same value is obtained by inverse linear interpolation in Table 5 between $\sigma = 0.45$ and $\sigma = 0.50$. For additional accuracy, we make further calculations of the right side of the first equation of (6.3) and interpolate as follows:

σ	ω	$J(n, \omega)$
0.4710	1.24837	0.6286
0.4703	1.24755	0.6277
0.4700	1.24720	0.6273

Here $J(n, \omega)$ signifies the right side of the first equation of (6.3). As a final estimate, we have $\hat{\sigma} = 0.4703$, which is subsequently rounded off to 0.470. We use the second and third equations of (6.3) to calculate $\hat{\gamma} = 0.171$ and $\hat{\beta} = 0.226$. It follows from (6.5) that $\hat{\mu} = -1.487$. Of course, $\hat{\omega} = 1.247$.

A FORTRAN program (Cohen and Whitten, 1980) was used to calculate modified moment estimates that agree with those presented above. This program was also used to calculate moment and maximum likelihood estimates. For comparisons, the three sets of estimates are listed below.

Estimates for Example 1

Estimator	Estimates							
	$\hat{\gamma}$	$\hat{\mu}$	$\hat{\sigma}$	$\hat{\omega}$	$\hat{\beta}$	$\hat{\alpha}_3$	$\hat{E}(X)$	$\sqrt{\hat{V}(X)}$
MME	0.171	−1.487	0.470	1.247	0.226	1.61	0.423	0.125
ME	0.056	−1.058	0.333	1.117	0.347	1.07	0.423	0.125
LMLE	0.185	−1.561	0.507	1.293	0.210	1.78	0.424	0.129

Asymptotic standard deviations of the estimates calculated using (5.4) with the LMLE estimates $\hat{\sigma} = 0.507$ and $\hat{\beta} = 0.210$ are approximately $\sigma_{\hat{\gamma}} \doteq 0.063$, and $\sigma_{\hat{\mu}} \doteq 0.358$, $\sigma_{\hat{\beta}} \doteq 0.075$, and $\sigma_{\hat{\sigma}} = 0.190$. It is to be observed that differences between the MME and LMLE are quite small. Comparable differences between estimators have been observed by the author in practically all applications where the MME were calculated.

Example 2 This example is taken from McCool (1974). The data consists of fatigue life in hours of ten bearings of a certain type. Listed in order of magnitude, these are:

152.7	172.0	172.5	173.3	193.0
204.7	216.5	234.9	262.6	422.6

In summary, $n = 10$, $\bar{x} = 220.4804$, $s^2 = 6147.438$, $s = 78.406$, $a_3 = 1.8635835$, $x_1 = 152.7$ and $s^2/(\bar{x} - x_1)^2 = 1.33809$. To calculate the MME, we enter the graphs of Figure 2 with $n = 10$, and $s^2/(\bar{x} - x_1)^2$ rounded off to 1.34 to read $\sigma_1 = 0.76$. Interpolation in Table 5 between entries for $\sigma = 0.75$ and $\sigma = 0.80$ also yields this same value. By calculating additional values of $J(\omega)$ for $\sigma = 0.760$ and 0.770 and then interpolating as was done with example 1, we obtain as a final estimate $\hat{\sigma} = 0.764$. We then use the last two equations of (6.3) to calculate $\hat{\beta} = 65.94$ and $\hat{\gamma} = 132.3$. Computer calculations using the FORTRAN program gave these same values for the MME, and also produced moment estimates. The LMLE failed to produce estimates from this sample. Failure of the LMLE is not an uncommon occurrence for small samples. Sometimes failures occur even for large samples. This limitation of the LMLE is one of the principal reasons for favoring the MME. The moment and the modified moment estimates from example 2 are displayed below for comparison.

Estimates for Example 2

Estimator	Estimates							
	$\hat{\gamma}$	$\hat{\mu}$	$\hat{\sigma}$	$\hat{\omega}$	$\hat{\beta}$	$\hat{\alpha}_3$	$\hat{E}(X)$	$\sqrt{\hat{V}(X)}$
MME	132.3	4.19	0.764	1.79	65.94	3.37	220.5	78.41
ME	81.9	4.79	0.524	1.32	120.65	1.86	220.5	78.41

In addition to the Cohen-Whitten FORTRAN program, which was employed in connection with the two illustrative examples presented here, attention is invited to a FORTRAN program available from Mathstat Associates, P.O. Box 723, Cranford, N.J. 07016. According to their announcement, this program (designated as LOGNPF) computes local maximum likelihood estimates of lognormal parameters by employing the interior penalty function routine described by Wingo (1975).

8. ESTIMATION FROM GROUPED DATA

When data from complete (unrestricted) samples are grouped, the estimation procedure is essentially the same as for ungrouped data. For *local maximum likelihood estimators*, the only change involves the summations in estimating equations (5.1, 5.2, 5.3). For example, $\sum_1^n \ln(x_1 - \hat{\gamma})$ would be replaced by $\sum_1^k f_j \ln(x_j - \hat{\gamma})$ where f_j and x_j are the number of observations and the mid-point respectively of the jth class (group), k is the number of classes, and $n = \sum_1^k f_j$. Corresponding changes would be made in the other summations. Unfortunately grouping errors might be rather large unless grouping intervals are quite small.

Grouping errors are less likely to effect the *modified estimators* $\bar{x} = \sum_1^k f_j x_j/n$ and $s^2 = [\sum_1^k f_j(x_j - \bar{x})^2/(n-1)] - (h^2/12)$ where $(h^2/12)$ is Sheppard's grouping correction and h is the class width. The only problem involved might concern determination of x_1, the smallest sample observation. If x_1 is recorded, then of course it should be used. Otherwise, it is at least known that $b_1 \leq x_1 \leq b_2$ where b_1 is the lower boundary of the smallest class and b_2 is the upper boundary. Therefore, we need to choose a value of x_1 between b_1 and b_2. In most instances, small differences in x_1 will have little effect on the estimates. The standard Chi-Square test or the Kolmogorov-Smirnov test for goodness of fit can be employed to determine whether or not expected frequencies based on the calculated estimates are in satisfactory agreement with observed frequencies.

With $s^2/(\bar{x} - x_1)^2$ calculated from the grouped data, equations (6.3) are employed to obtain estimates of σ, β, and γ. Computational procedures, including use of Table 5 and/or Figure 2, are the same as those followed for ungrouped data.

ACKNOWLEDGMENTS

The assistance of Dr. Betty Jones Whitten in computing entries for Tables 1, 2, 3, and 5, and for performing numerous other computations is gratefuly acknowledged. Appreciation is also expressed to the Editors and Publishers of the *Journal of Quality Technology* for permission to reproduce the chart of Figure 2, portions of Table 5, plus other selected text from Cohen, Whitten, and Ding (1985).

REFERENCES

Abramowitz, M. and Stegun, I.A. (1964). *Handbook of Mathematical Functions with Formulas, Graphs, and Mathematical Tables*, National Bureau of Standards Applied Mathematics Series 55. U.S. Government Printing Office, Washington, D.C. 20402.

Aitchison, J. and Brown, J.A.C. (1957), *The Lognormal Distribution*, London: Cambridge University Press.

Antle, C.E. (1985). Lognormal distribution, *Encyclopedia of Statistical Sciences*, *5*, 134–136. Editors: Kotz, S., Johnson, N.L., and Read, C.B. John Wiley & Sons, New York.

Box, G.E.P. and Cox, D.R., (1964). An analysis of transformations. *J. Roy. Statistic. Soc.*, *B*, *26*, 211-251.

Bradu, D. and Mundlak, Y., (1970). Estimation in lognormal linear models models, *J. Amer. Statist. Assoc.*, *65*, 198–211.

Calitz, F. (1973). Maximum likelihood estimation of the parameters of the three-parameter lognormal distribution—a reconsideration, *Austral. J. Statist.* *3*, 185–190.

Cohen, A.C. (1951). Estimating parameters of logarithmic-normal distributions by maximum likelihood, *J. Amer. Statist. Assoc.*, *46*, 206–212.

Cohen, A.C. and Whitten, B.J. (1980). Estimation in the three-parameter lognormal distribution, *J. Amer. Statist. Assoc.*, *75*, 399–404.

Cohen, A.C., Whitten, B.J. and Ding, Y. (1985). Modified moment estimation for the three-parameter lognormal distribution, *J. Qual. Tech.*, *17*, 92–99.

Dumonceaux, R. and Antle, C.E. (1973). Discrimination between the lognormal and the Weibull distributions, *Technometrics*, *15*, 923–926.

Evans, I.G. and Shaban, S.A. (1974). A note on estimation in lognormal models, *J. Amer. Statist. Assoc.*, *69*, 779–781.

Finney, D.J. (1941). On the distribution of a variate whose logarithm is normally distributed, *J. Statist. Soc.*, *Ser B*, *7*, 155–161.

Giesbrecht, F. and Kempthorne, O. (1976). Maximum likelihood estimation in the three-parameter lognormal distribution, *J. Roy. Statist. Soc.*, *Ser. B*, *38*, 257–264.

Harter, H.L. (1961). Expected values of normal order statistics, *Biometrika*, *48*, 151–166.

Harter, H.L. (1969). Order statistics and their use in testing and estimation Vol. 2. Table C1, 426–456, Aerospace Res. Lab., Wright-Patterson A.F. Base, Ohio.

Harter, H.L. and Moore, A.H. (1966). Local-maximum-likelihood estimation of the parameters of three-parameter lognormal populations from complete and censored samples, *J. Amer. Statist. Assoc.*, *61*, 842–851.

Heyde, C.C. (1963). On a property of the lognormal distribution, *J. Roy. Statist. Soc.*, *Ser. B*, *25*, 392–393.

Hill, B.M. (1963). The three-parameter lognormal distribution and Bayesian analysis of a point-source epidemic, *J. Amer. Statist. Assoc.*, *58*, 72–84.

Hoshi, K., Stedinger, J.R., and Burges, J. (1984). Estimation of log-normal quantiles: Monte Carlo results and first-order approximations, *J. Hydrology*, *71*, 1–30.

Johnson, N.L. and Kotz, S. (1970). *Continuous Univariate Distributions-1*, Houghton Mifflin, Boston.

Kane, V.E. (1982). Standard and goodness-of-fit parameter estimation methods for the three-parameter lognormal distribution, *Commun. Stat. Theor. Meth.*, *11*, 1935–1957.

Lambert, J.A. (1964). Estimation of parameters in the three-parameter lognormal distribution, *Austral. J. Statist.*, *6*, 29–32.

Lawrence, R.J. (1979). The lognormal as Inter-Event Time Distribution, *Unpublished manuscript.*

Lawrence, R.J. (1980). The lognormal distribution of buying frequency rates, *J. Marketing Res.*, *17*, 212–220.

Lawrence, R.J. (1984).The lognormal distribution of the duration of strikes, *J. Roy. Statist. Soc. Ser.*, *A*, *147*, 464–483.

McCool, J.I. (1974). Inferential techniques for Weibull populations, *Aerospace Research Laboratories Report ARL TR 74-0180, Wright-Patterson AFB, OH.*

Monlezum, C.J., Antle, C.E., and Klimko, L.A. (1975). Unpublished manuscript (concerning maximum likelihood estimation of parameter in the lognormal model).

Munro, A.H. and Wixley, R.A.J. (1970). Estimators based on order statistics of small samples from a three-parameter lognormal distribution, *J. Amer. Statist. Assoc.*, *65*, 212–225.

Nelson, W.B. (1982). *Applied Life Data Analysis*, John Wiley & Sons, New York.

Rukhin, A.L. (1984). Improved estimation in lognormal models, T.R. 84–38 Dept. of Statist., Purdue Univ., West Lafayette, Indiana.

Stedinger, J.R. (1980). Fitting lognormal distributions to hydrologic data. *Water Resources Research*, *16*, 481–490.

Tiku, M.L. (1968). Estimating the parameters of log-normal distribution from censored samples, *J. Amer. Statist. Assoc.*, *63*, 134–140.

Wilson, Edwin B. and Worcester, J. (1945). The normal logarithmic transform, *Rev. Econ. Statist.*, 27, 17–22.

Wingo, D.R. (1975). The use of interior penalty functions to overcome log-normal distribution parameter estimation anomalies, *J. Statist. Comp. Simul.*, *4*, 49–61.

Wingo, D.R. (1976). Moving truncations barrier-function methods for estimation in three-parameter lognormal models, *Commun. Statist.-Simula. Computa.*, *B5*, 65–80.

Yuan, P.T. (1933). On the logarithmic frequency distributions and the semi-logarithmic correlation surface, *Ann. Math. Statist.*, *4*, 30–74.

5

Censored, Truncated, and Grouped Estimation

A. CLIFFORD COHEN Department of Statistics, University of Georgia, Athens, Georgia

1. INTRODUCTION

Censored samples occur frequently in life, fatigue, and reaction time tests where individual observations are time (magnitude) ordered and where tests are terminated with one or more survivors. Complete life spans of the survivors thus remain unknown except for the information that they exceed the time (point) of censoring. A sample of the type described is said to be singly censored on the right. A typical right-censored sample in which a total of N items are placed on test would consist of n ordered observations $\{x_i\}$, $i = 1, \ldots, n$, plus the information that the life spans of $r = N - n$ of the sample specimens exceeded the time (point) of censoring, where X designates the random variable (in this instance the life span or time of failure). In generating censored samples, tests might be terminated at a fixed time T, with a random number of survivors, r, or they might be terminated at a random time x_n with a fixed number of survivors r. Samples produced by fixed time censoring are designated as Type I samples while those produced by fixed survivor number censoring are designated as Type II samples. Truncated samples or rather samples from a truncated distribution occur when sample selection and/or observation and measurement is not possible in certain regions (usually in the distribution tails) of the sample space. Truncated samples might be considered as a special case of censored samples in which the number of censored observations is unknown.

Truncated and censored samples from the normal, exponential, gamma, lognormal, Weibull and various other distributions have received attention from numerous writers beginning with Galton (1897), and including Pearson and Lee (1908), Fisher (1931), Bliss and Stevens (1937), Hald (1949), Ipsen (1949), Cohen (1950, 1959, 1961, 1963, 1965, 1975, 1976), Gupta (1952), Halperin (1952), Herd (1957), Roberts (1962a, 1962b), Harter and Moore (1966), Tiku (1968), Gajjar and Khatri (1969), Ringer and Springle (1972), Lemon (1975) and many others.

This chapter is primarily concerned with estimation in the three-parameter lognormal distribution from samples that are singly censored or singly truncated on the right. Consideration is also given to samples that are truncated or censored on the left as well as to doubly censored, doubly truncated, and progressively censored samples. Brief mention is made of estimation in the two-parameter distribution as a special case of the three-parameter distribution.

2. MAXIMUM LIKELIHOOD ESTIMATION

The likelihood function for a Type I sample that is singly censored on the right at $X = T$ from a distribution with pdf $f(x;\theta_1,\theta_2,\theta_3)$ and c.d.f. $F(x;\theta_1,\theta_2,\theta_3)$ is

$$L(S) = C\prod_{1}^{n} f(x_1;\theta_1,\theta_2,\theta_3)\cdot[1-F(T)]^r \tag{2.1}$$

where S designates the censored sample $\{x_i\}$, n, r, T, and C is a constant that depends only on n and r. The corresponding function of a Type II sample differs from (2.1) only in that T is replaced by the nth order statistic x_n.

Our concern here is with the three-parameter lognormal distribution whose pdf as given in Chapter 4, equation (1.1), is

$$f(x;\gamma,\mu,\sigma) = \frac{1}{\sigma\sqrt{2\pi}(x-\gamma)}\exp -\frac{1}{2\sigma^2}[\ln(x-\gamma)-\mu]^2$$

$$\gamma < x < \infty \qquad \sigma > 0 \tag{2.2}$$

zero otherwise.

When regularity conditions do not present problems, maximum likelihood estimates are obtained by taking logarithms of (2.1), differentiating with respect to the parameters, equating to zero and solving the resulting equations simultaneously. For the lognormal distribution with p.d.f. (2.2), $L(S) \to \infty$ as $\gamma \to x_1$ and a local maximum of the likelihood function is the best that can be achieved by following the procedure described here. Fortunately in most instances these local maximum likelihood estimates (LMLE) are reasonable and useful.

Estimating equations for the LMLE $\hat{\mu}$, $\hat{\sigma}$, and $\hat{\gamma}$ given by Cohen (1976) are

$$\sum_{1}^{n} [\ln(x_i - \gamma) - \mu] + r\sigma Z = 0$$

$$\sum_{1}^{n} [\ln(x_i - \gamma) - \mu]^2 - \sigma^2 [n - \xi r Z] = 0 \tag{2.3}$$

$$\sum_{1}^{n} \left[\frac{\ln(x_i - \gamma) - \mu}{x_i - \gamma}\right] + \sigma^2 \sum_{1}^{n} \left(\frac{1}{x_i - \gamma}\right) + \frac{r\sigma Z}{T - \gamma} = 0$$

where

$$\xi = \frac{\ln(T - \gamma) - \mu}{\sigma} \quad \text{and} \quad Z = Z(\xi) = \frac{\phi(\xi)}{1 - \Phi(\xi)} \tag{2.4}$$

and where $\phi(\ )$ and $\Phi(\ )$ are the pdf and the cdf of the standard normal distribution (0,1). Note that Z is the hazard function of this distribution.

A straightforward "trial and error" iterative procedure for simultaneously solving the three equations of (2.3) was outlined in the 1976 paper. However these calculations can be substantially simplified by taking advantage of the simple relation that exists between the normal and the lognormal distribution.

If the random variable X has a lognormal distribution (γ, μ, σ) then $Y = \ln(X - \gamma)$ has a normal distribution (μ, σ^2). Consequently if we are analyzing sample data from a lognormal distribution, we need only make the transformation $\{y_i\} = \{\ln(x_i - \gamma)\}$ and the transformed sample becomes a normal sample to be analyzed using applicable normal theory. When γ is known, the transformation and any subsequent estimation or analysis is quite simple and further comments seem unnecessary. The situation is somewhat more complicated when γ is unknown and must be estimated from sample data. In this case, we begin with a first approximation $\gamma_1 (< x_1)$ and calculate conditional estimates $\mu(\gamma_1)$ and $\sigma(\gamma_1)$. These conditional estimates can then be "tested" by imposing a third requirement or by substitution in a third estimating equation. Giesbrecht and Kempthorne (1976) employed this approach in estimating lognormal parameters from complete (uncensored) samples by evaluating the likelihood function $L[\gamma_i, \mu(\gamma_i), \sigma(\gamma_i)]$ for several values γ_i in the vicinity of a maximum. The value of γ which results in maximum L was then chosen as $\hat{\gamma}$. Although they considered only complete samples, the procedure is also applicable for truncated and censored samples.

For complete samples $\hat{\mu}(\gamma_1) = \frac{1}{n}\sum_1^n \ln(x_i - \gamma_1) = \bar{y}(\gamma_1)$ and $\hat{\sigma}^2(\gamma_1) = \frac{1}{n-1}\sum_1^n [\ln(x_i - \gamma_1) - \hat{\mu}(\gamma_1)]^2 = s_y^2(\gamma_1)$. The third requirement could be maximizing L as was done by Giesbrecht and Kempthorne or it could be satisfying the equation $\partial \ln L/\partial\gamma = 0$; i.e., the third equation of (2.3) with $r = 0$. If $\partial \ln L/\partial\gamma \,|_{\gamma=\gamma_1} = 0$, our task is finished and $\hat{\gamma} = \gamma_1$. Otherwise, we select a second approximation γ_2 and repeat the calculations. We continue until we find two values γ_i and γ_j in a sufficiently narrow interval and such that $(\partial \ln L/\partial\gamma)(\gamma_i) \lesseqgtr 0 \lesseqgtr (\partial \ln L/\partial\gamma)(\gamma_j)$. We then interpolate for the final estimates $\hat{\gamma}$, $\hat{\mu}$, $\hat{\sigma}$. Both the Giesbrecht and Kempthorne procedure and that described above produce local maximum likelihood estimates.

3. TRUNCATED AND CENSORED SAMPLES

For truncated and censored samples, the estimating procedure is quite similar to that described in the preceding paragraphs for complete samples. The two cases differ only in the estimation of $\mu(\gamma)$ and $\sigma(\gamma)$. If a sample of size N from a lognormal distribution (γ, μ, σ) is censored on the right at $X = T$ with r survivors and thus includes the complete ordered observations $\{x_i\}$, $i = 1, \ldots, n$, the transformed sample from a normal distribution (μ, σ) is censored on the right at $y_T = \ln(T - \gamma)$ with r survivors and includes the transformed observations $\{y_i\}$, $i = 1, \ldots,, n$, where $y_i = \ln(x_i - \gamma)$. Corresponding statements can be made for other types of censoring and for truncation. In these cases, conditional estimates, $\hat{\mu}(\gamma)$ and $\hat{\sigma}^2(\gamma)$ can be calculated from the transformed sample using estimators derived by Cohen (1959) for truncated and censored samples from the normal distribution. In the notation employed here, these estimators conditioned on γ can be written for both left and right truncated and censored samples as

Estimators for Singly Truncated Samples

$$\begin{aligned} \hat{\mu}(\hat{\gamma}) &= \bar{y} + \theta(\hat{\gamma})\,[y_T - \bar{y}] \\ \hat{\sigma}^2(\hat{\gamma}) &= s_y^2 + \theta(\hat{\gamma})\,[y_T - \bar{y}]^2 \end{aligned} \tag{3.1}$$

where $\theta(\gamma)$ is defined in (3.5) below.

Estimators for Type I Singly Censored Samples

$$\begin{aligned}\hat{\mu}(\hat{\gamma}) &= \bar{y} + \lambda(\hat{\gamma})\,[y_T - \bar{y}] \\ \hat{\sigma}^2(\hat{\gamma}) &= s_y^2 + \lambda(\hat{\gamma})\,[y_T - \bar{y}]^2\end{aligned} \tag{3.2}$$

where $\lambda(\ )$ is defined in (3.6) below.

Estimators for Type II Singly Censored Samples

$$\begin{aligned}\hat{\mu}(\hat{\gamma}) &= \bar{y} + \lambda(\hat{\gamma})\,[y_n - \bar{y}] \\ \hat{\sigma}^2(\hat{\gamma}) &= s_y^2 + \lambda(\hat{\gamma})\,[y_n - \bar{y}]^2\end{aligned} \tag{3.3}$$

In each of the above equations, (3.1, 3.2, 3.3), $\bar{y}$ and s_y^2 are the conditional mean and variance respectively based on the n complete observations of the transformed normal sample,

$$\bar{y}(\gamma) = \frac{1}{n}\sum_1^n y_i \qquad s_y^2(\gamma) = \frac{1}{(n-1)}\sum_1^n [y_i - \bar{y}]^2 \tag{3.4}$$

The auxiliary estimating functions θ and λ, here written as $\theta(\gamma)$ and $\lambda(\gamma)$ to signify their dependence on the threshold parameter γ, are defined as

$$\theta(\gamma) = \frac{Z(\xi)}{Z(\xi) - \xi} \quad \text{where } Z(\xi) = \frac{\phi(\xi)}{1 - \Phi(\xi)} \tag{3.5}$$

and with $h = r/N$

$$\lambda(\gamma) = \frac{Y(h, \xi)}{Y(h, \xi) - \xi} \quad \text{where } Y(h, \xi) = \frac{r}{n} Z(-\xi) \tag{3.6}$$

In the above formulas, $\xi = \xi(\gamma) = [y_T - \mu(\gamma)]/\sigma(\gamma)$ for truncated and Type I censored samples. In Type II right censored samples, y_T is replaced with y_n where $y_n = \ln(x_n - \gamma)$. In left censored Type II samples, y_T is replaced with y_{r+1}. Although original derivations [Cohen (1959)] were based on left truncated and left censored samples, as a result of symmetry in the normal distribution, the functions $\theta(\ )$ and $\lambda(\ )$ are also applicable to right restricted samples. In the interest of a simpler notation, the symbol (γ) is subsequently dropped except where clarity of expression demands that it be included.

The function $Y(h, \xi)$ introduced here should not be confused with the random variable $Y = \ln(X - \gamma)$, introduced earlier and used throughout chapters 4 and 5.

Maximum likelihood estimators of μ and σ^2 require that

$$\hat{\alpha} = \alpha(\hat{\xi}) = \frac{s_y^2}{(y_T - \bar{y})^2} \tag{3.7}$$

in truncated and Type I censored samples, where

$$\alpha(\xi) = \frac{1 - Z(Z - \xi)}{(Z - \xi)^2} \tag{3.8}$$

for truncated samples, and where

$$\alpha(\xi) = \alpha(h, \xi) = \frac{1 - Y(Y - \xi)}{(Y - \xi)^2} \tag{3.9}$$

for both Type I and Type II censored samples. For Type II censored samples, y_T in equation (3.7) is replaced with y_n, where $y_n = \ln(x_n - \gamma)$.

In order to calculate estimates using equations (3.1), (3.2) or (3.3) in practical applications, it is necessary first to calculate $\bar{y}$, s_y^2 and either $s_y^2/(y_T - \bar{y})^2$ or $s_y^2/(y_n - \bar{y})^2$ and with N, r, n, and either y_T or y_n also available from the sample data (conditioned on γ), to calculate the applicable auxiliary function θ or λ. Tables to facilitate this latter calculation were provided by Cohen (1959) and expanded by Cohen (1961). They were further expanded by Cooley and Cohen (1970). In each of these tables, the argument $\hat{\alpha}$ is limited to values ≤ 1. A more comprehensive tabulation with values of $\hat{\alpha}$ as large as 10 has been provided by Schmee and Nelson (1977). An abridged table from this source with $\hat{\alpha} = 0(.1)\ 1.0\ (1.0)\ 10$, and with $r/N = 0.1(0.1)0.9$ is given by Schmee, Gladstein and Nelson (1985). When tables are not available, it is relatively easy to calculate required values of λ from the defining relations of (3.5), (3.6), and (3.9).

Although θ and λ as defined in equations (3.5) and (3.6) are functions of ξ they are tabulated as functions of α which in turn is a function of ξ as defined in equations (3.8) and (3.9). This change of argument eliminates the necessity for first calculating $\hat{\xi}$ in order subsequently to calculate $\hat{\theta}$ or $\hat{\lambda}$. It is necessary only that we enter the tables with $\alpha = s_y^2/(y_T - \bar{y})^2$ or with $\alpha = s_y^2/(y_n - \bar{y})^2$ as applicable and read the corresponding value of $\hat{\theta}$ or $\hat{\lambda}$. Of course we enter with the appropriate value of h for censored samples, where $h = r/N$.

Tables of $\theta(\alpha)$ and $\lambda(h, \alpha)$ from Cohen (1961) are included here as Table 1 and Table 2 to aid in the calculation of estimates. Graphs of θ and λ are included in Figures 1, 2, and 3 respectively.

Table 1 Auxiliary Estimation Function θ for Singly Truncated Samples from the Normal Distribution

α	.000	.001	.002	.003	.004	.005	.006	.007	.008	.009	α
0.05	.00000	.00000	.00000	.00001	.00001	.00001	.00001	.00001	.00002	.00002	0.05
0.06	.00002	.00003	.00003	.00003	.00004	.00004	.00005	.00006	.00007	.00007	0.06
0.07	.00008	.00009	.00010	.00011	.00013	.00014	.00016	.00017	.00019	.00020	0.07
0.08	.00022	.00024	.00026	.00028	.00031	.00033	.00036	.00039	.00042	.00045	0.08
0.09	.00048	.00051	.00055	.00059	.00063	.00067	.00071	.00075	.00080	.00085	0.09
0.10	.00090	.00095	.00101	.00106	.00112	.00118	.00125	.00131	.00138	.00145	0.10
0.11	.00153	.00160	.00168	.00176	.00184	.00193	.00202	.00211	.00220	.00230	0.11
0.12	.00240	.00250	.00261	.00272	.00283	.00294	.00305	.00317	.00330	.00342	0.12
0.13	.00355	.00369	.00382	.00396	.00410	.00425	.00440	.00455	.00470	.00486	0.13
0.14	.00503	.00519	.00536	.00553	.00571	.00589	.00608	.00627	.00646	.00665	0.14
0.15	.00685	.00705	.00726	.00747	.00769	.00791	.00813	.00835	.00858	.00882	0.15
0.16	.00906	.00930	.00955	.00980	.01006	.01032	.01058	.01085	.01112	.01140	0.16
0.17	.01168	.01197	.01226	.01256	.01286	.01316	.01347	.01378	.01410	.01443	0.17
0.18	.01476	.01509	.01543	.01577	.01611	.01646	.01682	.01718	.01755	.01792	0.18
0.19	.01830	.01868	.01907	.01946	.01986	.02026	.02067	.02108	.02150	.02193	0.19
0.20	.02236	.02279	.02323	.02368	.02413	.02458	.02504	.02551	.02599	.02647	0.20
0.21	.02695	.02744	.02794	.02844	.02895	.02946	.02998	.03050	.03103	.03157	0.21
0.22	.03211	.03266	.03322	.03378	.03435	.03492	.03550	.03609	.03668	.03728	0.22
0.23	.03788	.03849	.03911	.03973	.04036	.04100	.04165	.04230	.04296	.04362	0.23
0.24	.04429	.04497	.04565	.04634	.04704	.04774	.04845	.04917	.04989	.05062	0.24
0.25	.05136	.05211	.05286	.05362	.05439	.05516	.05594	.05673	.05753	.05834	0.25
0.26	.05915	.05997	.06080	.06163	.06247	.06332	.06418	.06504	.06591	.06679	0.26
0.27	.06768	.06858	.06948	.07039	.07131	.07224	.07317	.07412	.07507	.07603	0.27
0.28	.07700	.07797	.07896	.07995	.08095	.08196	.08298	.08401	.08504	.08609	0.28
0.29	.08714	.08820	.08927	.09035	.09144	.09254	.09364	.09476	.09588	.09701	0.29
0.30	.09815	.09930	.10046	.10163	.10281	.10400	.10520	.10641	.10762	.10885	0.30
0.31	.1101	.1113	.1126	.1138	.1151	.1164	.1177	.1190	.1203	.1216	0.31
0.32	.1230	.1243	.1257	.1270	.1284	.1298	.1312	.1326	.1340	.1355	0.32
0.33	.1369	.1383	.1398	.1413	.1428	.1443	.1458	.1473	.1488	.1503	0.33
0.34	.1519	.1534	.1550	.1566	.1582	.1598	.1614	.1630	.1647	.1663	0.34
0.35	.1680	.1697	.1714	.1731	.1748	.1765	.1782	.1800	.1817	.1835	0.35
0.36	.1853	.1871	.1889	.1907	.1926	.1944	.1963	.1982	.2001	.2020	0.36
0.37	.2039	.2058	.2077	.2097	.2117	.2136	.2156	.2176	.2197	.2217	0.37
0.38	.2238	.2258	.2279	.2300	.2321	.2342	.2364	.2385	.2407	.2429	0.38
0.39	.2451	.2473	.2495	.2517	.2540	.2562	.2585	.2608	.2631	.2655	0.39
0.40	.2678	.2702	.2726	.2750	.2774	.2798	.2822	.2847	.2871	.2896	0.40
0.41	.2921	.2947	.2972	.2998	.3023	.3049	.3075	.3102	.3128	.3155	0.41
0.42	.3181	.3208	.3235	.3263	.3290	.3318	.3346	.3374	.3402	.3430	0.42
0.43	.3459	.3487	.3516	.3545	.3575	.3604	.3634	.3664	.3694	.3724	0.43
0.44	.3755	.3785	.3816	.3847	.3878	.3910	.3941	.3973	.4005	.4038	0.44
0.45	.4070	.4103	.4136	.4169	.4202	.4236	.4269	.4303	.4338	.4372	0.45
0.46	.4407	.4442	.4477	.4512	.4547	.4583	.4619	.4655	.4692	.4728	0.46
0.47	.4765	.4802	.4840	.4877	.4915	.4953	.4992	.5030	.5069	.5108	0.47
0.48	.5148	.5187	.5227	.5267	.5307	.5348	.5389	.5430	.5471	.5513	0.48
0.49	.5555	.5597	.5639	.5682	.5725	.5768	.5812	.5856	.5900	.5944	0.49
0.50	.5989	.6034	.6079	.6124	.6170	.6216	.6263	.6309	.6356	.6404	0.50
0.51	.6451	.6499	.6547	.6596	.6645	.6694	.6743	.6793	.6843	.6893	0.51
0.52	.6944	.6995	.7046	.7098	.7150	.7202	.7255	.7308	.7361	.7415	0.52
0.53	.7469	.7524	.7578	.7633	.7689	.7745	.7801	.7857	.7914	.7972	0.53
0.54	.8029	.8087	.8146	.8204	.8263	.8323	.8383	.8443	.8504	.8565	0.54
0.55	.8627	.8689	.8751	.8813	.8876	.8940	.9004	.9068	.9133	.9198	0.55
0.56	.9264	.9330	.9396	.9463	.9530	.9598	.9666	.9735	.9804	.9874	0.56
0.57	.9944	1.001	1.009	1.016	1.023	1.030	1.037	1.045	1.052	1.060	0.57
0.58	1.067	1.075	1.082	1.090	1.097	1.105	1.113	1.121	1.129	1.137	0.58
0.59	1.145	1.153	1.161	1.169	1.177	1.185	1.194	1.202	1.211	1.219	0.59
0.60	1.228	1.236	1.245	1.254	1.262	1.271	1.280	1.289	1.298	1.307	0.60
0.61	1.316	1.326	1.335	1.344	1.353	1.363	1.373	1.382	1.392	1.402	0.61
0.62	1.411	1.421	1.431	1.441	1.451	1.461	1.472	1.482	1.492	1.503	0.62
0.63	1.513	1.524	1.534	1.545	1.556	1.567	1.578	1.589	1.600	1.611	0.63
0.64	1.622	1.634	1.645	1.657	1.668	1.680	1.692	1.704	1.716	1.728	0.64
0.65	1.740	1.752	1.764	1.777	1.789	1.802	1.814	1.827	1.840	1.853	0.65
0.66	1.866	1.879	1.892	1.905	1.919	1.932	1.946	1.960	1.974	1.988	0.66
0.67	2.002	2.016	2.030	2.044	2.059	2.073	2.088	2.103	2.118	2.133	0.67
0.68	2.148	2.163	2.179	2.194	2.210	2.225	2.241	2.257	2.273	2.290	0.68
0.69	2.306	2.322	2.339	2.356	2.373	2.390	2.407	2.424	2.441	2.459	0.69
0.70	2.477	2.495	2.512	2.531	2.549	2.567	2.586	2.605	2.623	2.643	0.70
0.71	2.662	2.681	2.701	2.720	2.740	2.760	2.780	2.800	2.821	2.842	0.71
0.72	2.863	2.884	2.905	2.926	2.948	2.969	2.991	3.013	3.036	3.058	0.72
0.73	3.081	3.104	3.127	3.150	3.173	3.197	3.221	3.245	3.270	3.294	0.73
0.74	3.319	3.344	3.369	3.394	3.420	3.446	3.472	3.498	3.525	3.552	0.74
0.75	3.579	3.606	3.634	3.662	3.690	3.718	3.747	3.776	3.805	3.834	0.75
0.76	3.864	3.894	3.924	3.955	3.986	4.017	4.048	4.080	4.112	4.144	0.76
0.77	4.177	4.210	4.243	4.277	4.311	4.345	4.380	4.415	4.450	4.486	0.77
0.78	4.52	4.56	4.60	4.63	4.67	4.71	4.75	4.79	4.82	4.86	0.78
0.79	4.90	4.94	4.99	5.03	5.07	5.11	5.15	5.20	5.24	5.28	0.79
0.80	5.33	5.37	5.42	5.46	5.51	5.56	5.61	5.65	5.70	5.75	0.80
0.81	5.80	5.85	5.90	5.95	6.01	6.06	6.11	6.17	6.22	6.28	0.81
0.82	6.33	6.39	6.45	6.50	6.56	6.62	6.68	6.74	6.81	6.87	0.82
0.83	6.93	7.00	7.06	7.13	7.19	7.26	7.33	7.40	7.47	7.54	0.83
0.84	7.61	7.68	7.76	7.83	7.91	7.98	8.06	8.14	8.22	8.30	0.84
0.85	8.39	8.47	8.55	8.64	8.73	8.82	8.91	9.00	9.09	9.18	0.85

$\alpha = s_y^2/(\bar{y} - y_T)^2$

Table 2 Auxiliary Estimation Function λ for Singly Censored Samples from the Normal Distribution

α \ h	.01	.02	.03	.04	.05	.06	.07	.08	.09	.10	.15	.20	h / α
.00	.010100	.020400	.030902	.041583	.052507	.063627	.074953	.086488	.09824	.11020	.17342	.24268	.00
.05	.010551	.021294	.032225	.043350	.054670	.066189	.077909	.089834	.10197	.11431	.17935	.25033	.05
.10	.010950	.022082	.033398	.044902	.056596	.068483	.080568	.092852	.10534	.11804	.18479	.25741	.10
.15	.011310	.022798	.034466	.046318	.058356	.070586	.083009	.095629	.10845	.12148	.18985	.26405	.15
.20	.011642	.023459	.035453	.047629	.059990	.072539	.085280	.098216	.11135	.12469	.19460	.27031	.20
.25	.011952	.024076	.036377	.048858	.061522	.074372	.087413	.10065	.11408	.12772	.19910	.27626	.25
.30	.012243	.024658	.037249	.050018	.062969	.076106	.089433	.10295	.11667	.13059	.20338	.28193	.30
.35	.012520	.025211	.038077	.051120	.064345	.077756	.091355	.10515	.11914	.13333	.20747	.28737	.35
.40	.012784	.025738	.038866	.052173	.065660	.079332	.093193	.10725	.12150	.13595	.21139	.29260	.40
.45	.013036	.026243	.039624	.053182	.066921	.080845	.094958	.10926	.12377	.13847	.21517	.29765	.45
.50	.013279	.026728	.040352	.054153	.068135	.082301	.096657	.11121	.12595	.14090	.21882	.30253	.50
.55	.013513	.027196	.041054	.055089	.069306	.083708	.098298	.11308	.12806	.14325	.22235	.30725	.55
.60	.013739	.027649	.041733	.055995	.070439	.085068	.099887	.11490	.13011	.14552	.22578	.31184	.60
.65	.013958	.028087	.042391	.056874	.071538	.086388	.10143	.11666	.13209	.14773	.22910	.31630	.65
.70	.014171	.028513	.043030	.057726	.072605	.087670	.10292	.11837	.13402	.14987	.23234	.32065	.70
.75	.014378	.028927	.043652	.058556	.073643	.088917	.10438	.12004	.13590	.15196	.23550	.32489	.75
.80	.014579	.029330	.044258	.059364	.074655	.090133	.10580	.12167	.13773	.15400	.23858	.32903	.80
.85	.014775	.029723	.044848	.060153	.075642	.091319	.10719	.12325	.13952	.15599	.24158	.33307	.85
.90	.014967	.030107	.045425	.060923	.076606	.092477	.10854	.12480	.14126	.15793	.24452	.33703	.90
.95	.015154	.030483	.045989	.061676	.077549	.093611	.10987	.12632	.14297	.15983	.24740	.34091	.95
1.00	.015338	.030850	.046540	.062413	.078471	.094720	.11116	.12780	.14465	.16170	.25022	.34471	1.00

α \ h	.25	.30	.35	.40	.45	.50	.55	.60	.65	.70	.80	.90	h / α
.00	.31862	.4021	.4941	.5961	.7096	.8368	.9808	1.145	1.336	1.561	2.176	3.283	.00
.05	.32793	.4130	.5066	.6101	.7252	.8540	.9994	1.166	1.358	1.585	2.203	3.314	.05
.10	.33662	.4233	.5184	.6234	.7400	.8703	1.017	1.185	1.379	1.608	2.229	3.345	.10
.15	.34480	.4330	.5296	.6361	.7542	.8860	1.035	1.204	1.400	1.630	2.255	3.376	.15
.20	.35255	.4422	.5403	.6483	.7678	.9012	1.051	1.222	1.419	1.651	2.280	3.405	.20
.25	.35993	.4510	.5506	.6600	.7810	.9158	1.067	1.240	1.439	1.672	2.305	3.435	.25
.30	.36700	.4595	.5604	.6713	.7937	.9300	1.083	1.257	1.457	1.693	2.329	3.464	.30
.35	.37379	.4676	.5699	.6821	.8060	.9437	1.098	1.274	1.476	1.713	2.353	3.492	.35
.40	.38033	.4755	.5791	.6927	.8179	.9570	1.113	1.290	1.494	1.732	2.376	3.520	.40
.45	.38665	.4831	.5880	.7029	.8295	.9700	1.127	1.306	1.511	1.751	2.399	3.547	.45
.50	.39276	.4904	.5967	.7129	.8408	.9826	1.141	1.321	1.528	1.770	2.421	3.575	.50
.55	.39870	.4976	.6051	.7225	.8517	.9950	1.155	1.337	1.545	1.788	2.443	3.601	.55
.60	.40447	.5045	.6133	.7320	.8625	1.007	1.169	1.351	1.561	1.806	2.465	3.628	.60
.65	.41008	.5114	.6213	.7412	.8729	1.019	1.182	1.366	1.577	1.824	2.486	3.654	.65
.70	.41555	.5180	.6291	.7502	.8832	1.030	1.195	1.380	1.593	1.841	2.507	3.679	.70
.75	.42090	.5245	.6367	.7590	.8932	1.042	1.207	1.394	1.608	1.858	2.528	3.705	.75
.80	.42612	.5308	.6441	.7676	.9031	1.053	1.220	1.408	1.624	1.875	2.548	3.730	.80
.85	.43122	.5370	.6515	.7761	.9127	1.064	1.232	1.422	1.639	1.892	2.568	3.754	.85
.90	.43622	.5430	.6586	.7844	.9222	1.074	1.244	1.435	1.653	1.908	2.588	3.779	.90
.95	.44112	.5490	.6656	.7925	.9314	1.085	1.255	1.448	1.668	1.924	2.607	3.803	.95
1.00	.44592	.5548	.6724	.8005	.9406	1.095	1.267	1.461	1.682	1.940	2.626	3.827	1.00

For all values $0 \le \alpha \le 1$ $\quad \lambda(0, \alpha) = 0$ $\quad n = r/N$

Calculations for the LMLE can be completed by finding a value $\hat{\gamma}$ such that $\mu(\hat{\gamma})$, $\sigma(\hat{\gamma})$, and $\hat{\gamma}$ satisfy the third equation of (2.3). Several cycles of calculation of $\mu(\gamma)$ and $\sigma(\gamma)$ using different values of γ will usually be necessary until we find two values of γ_i and γ_j in a sufficiently narrow interval such that $J(\gamma_i) \lesseqgtr 0 \lesseqgtr J(\gamma_j)$, where $J(\ )$ signifies the left side of the third equation of (2.3). We interpolate for final estimates. Alternately, we might resort to the Giesbrecht-Kempthorne procedure and maximize $L(\ )$ directly.

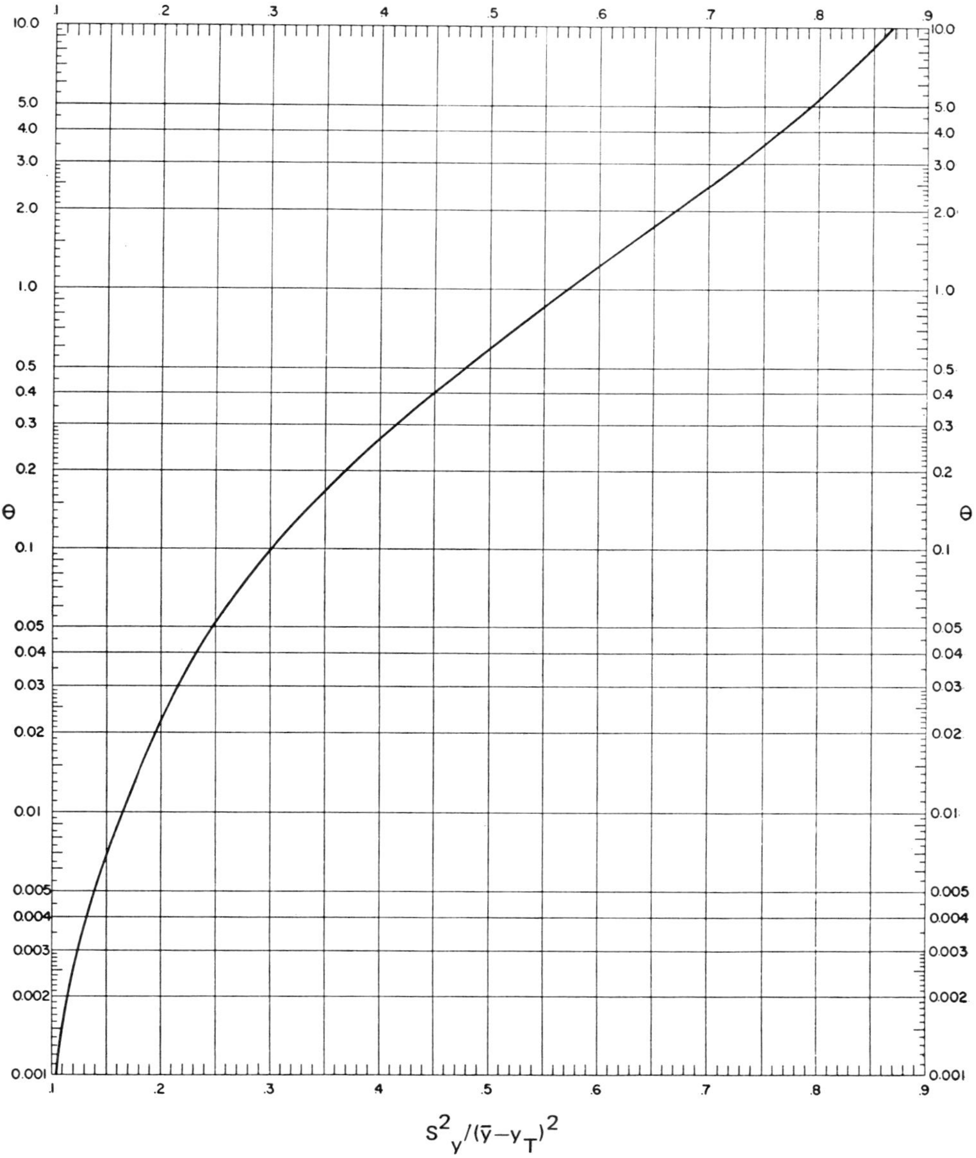

Figure 1 Estimation curve for singly truncated samples from the normal distribution.

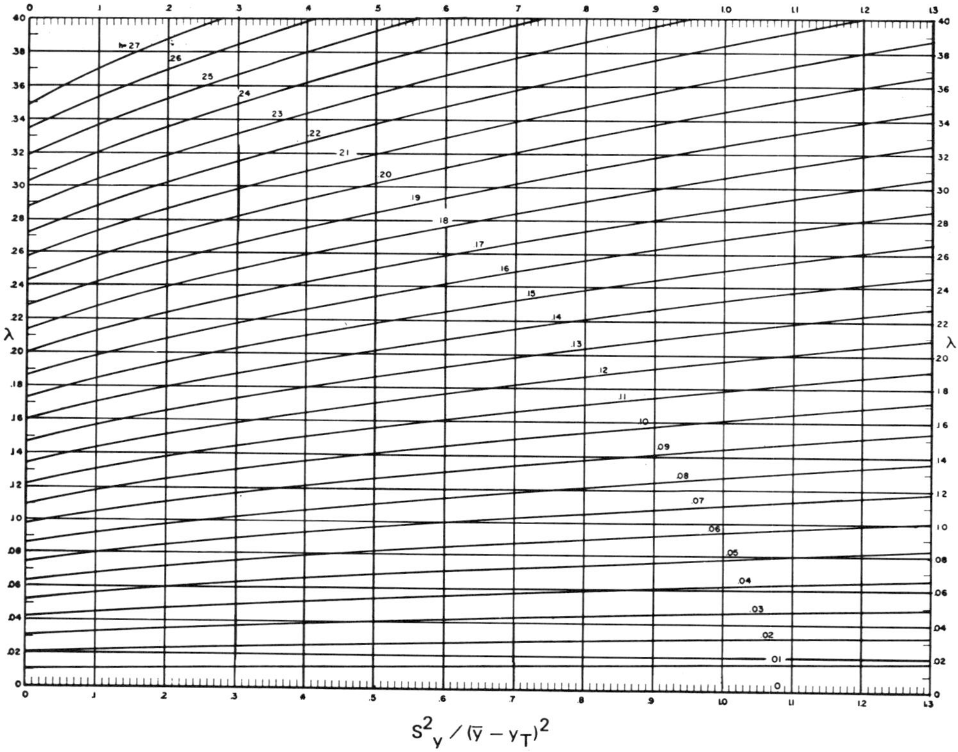

Figure 2 Estimation curves for singly censored normal samples $h = r/N = 0(.01)0.27$.

3.1 Modified Estimators

Modified estimators MMLE, which employ the first order statistic, are often to be preferred over the LMLE in the cases of truncation and censoring as well as in the case of complete samples.

Two choices as replacements for the third estimating equation of (2.3) will be considered. We might employ

$$E[F(X_1)] = F(x_1) \tag{3.10}$$

or alternately

$$E[\ln(X_1 - \gamma)] = \ln(x_1 - \gamma) \tag{3.11}$$

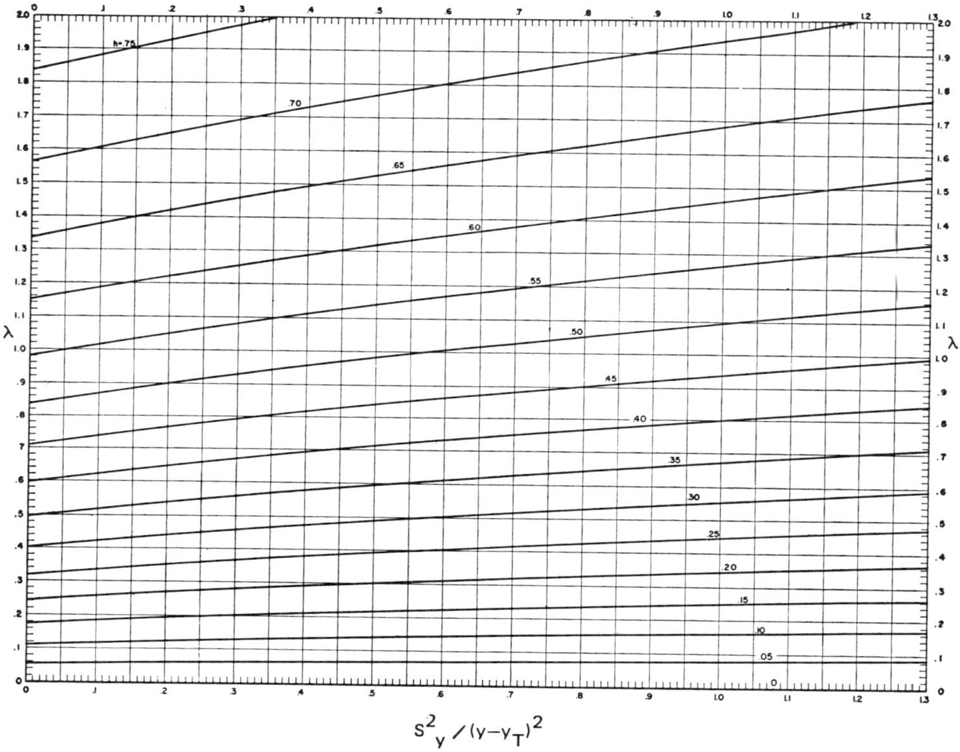

Figure 3 Estimation curves for singly censored normal samples $h = r/N = 0(.05)0.75$.

as described in Chapter 4 for complete samples.

Let us first consider equation (3.10). Since $E[F(X_1)] = \frac{1}{N+1}$, this equation becomes

$$F(x_1) = \frac{1}{N+1} \tag{3.12}$$

and since

$$F(x_1) = G(y_1) = \Phi(w_1)$$

where

$$w_1 = \Phi^{-1}\left(\frac{1}{N+1}\right) \tag{3.13}$$

it follows that

$$w_1 = \frac{\ln(x_1 - \gamma) - \mu}{\sigma}$$

and the resulting third estimating equation (3.12) becomes

$$\gamma + e^{\mu + \sigma w_1} = x_1 \tag{3.14}$$

where w_1 is defined by (3.13). In practical applications, w_1 can be obtained by inverse interpolation in a table of the cumulative standard normal distribution (0,1) as a function of N. Note that in this chapter, W is used to designate the standard unit whereas Z designates the hazard function.

For a given or assumed value of γ, $\hat{\mu}(\gamma)$ and $\hat{\sigma}(\gamma)$ are calculated as described for the LMLE. The final estimates $\hat{\gamma}$, $\hat{\mu}$, and $\hat{\sigma}$ must, of course, satisfy equation (3.14). This time we seek two values γ_i and γ_j in a sufficiently narrow interval such that

$$(\gamma_i + e^{u_i + \sigma_i w_1}) \lesseqgtr x_1 \lesseqgtr (\gamma_j + e^{\mu_j + \sigma_j w_1})$$

We subsequently interpolate for the final estimates as summarized below.

γ	$\mu(\gamma)^*$	$\sigma(\gamma)^*$	$w_1 = \Phi^{-1}\left(\frac{1}{N+1}\right)$	$\gamma + e^{\mu + w_1 \sigma}$
γ_i	μ_i	σ_i	w_1	$\gamma_i + e^{\mu_i + w_1 \sigma_i}$
$\underline{\hat{\gamma}}$	$\underline{\hat{\mu}}$	$\underline{\hat{\sigma}}$	w_1	$\underline{x_1}$
γ_j	μ_j	σ_j	w_1	$\gamma_j + e^{\mu_j + w_1 \sigma_j}$

*$\mu(\gamma)$ and $\sigma(\gamma)$ are calculated as previously described using the applicable equations (3.1), (3.2), or (3.3).

The alternate third estimating equation (3.11) leads to

$$\gamma + e^{\mu + \sigma E(W_{1,N})} = x_1 \tag{3.15}$$

where $E(W_{1,N})$ is the expected value of the first order statistic in a sample of size N from the standard normal distribution (0,1). As stated in Chapter 4, tables of these values have been provided by Harter (1961, 1969). Selected values from this source were included in Table 4 of Chapter 4. Note that estimating equation (3.15) differs from (3.14) only in that $w_1 = \Phi^{-1}[1/(N+1)]$ has been replaced by $E(W_{1,N})$. The following abbreviated table provides comparisons between $\Phi^{-1}[1/(N+1)]$ and $E(W_{1,N})$ for selected values of N.

N	$\Phi^{-1}\left(\frac{1}{N+1}\right)$	$E(W_{1,N})$
5	−0.9675	−1.16296
10	−1.34	−1.53875
20	−1.67	−1.86748
30	−1.85	−2.04276
40	−1.97	−2.16078
50	−2.06	−2.24907
75	−2.22	−2.40299
100	−2.33	−2.50759
200	−2.58	−2.74604
300	−2.72	−2.87777
400	−2.808	−2.96818

It is noted that $\Phi^{-1}[1/(N+1)] > E(W_{1,N})$ for all N, and that the difference ranges from 0.196 for $N = 5$ to 0.160 for $N = 400$.

4. PROGRESSIVELY CENSORED SAMPLES

In numerous practical situations, the initial censoring in a life test results only in withdrawal of a portion of the survivors. Those which remain on test continue to be observed until failure or until a subsequent stage of censoring. For sufficiently large samples, censoring might be progressive through several stages. Such samples arise naturally when certain specimens must be withdrawn prior to failure for use as test objects in related experimentation. In medical tests, samples of this type sometimes result when participants simply drop out and observation is discontinued. When test facilities are limited, and when prolonged tests are expensive, early censoring of a substantial number of sample specimens frees facilities for other tests while specimens which are allowed to continue on test, permit observation of at least some of the extreme sample values.

In progressive censoring as in single stage censoring, a distinction is to be made between censoring of Type I in which times of censoring are predetermined and Type II in which censoring occurs when the number of survivors drops to predetermined levels. Estimating equations for the two types are essentially identical, but there are minor differences in estimate variances. Progressive censoring of a more complex type in which both times of censoring and the number of specimens withdrawn are the result of random causes were considered by Sampford (1952) in connection with response-time studies involving animals.

Type II progressively censored samples from the normal distribution have been considered by Herd (1956, 1957, 1960), who referred to them as "multiple censored samples," and by Roberts (1962a, 1962b), who designated them as "hypercensored samples." Both Type I and Type II censoring in the normal, exponential, Weibull, lognormal, and gamma distributions have previously been considered by Cohen (1963, 1975, 1976, 1977).

4.1 The Sample

Let N designate the total sample size and n the number of sample specimens which fail and therefore result in complete life spans. Suppose that censoring occurs progressively in k stages at times $\{T_i\}$, $i = 1, 2, \ldots, k$, such that $T_{i+1} > T_i$, and that at the ith stage of censoring, r_i sample specimens selected randomly from the survivors at time T_i are removed (censored) from further observation. The sample data thus consist of the n complete life spans $\{x_j\}$, $j = 1, 2, \ldots, n$, plus the k censoring times T_i, $i = 1, \ldots, k$, plus the k values $\{r_i\}$, $i = 1, \ldots, k$. It follows that

$$N = n + \sum_{1}^{k} r_i$$

Note that the first order statistic x_1 is included in the sample data.

In Type I censoring, the T_i are fixed and the number of survivors at these times are random variables. In Type II censoring, the T_i coincide with failure times and are random variables, whereas the number of survivors at these times are fixed. For both types the r_i are fixed. In some cases $r_i = 1$ for all i.

The likelihood function, $L(S)$, where S is a k-stage Type I progressively censored sample, may be written as

$$L(S) = C \prod_{j=1}^{n} f(x_j) \prod_{i=1}^{k} [1 - F(T_i)]^{r_i} \tag{4.1}$$

where C is an ordering constant, f is a density function, and F is a distribution function. Note that the likelihood function of equation (2.1) for a singly censored sample is a special case of (4.1) with $k = 1$ and $r_k = r$.

For a Type II progressively censored sample, which is the type considered by Herd (1956, 1957), the likelihood function is

$$L(S) = \prod_{i=1}^{n} \{n_i^* f(x_i)[1 - F(x_i)]^{r_i}\} \tag{4.2}$$

where $n_i^* = N - \sum_{j=1}^{i-1} r_j - i + 1$.

In the special case of Type II progressive censoring where the total sample is partitioned into equal subgroups of $r + 1$ units each, and where immediately following the first failure in a subgroup, the remaining r units of the subgroup are withdrawn, the density function of these "smallest observations" is well known. [c.f. for example Hoel (1954, p. 304)]. The density of the smallest observation x_1 in a random sample of size $r + 1$ may be written as

$$h(x_1) = (r + 1) f(x_1)[1 - F(x_1)]^r$$

where $f(x)$ and $F(x)$ are the density and distribution functions respectively.

In practical applications, samples of this kind might arise from testing "throw away" units, each consisting of $r + 1$ identical components, where the failure of any single component means failure of the unit.

4.2 Local Maximum Likelihood Estimators

Estimating equations for the LMLE in the case of progressively censored samples from a normal distribution are

$$\begin{gathered}
\sum_1^n [\ln(x_j - \gamma) - \mu] + \sigma \sum_1^k r_i Z_i = 0 \\
\sum_1^n [\ln(x_j - \gamma) - \mu]^2 - n\sigma^2 + \sigma^2 \sum_1^k r_i \xi_i Z_i = 0 \\
\sum_1^n \left[\frac{\ln(x_j - \gamma) - \mu}{x_j - \gamma}\right] + \sigma^2 \sum_1^n \left(\frac{1}{x_j - \gamma}\right) + \sigma \sum_1^k \frac{r_i Z_i}{T_i - \gamma} = 0
\end{gathered} \tag{4.3}$$

where

$$\xi_i = \frac{\ln(T_i - \gamma) - \mu}{\sigma} \quad \text{and} \quad Z_i(\xi) = \frac{\phi(\xi_i)}{1 - \Phi(\xi_i)} \tag{4.4}$$

Note that equations (2.3) for singly censored samples become a special case of (4.3) where $k = 1$ and $r_k = r$.

We follow a procedure here that is essentially the same as that employed with singly censored samples. For a given or assumed value of γ, we make the normalizing transformation $Y = \ln(X - \gamma)$ and employ the normal estimating equations for $\mu(\gamma)$ and $\sigma^2(\gamma)$ derived by Cohen (1963). These follow from the first two equations of (4.3) as

$$\begin{aligned} \bar{y} &= \mu(\gamma) - \sigma(\gamma) \sum_1^k \frac{r_i}{n} Z_i \\ s_y^2 &= \sigma^2(\gamma) \left[1 - \sum_1^k \frac{r_i}{n} \xi_i Z_i - \left(\sum_1^k \frac{r_i}{n} Z_i \right)^2 \right] \end{aligned} \tag{4.5}$$

Standard iterative procedures may be employed to solve the pair of equations (4.5) simultaneously for the required estimates $\hat{\mu}$ and $\hat{\sigma}$. Newton's method, for example, will usually be satisfactory. This is a method of iteration based on Taylor series expansions of the estimating equations in the vicinity of their simultaneous solution. With μ_0 and σ_0 designating approximate solutions, let

$$\hat{\mu} = \mu_0 + b \qquad \text{and} \qquad \hat{\sigma} = \sigma_0 + d \tag{4.6}$$

where b and d are corrections to be determined by the iteration process. The symbol ($\hat{\ }$) is employed to distinguish estimates from the parameters being estimated. Using Taylor's theorem and neglecting powers of b and d above the first, we have

$$\begin{aligned} b\frac{\partial^2 \ln L}{\partial \mu_0^2} + d\frac{\partial^2 \ln L}{\partial \mu_0 \partial \sigma_0} &= -\frac{\partial \ln L}{\partial \mu_0} \\ b\frac{\partial^2 \ln L}{\partial \mu_0 \partial \sigma_0} + d\frac{\partial^2 \ln L}{\partial \sigma_0^2} &= -\frac{\partial \ln L}{\partial \sigma_0} \end{aligned} \tag{4.7}$$

Corrections b and d are then obtained by solving these two equations simultaneously. The right hand members are given in (4.8).

$$\begin{aligned}\frac{\partial \ln L}{\partial \mu} &= \frac{n}{\sigma}\left[\frac{(\bar{x}-\mu)}{\sigma} + \sum_{1}^{k} \frac{r_i}{n} Z_i\right] \\ \frac{\partial \ln L}{\partial \sigma} &= \frac{n}{\sigma}\left[\frac{s^2 + (\bar{x}-\mu)^2}{\sigma^2} - 1 + \sum_{1}^{k} \frac{r_i}{n} \xi_i Z_i\right]\end{aligned} \tag{4.8}$$

For the coefficients on the left side, we differentiate a second time to obtain

$$\begin{aligned}\frac{\partial^2 \ln L}{\partial \mu^2} &= -\frac{n}{\sigma^2}\left[1 + \sum_{1}^{k} \frac{r_i}{n} A_i\right] \\ \frac{\partial^2 \ln L}{\partial \mu \partial \sigma} &= -\frac{n}{\sigma^2}\left[\frac{2(\bar{x}-\mu)}{\sigma} + \sum_{1}^{k} \frac{r_i}{n} B_i\right] \\ \frac{\partial^2 \ln L}{\partial \sigma^2} &= -\frac{n}{\sigma^2}\left[\frac{3\left\{s^2 + (\bar{x}-\mu)^2\right\}}{\sigma^2} - 1 + \sum_{1}^{k} \frac{r_i}{n} C_i\right]\end{aligned} \tag{4.9}$$

where

$$\begin{aligned}A_i &= Z_i(Z_i - \xi_i) \\ B_i &= Z_i + \xi_i A_i \\ C_i &= \xi_i(Z_i + B_i)\end{aligned} \tag{4.10}$$

The functions Z, ξZ, A, B, and C are given in Table 3 for values of the argument $\xi = -1.0(0.5)3.5$. The number of repetitions of the above iterative process required for a given degree of accuracy will depend largely on the initial approximations. In many instances, satisfactory first approximations μ_0 and σ_0 will be available from the sample percentiles. The pattern of censoring in specific examples will of course determine which percentiles are available, but in general, a suitable first approximation to $\hat{\mu}$ may be calculated as $\mu_0 = [P_i + P_{100-i}]/2$, where P_i is the ith percentile. It is desirable though not necessary to restrict i to the interval $25 \le i \le 50$.

In some circumstances, it might be desirable to employ the probit technique discussed in a later Section of this paper to obtain more accurate

Table 3 The Functions Z, ξZ, A, B, and C for Progressively Censored Normal Samples

ξ	Z	ξZ	$A = Z(Z - \xi)$	$B = Z + \xi A$	$C = \xi(Z + B)$
−1.0	0.28760	−0.28760	0.37031	−0.083	−0.205
−0.5	0.50916	−0.25458	.51382	0.252	−0.381
0	0.79788	0	.63662	0.798	0
0.5	1.14108	0.57054	.73152	1.507	1.324
1.0	1.52514	1.52514	.80090	2.326	3.851
1.1	1.60580	1.76638	.81221	2.499	4.515
1.2	1.68755	2.02506	.82277	2.675	5.235
1.3	1.77033	2.30143	.83263	2.853	6.010
1.4	1.85406	2.59568	.84185	3.033	6.841
1.5	1.93868	2.90802	.85045	3.214	7.730
1.6	2.02413	3.23861	.85849	3.398	8.675
1.7	2.11036	3.58761	.86600	3.583	9.678
1.8	2.19731	3.95516	.87302	3.769	10.739
1.9	2.28495	4.34140	.87958	3.956	11.858
2.0	2.37322	4.74644	.88572	4.145	13.036
2.5	2.82274	7.05685	.91103	5.100	19.808
3.0	3.28310	9.84930	.92944	6.071	28.064
3.5	3.75137	13.12980	.94307	7.052	37.812

first approximations and thereby reduce the number of iterations necessary in calculating maximum likelihood estimates.

Herd (1956) employed an iterative procedure for solving the estimating equations which in some respects is simpler than the traditional method of Newton. However, when close first approximations are available, Newton's method is quite satisfactory.

4.3 The Third Estimating Equation

Calculations can be completed by finding a value $\hat{\gamma}$ such that $\mu(\hat{\gamma})$ and $\sigma^2(\hat{\gamma})$ (obtained from equations (4.5)) together with $\hat{\gamma}$ will also satisfy the third equation of (4.3). Several cycles of calculation with different values of γ might be necessary in order to find two values γ_i and γ_j in a sufficiently narrow interval such that $Q(\gamma_i) \lesssim 0 \lesssim Q(\gamma_j)$, where $Q(\gamma)$ signifies the left side of the third equation of (4.3). Linear interpolation can then provide the final estimates.

4.4 Modified Estimators

Modified estimators can be calculated with $\hat{\mu}(\gamma)$ and $\hat{\sigma}(\gamma)$ calculated as described in the preceding section for the LMLE, and with $\hat{\gamma}$ calculated as described for singly censored samples. Accordingly, we seek values γ_i and γ_j in a sufficiently narrow interval such that either

$$\left[\gamma_i + e^{\mu(\gamma_i)+w_1\sigma(\gamma_i)}\right] \lesseqgtr x_1 \lesseqgtr \left[\gamma_j + e^{\mu(\gamma_j)+w_1\sigma(\gamma_j)}\right] \tag{4.11}$$

or

$$\left[\gamma_i + e^{\mu(\gamma_i)+\sigma(\gamma_i)\cdot E(W_{1,N})}\right] \lesseqgtr x_1 \lesseqgtr \left[\gamma_j + e^{\mu(\gamma_j)+\sigma(\gamma_j)\cdot E(W_{1,N})}\right] \tag{4.12}$$

Final estimates then follow by linear interpolation between γ_i and γ_j.

4.5 First Approximations to $\hat{\mu}$ and $\hat{\sigma}$ Using Probits

This section is concerned with the use of probits in obtaining first approximations to $\hat{\mu}$ and $\hat{\sigma}$ when samples from a normal population are progressively censored as described in Section 4.1. For convenience, here we assume that our sample data are grouped into k classes with boundaries T_0, T_1, $\ldots$, T_k. Let f_i designate the number of failures observed in the ith class, i.e., during the time interval $T_{i-1} \leq x \leq T_i$. As in Section 4.1, let r_i designate the number of specimens withdrawn (censored) at time T_i and assume that censoring can occur only at one of the class boundaries T_i, but does not necessarily occur at each of these boundaries. As in Sub-Section 4.1, $N = n + \sum_1^k r_i$, where N is the total sample size and $n = \sum_1^k f_i$. It is to be noted that some of the r_i may be zeros.

Where f is the density of X, the conditional density subject to the restriction that $x > T_i$ may be written as

$$f(x \mid x > T_i) = \frac{f(x)}{1 - F_i} \tag{4.13}$$

It follows that expected frequencies in progressively censored samples of the type under consideration are

$$\begin{aligned} E(f_1) &= N\int_{T_0}^{T_1} f(x)dx = Np_1 \\ E(f_i) &= \left[N - \sum_1^{i-1} \frac{r_j}{1 - F_j}\right] p_i, i = 2, 3, \ldots, k \end{aligned} \tag{4.14}$$

where

$$p_i = \int_{T_{i-1}}^{T_i} f(x)dx$$

Accordingly, estimates of p_i may be obtained as

$$p_1^* = \frac{f_1}{N}$$

$$p_i^* = f_i \Bigg/ \left[N - \sum_{1}^{i-1} \frac{r_j}{1 - F_j^*}\right] \qquad i = 2, 3, \ldots, k \tag{4.15}$$

with estimates of F_i given as

$$F_1^* = p_1^* \qquad F_i^* = F_{i-1}^* + p_i^* \cdot i = 2, 3, \ldots, k \tag{4.16}$$

The stars (*) serve to distinguish estimates from parameters.

The F_i^* are transformed into probits using a standard probit table or using ordinary tables of normal curve areas, where the probit estimate y^* is given as

$$y_i^* = 5 + \xi_i^* \tag{4.17}$$

where ξ_i^* is determined from the relation

$$F_i^* = \int_{-\infty}^{\xi_i^*} \phi(t)dt \tag{4.18}$$

The $k-1$ pairs of values (T_i, y_i^*), $1 \leq i \leq k-1$, should lie approximately on the probit regression line. This is a straight line, and its equation may be written in the form

$$y = 5 + \frac{1}{\sigma}(x - \mu) \tag{4.19}$$

Thus μ_0 is the value of x corresponding to the probit value of 5, and σ_0 is the reciprocal of the slope of the probit regression line.

For purposes of most analyses, a graphic fit of the probit regression line will ordinarily be satisfactory. A weighted least square fit would, of course, give more accurate results, but the computational effort involved would be considerably greater.

When a high degree of accuracy is not required, the probit estimates obtained from a regression line fitted graphically by eye might be satisfactory without further refinement. In some cases, estimates obtained from a probit regression line fitted by the method of least squares might even be preferred over estimates obtained by the methods of Section 4.2.

5. DOUBLY CENSORED AND DOUBLY TRUNCATED SAMPLES

The likelihood function for a Type I sample that is *doubly censored* (left and right) from a lognormal distribution is written as

$$L(S) = C[F(T_1)]^{r_1} \prod_{i=1}^{n} f(x_i; \gamma, \mu, \sigma) \cdot [1 - F(T_2)]^{r_2} \tag{5.1}$$

where T_1 and T_2 are the times of censoring, r_1 is the number of items censored from the left and r_2 is the number censored from the right. For Type II censoring, T_1 would be replaced with x_{r_1+1} and T_2 with x_{n+r_1}. For both types $N = n + r_1 + r_2$.

For a *doubly truncated sample*, $N = n$ and the likelihood function is

$$L(S) = [F(T_2) - F(T_1)]^{-n} \prod_{i=1}^{n} f(x_i; \gamma, \mu, \sigma) \tag{5.2}$$

Estimating equations for *doubly censored* samples obtained by differentiating and equating to zero the logarithm of (5.1) with respect to μ, σ, and γ in turn, where $f(x; \gamma, \mu, \sigma)$ is given by equation (2.2), are

$$\frac{1}{n} \sum_{1}^{n} [\ln(x_i - \gamma) - \mu] - \sigma(Y_1 - Y_2) = 0$$

$$\frac{1}{n} \sum_{1}^{n} [\ln(x_i - \gamma) - \mu]^2 - \sigma^2(1 + \xi_1 Y_1 - \xi_2 Y_2) = 0 \tag{5.3}$$

$$\frac{1}{n} \sum_{1}^{n} \left[\frac{\ln(x_i - \gamma) - \mu}{x_i - \gamma}\right] + \frac{\sigma^2}{n} \sum_{1}^{n} \left(\frac{1}{x_i - \gamma}\right) - \sigma \left(\frac{Y_1}{T_1 - \gamma} - \frac{Y_2}{T_2 - \gamma}\right) = 0$$

where

$$Y_1 = \frac{r_1}{n}\left(\frac{\phi(\xi_1)}{\Phi(\xi_1)}\right) \qquad Y_2 = \frac{r_2}{n}\left(\frac{\phi(\xi_2)}{1-\Phi(\xi_2)}\right)$$
$$\xi_1 = \frac{\ln(T_1-\gamma)-\mu}{\sigma} \quad \text{and} \quad \xi_2 = \frac{\ln(T_2-\gamma)-\mu}{\sigma} \tag{5.4}$$

Corresponding equations for *doubly truncated* samples are

$$\frac{1}{n}\sum_1^n [\ln(x_i-\gamma)-\mu] - \sigma(\overline{Z}_1 - \overline{Z}_2) = 0$$
$$\frac{1}{n}\sum_1^n [\ln(x_i-\gamma)-\mu]^2 - \sigma^2(1+\xi_1\overline{Z}_1 - \xi_2\overline{Z}_2) = 0 \tag{5.5}$$
$$\frac{1}{n}\sum_1^n \left[\frac{\ln(x_1-\gamma)-\mu}{x_i-\gamma}\right] + \frac{\sigma^2}{n}\sum_1^n\left(\frac{1}{x_i-\gamma}\right) - \sigma\left[\frac{\overline{Z}_1}{T_1-\gamma} - \frac{\overline{Z}_2}{T_2-\gamma}\right] = 0$$

where

$$\overline{Z}_i = \frac{\phi(\xi_i)}{\Phi(\xi_2)-\Phi(\xi_1)} \qquad i = 1,2 \tag{5.6}$$

When we make the substitution $y_i = \ln(x_i - \gamma)$, the first two equations of (5.3) and the first two equations of (5.5) reduce to expressions previously derived by Cohen (1957) for doubly censored and for doubly truncated samples from the normal distribution (μ, σ^2). Accordingly, the estimation procedure is similiar to that employed with singly censored and singly truncated samples. We select a trial value γ_1 (i.e., a first approximation) and make the transformation $y_i = \ln(x_i - \gamma_1)$ to render the sample normal (μ, σ). We then calculate $\hat{\mu}(\gamma_1)$ and $\hat{\sigma}(\gamma_1)$ in accordance with the procedure outlined in the 1957 paper. As an aid in the calculation of $\hat{\mu}(\gamma_i)$ and $\hat{\sigma}(\gamma_i)$, for doubly truncated samples, Figure 4 has been reproduced here from the 1957 paper. We then substitute these conditional estimates into the third equation of (5.3) or (5.5) as is appropriate. If the third estimating equation is satisfied, no further calculations are needed and $\hat{\gamma} = \gamma_1$, $\hat{\mu} = \mu(\gamma_1)$, $\hat{\sigma} = \sigma(\gamma_1)$. Otherwise we select a new value of γ and continue through additional cycles of computation until we find two values γ_i and γ_j in a sufficiently narrow interval such that the third estimating equation is satisfied for some value $\gamma_i \lesssim \hat{\gamma} \lesssim \gamma_j$. The final estimate is obtained by linear interpolation.

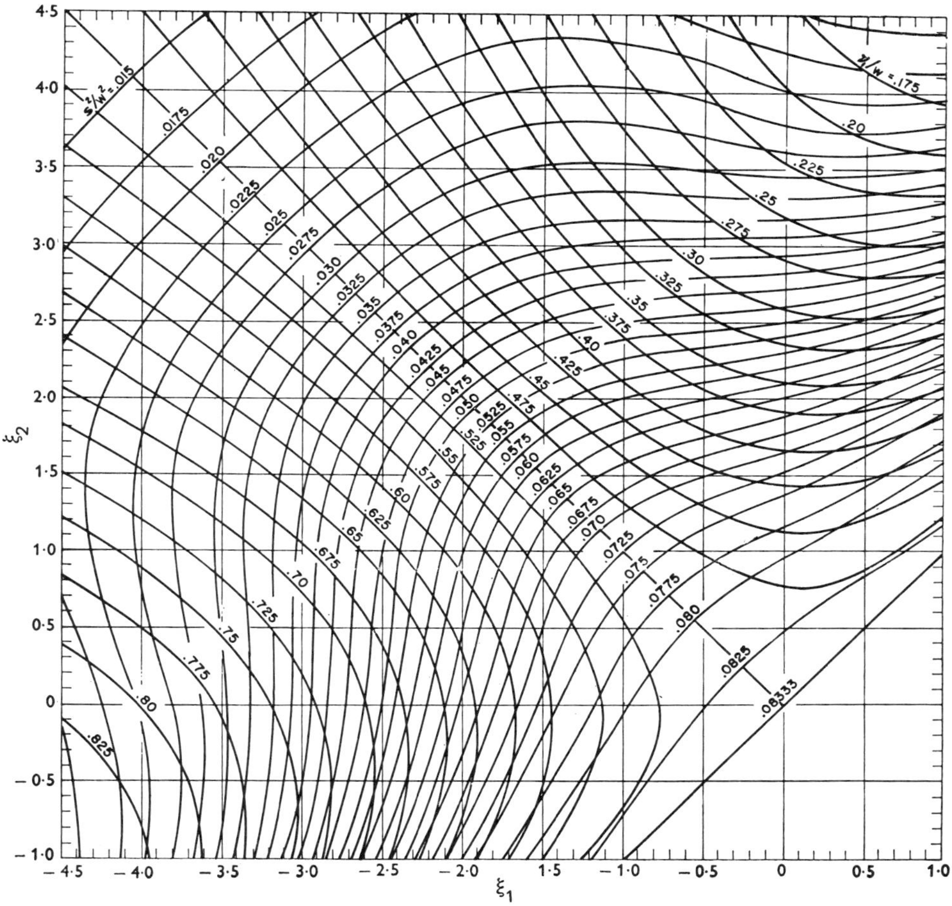

Figure 4 Estimation curves for doubly truncated normal samples. (1) Locate curve corresponding to sample value of ν_1/w. (2) Follow this curve to its intersection with curve corresponding to sample value of s_y^2/w^2. (3) Read coordinates of intersection. These are estimates $\hat{\xi}_1$ and $\hat{\xi}_2$. $\nu_1 = \bar{y} - y_{T1}$, $w = y_{T2} - y_{T1}$, $\xi_1 = (y_{T1} - \mu)/\sigma$, $\xi_2 = (y_{T2} - \mu)/\sigma$, $\hat{\sigma} = w/(\hat{\xi}_2 - \hat{\xi}_1)$, $\hat{\mu} = y_{T1} - \hat{\sigma}\hat{\xi}_1$. Reproduced with permission of the Biometrika Trustees from *Biometrika* 44 (1957) p. 227.

6. ESTIMATE VARIANCES AND COVARIANCES

The asymptotic variance-covariance matrix of the estimators $\hat{\mu}$, $\hat{\sigma}$, and $\hat{\gamma}$ is obtained by inverting the information matrix, in which elements are negatives of expected values of the second partial derivatives of the logarithm of the likelihood function. For sufficiently large samples, these expected values can be approximated by substituting the estimates obtained from a given sample directly into the partial derivatives.

Progressively Censored Samples. Second partial derivatives for progressively censored samples as given by Cohen (1976) are

$$\begin{aligned}
\frac{\partial^2 \ln L}{\partial \mu^2} &= \frac{-n}{\sigma^2} - \frac{1}{\sigma^2}\sum_1^k r_j Z_j (Z_j - \xi_j) \\
\frac{\partial^2 \ln L}{\partial \sigma^2} &= \frac{n}{\sigma^2} - \frac{3}{\sigma^4}\sum_1^n [\ln(x_i - \gamma) - \mu]^2 - \frac{1}{\sigma^2}\sum_1^k r_j \xi_j Z_j \left[2 + \xi_j (Z_j - \xi_j)\right] \\
\frac{\partial^2 \ln L}{\partial \gamma^2} &= \frac{1}{\sigma^2}\sum_1^n \frac{\left[\ln(x_i - \gamma) - \mu - 1 + \sigma^2\right]}{(x_i - \gamma)^2} + \sum_1^k \frac{r_j Z_j \left[\sigma - (Z_j - \xi_j)\right]}{(T_j - \gamma)^2} \\
\frac{\partial^2 \ln L}{\partial \mu \partial \gamma} &= \frac{\partial^2 \ln L}{\partial \gamma \partial \mu} = \frac{1}{\sigma^2}\sum_1^n \left(\frac{1}{x_i - \gamma}\right) - \frac{1}{\sigma^2}\sum_1^k \frac{r_j Z_j (Z_j - \xi_j)}{(T_j - \gamma)} \\
\frac{\partial^2 \ln L}{\partial \sigma \partial \gamma} &= \frac{\partial^2 \ln L}{\partial \gamma \partial \sigma} = -\frac{2}{\sigma^3}\sum_1^n \frac{[\ln(x_i - \gamma) - \mu]}{(x_i - \gamma)} \\
&\quad - \frac{1}{\sigma^2}\sum_1^k \frac{r_j Z_j \left[1 + \xi_j (Z_j - \xi_j)\right]}{(T_j - \gamma)} \\
\frac{\partial^2 \ln L}{\partial \mu \partial \sigma} &= \frac{\partial^2 \ln L}{\partial \sigma \partial \mu} = -\frac{2}{\sigma^3}\sum_1^n [\ln(x_i - \gamma) - \mu] \\
&\quad - \frac{1}{\sigma^2}\sum_1^k r_j Z_j \left[1 + \xi_j (Z_j - \xi_j)\right]
\end{aligned} \tag{6.1}$$

Since the estimators $\hat{\mu}$, $\hat{\sigma}$, and $\hat{\gamma}$ are local rather than global maximum likelihood estimators, the applicability of the variance-covariance matrix obtained here, might be open to question. However, a Monte Carlo study by Norgaard (1975) indicates that the approximate asymptotic variances and covariances obtained here should be considered satisfactory when $n \geq 50$, although they might be misleading as measures of sampling error for small samples. Norgaard's results are consistent with results of an earlier Monte Carlo study by Harter and Moore (1966) in connection with singly and doubly censored samples. It is also to be noted that Norgaard's study indicates that variances and covariances of the MMLE are approximately equal to corresponding measures of the MLE.

Singly Censored Samples. Second partial derivatives for samples that are singly censored on the right are a special case of (6.1) in which $k = 1$. As previously mentioned, when γ is known, the estimation problem is reduced to that of estimating μ and σ from a singly censored normal distribution. In that case it is relatively easy to obtain expected values of the second partial derivatives. Accordingly the asymptotic variances and covariances as given by Cohen (1959, 1961) are applicable. These may be expressed as

$$\begin{aligned} V(\hat{\mu}) &= \frac{\sigma^2}{N}\mu_{11}(\xi) \qquad & \mathrm{Cov}(\hat{\mu},\hat{\sigma}) &= \frac{\sigma^2}{N}\mu_{12}(\xi) \\ V(\hat{\sigma}) &= \frac{\sigma^2}{N}\mu_{22}(\xi) \qquad & \rho(\hat{\mu},\hat{\sigma}) &= \frac{\mu_{12}}{\sqrt{\mu_{11}\mu_{22}}} \end{aligned} \tag{6.2}$$

where

$$\begin{aligned} \mu_{11} &= \frac{\phi_{22}}{\Phi(\xi)\left[\phi_{11}\phi_{22} - \phi_{12}^2\right]} \\ \mu_{22} &= \frac{\phi_{11}}{\Phi(\xi)\left[\phi_{11}\phi_{22} - \phi_{12}^2\right]} \\ \mu_{12} &= \frac{-\phi_{12}}{\Phi(\xi)\left[\phi_{11}\phi_{22} - \phi_{12}^2\right]} \end{aligned} \tag{6.3}$$

and where the ϕ_{ij} are given below for samples that are singly truncated or censored on the left.

Truncated Samples

$$\begin{aligned} \phi_{11}(\xi) &= 1 - Z(\xi)\left[Z(\xi) - \xi\right] \\ \phi_{12}(\xi) &= Z(\xi)\left\{1 - \xi\left[Z(\xi) - \xi\right]\right\} \\ \phi_{22}(\xi) &= 2 + \xi\phi_{12}(\xi) \end{aligned}$$

Type I Censored Samples

$$\begin{aligned} \phi_{11}(\xi) &= 1 + Z(\xi)\left[Z(-\xi) + \xi\right] \\ \phi_{12}(\xi) &= Z(\xi)\left\{1 + \xi\left[Z(-\xi) + \xi\right]\right\} \\ \phi_{22}(\xi) &= 2 + \xi\phi_{12}(\xi) \end{aligned}$$

Type II Censored Samples

$$\begin{aligned} \phi_{11}\left(\frac{r}{N},\xi\right) &= 1 + Y\left(\frac{r}{N},\xi\right)\left[Z(-\xi) + \xi\right] \\ \phi_{12}\left(\frac{r}{N},\xi\right) &= Y\left(\frac{r}{N},\xi\right)\left\{1 + \xi[Z(-\xi) + \xi]\right\} \\ \phi_{22}\left(\frac{r}{N},\xi\right) &= 2 + \xi\phi_{12}\left(\frac{r}{N},\xi\right) \end{aligned} \tag{6.4}$$

It is to be noted that as $N \to \infty$, the ϕ_{ij} for Type II censored samples approach the ϕ_{ij} for Type I censored samples. Likewise as $N \to \infty$, then $[1 - \Phi(\hat{\xi})] \to \lim_{N\to\infty}(n/N)$. Therefore, limiting values of estimate

Table 4 Variance-Covariance Factors for Singly Truncated and Singly Censored Samples from the Normal Distribution*

η	For Truncated Samples				For Censored Samples				Percent Rest.	η
	μ_{11}	μ_{12}	μ_{22}	ρ	μ_{11}	μ_{12}	μ_{22}	ρ		
-4.0	1.00054	-.001143	.502287	-.001613	1.00000	-.000006	.500030	-.000001	0.00	-4.0
-3.5	1.00313	-.005922	.510366	-.008277	1.00001	-.000052	.500208	-.000074	0.02	-3.5
-3.0	1.01460	-.024153	.536283	-.032744	1.00010	-.000335	.501180	-.000473	0.13	-3.0
-2.5	1.05738	-.081051	.602029	-.101586	1.00056	-.001712	.505280	-.002407	0.62	-2.5
-2.4	1.07437	-.101368	.622786	-.123924	1.00078	-.002312	.506935	-.003247	0.82	-2.4
-2.3	1.09604	-.126136	.646862	-.149803	1.00107	-.003099	.509030	-.004341	1.07	-2.3
-2.2	1.12365	-.156229	.674663	-.179434	1.00147	-.004121	.511658	-.005757	1.39	-2.2
-2.1	1.15880	-.192688	.706637	-.212937	1.00200	-.005438	.514926	-.007571	1.79	-2.1
-2.0	1.20350	-.236743	.743283	-.250310	1.00270	-.007123	.518960	-.009875	2.28	-2.0
-1.9	1.26030	-.289860	.785158	-.291388	1.00363	-.009266	.523899	-.012778	2.87	-1.9
-1.8	1.33246	-.353771	.832880	-.335818	1.00485	-.011971	.529899	-.016405	3.59	-1.8
-1.7	1.42405	-.430531	.887141	-.383041	1.00645	-.015368	.537141	-.020901	4.46	-1.7
-1.6	1.54024	-.522564	.948713	-.432293	1.00852	-.019610	.545827	-.026431	5.48	-1.6
-1.5	1.68750	-.632733	1.01846	-.482644	1.01120	-.024884	.556186	-.033181	6.68	-1.5
-1.4	1.87398	-.764405	1.09734	-.533054	1.01467	-.031410	.568471	-.041358	8.08	-1.4
-1.3	2.10982	-.921533	1.18642	-.582464	1.01914	-.039460	.582981	-.051193	9.68	-1.3
-1.2	2.40764	-1.10874	1.28690	-.629889	1.02488	-.049355	.600046	-.062937	11.51	-1.2
-1.1	2.78311	-1.33145	1.40009	-.674498	1.03224	-.061491	.620049	-.076861	13.57	-1.1
-1.0	3.25557	-1.59594	1.52746	-.715676	1.04168	-.076345	.643438	-.093252	15.87	-1.0
-0.9	3.84879	-1.90952	1.67064	-.753044	1.05376	-.094501	.670724	-.112407	18.41	-0.9
-0.8	4.59189	-2.28066	1.83140	-.786452	1.06923	-.116674	.702513	-.134620	21.19	-0.8
-0.7	5.52036	-2.71911	2.01172	-.815942	1.08904	-.143744	.739515	-.160175	24.20	-0.7
-0.6	6.67730	-3.23612	2.21376	-.841703	1.11442	-.176798	.782574	-.189317	27.43	-0.6
-0.5	8.11482	-3.84458	2.43990	-.864019	1.14696	-.217183	.832691	-.222233	30.85	-0.5
-0.4	9.89562	-4.55921	2.69271	-.883229	1.18876	-.266577	.891077	-.259011	34.46	-0.4
-0.3	12.0949	-5.39683	2.97504	-.899688	1.24252	-.327080	.959181	-.299607	38.21	-0.3
-0.2	14.8023	-6.37653	3.28997	-.913744	1.31180	-.401326	1.03877	-.343800	42.07	-0.2
-0.1	18.1244	-7.51996	3.64083	-.925727	1.40127	-.492641	1.13198	-.391156	46.02	-0.1
0.0	22.1875	-8.85155	4.03126	-.935932	1.51709	-.605233	1.24145	-.441013	50.00	0.0
0.1	27.1403	-10.3988	4.46517	-.944623	1.66743	-.744459	1.37042	-.492483	53.98	0.1
0.2	33.1573	-12.1927	4.94678	-.952028	1.86310	-.917165	1.52288	-.544498	57.93	0.2
0.3	40.4428	-14.2679	5.48068	-.958345	2.11857	-1.13214	1.70381	-.595891	61.79	0.3
0.4	49.2342	-16.6628	6.07169	-.963742	2.45318	-1.40071	1.91942	-.645504	65.54	0.4
0.5	59.8081	-19.4208	6.72512	-.968361	2.89293	-1.73757	2.17751	-.692299	69.15	0.5
0.6	72.4834	-22.5896	7.44658	-.972322	3.47293	-2.16185	2.48793	-.735459	72.57	0.6
0.7	87.6276	-26.2220	8.24204	-.975727	4.24075	-2.69858	2.86318	-.774443	75.80	0.7
0.8	105.66	-30.376	9.1178	-.97866	5.2612	-3.3807	3.3192	-.80899	78.81	0.8
0.9	127.07	-35.117	10.081	-.98119	6.6229	-4.2517	3.8765	-.83912	81.59	0.9
1.0	152.40	-40.515	11.138	-.98338	8.4477	-5.3696	4.5614	-.86502	84.13	1.0
1.1	182.29	-46.650	12.298	-.98529	10.903	-6.8116	5.4082	-.88703	86.43	1.1
1.2	217.42	-53.601	13.567	-.98694	14.224	-8.6818	6.4616	-.90557	88.49	1.2
1.3	258.61	-61.465	14.954	-.98838	18.735	-11.121	7.7804	-.92109	90.32	1.3
1.4	306.78	-70.347	16.471	-.98964	24.892	-14.319	9.4423	-.93401	91.92	1.4
1.5	362.91	-80.350	18.124	-.99074	33.339	-18.539	11.550	-.94473	93.32	1.5
1.6	428.11	-91.586	19.922	-.99171	44.986	-24.139	14.243	-.95361	94.52	1.6
1.7	503.57	-104.17	21.874	-.99256	61.132	-31.616	17.706	-.96097	95.54	1.7
1.8	591.03	-118.31	24.003	-.99332	83.638	-41.664	22.193	-.96706	96.41	1.8
1.9	691.78	-134.10	26.311	-.99398	115.19	-55.252	28.046	-.97211	97.13	1.9
2.0	807.71	-151.73	28.813	-.99457	159.66	-73.750	35.740	-.97630	97.72	2.0
2.1	940.38	-171.30	31.511	-.99509	222.74	-99.100	45.930	-.97979	98.21	2.1
2.2	1091.4	-192.92	34.405	-.99555	312.73	-134.08	59.526	-.98270	98.61	2.2
2.3	1265.4	-217.17	37.575	-.99596	441.92	-182.68	77.810	-.98514	98.93	2.3
2.4	1458.6	-243.23	40.858	-.99632	628.58	-250.68	102.59	-.98718	99.18	2.4
2.5	1677.8	-271.99	44.392	-.99665	899.99	-346.53	136.44	-.98890	99.38	2.5

*When truncation of type I censoring occurs on the left, entries in this table corresponding to $\eta = \xi$ are applicable. For right truncated or type I right censored samples, read entries corresponding to $\eta = -\xi$, but delete negative signs from μ_{12} and ρ. For both type II left censored and type II right censored samples, read entries corresponding to Percent Restriction $= 100h$, but for right censoring delete negative signs from μ_{12} and ρ.

variances and covariances for both types of censored samples in this sense become equal.

The variances and covariances of estimators based on samples that are restricted on the left may be calculated by substituting appropriate values of $\phi_{ij}(\hat{\xi})$ from (6.4) into (6.3), where $\hat{\xi} = (y_T - \hat{\mu})/\hat{\sigma}$ or $\hat{\xi} = (y_n - \hat{\mu})/\hat{\sigma}$ as applicable. For samples that are restricted on the right, calculations are the same except that $\phi_{ij}(-\hat{\xi})$ from (6.4) are substituted into (6.3).

In order to facilitate the calculation of variances and covariances using (6.2), Table 4 of μ_{ij}, i, $j = 1, 2$, is reproduced here from Table 3 of the writer's 1961 paper.

Variances and covariances for the general case where γ must also be estimated are of course somewhat larger than those calculated from (6.2). However in view of the ease with which (6.2) can be evaluated, it seems worthwhile to do so as a rough check on calculations based on (6.1).

7. ILLUSTRATIVE EXAMPLES

To illustrate the practical application of the *probit technique* in estimating the mean and variance of a normal distribution from a progressively censored sample, two examples have been selected. The first pertains to a life test conducted on an electronic component and the second to a life test on certain biological specimens in a stress environment.

Example 1. Life Test of an Electronic Component A sample of 300 items was placed on test. After 1650 hours, 50 of the survivors were withdrawn and at the end of 1735 hours, the test was terminated with 95 of the original specimens surviving. The life span in hours was recorded for each specimen which failed. In the notation used here, data for this sample are summarized as follows: $N = 300$, $f_1 = 120$, $f_2 = 35$, $n = 155$, $r_1 = 50$, $r_2 = 95$, $T_1 = 1650$, $T_2 = 1735$, $\bar{x} = 1544.8$, and $s^2 = 17{,}022$.

To obtain first approximations to $\hat{\mu}$ and $\hat{\sigma}$, we estimate F_1 and F_2 using the methods of Section 4 and interpolate linearly between the points (T_1, y_1^*) and (T_2, y_2^*) which determine a line that approximates the probit regression line. From equations (4.15) and (4.16), we have

$$p_1^* = F_1^* = 120/300 = 0.4000$$

$$p_2^* = 35/[300 - 50/(1 - 0.4000)] = 0.1615$$

$$F_2^* = 0.4000 + 0.1615 = 0.5615$$

The linear interpolation with F_1^* and F_2^* transformed into probits is summarized below.

T_1	F_i^*	y_i^* (probits)
1650	0.4000	4.747
1703	0.5000	5.000
1735	0.5615	5.155

First approximations to the required estimates are thus

$$\mu_0 = 1703 \qquad \sigma_0 = \frac{1735 - 1650}{5.155 - 4.747} = 208$$

With the partials of (4.8) and (4.9) evaluated for these approximations to $\hat{\mu}$ and $\hat{\sigma}$, ($\hat{\xi}_1 = -0.255, \hat{\xi}_2 = 0.154$), equations (4.7) for this example become

$$-0.00572b + 0.00268d = 0.00188$$

$$0.00268b - 0.00717d = -0.00285$$

On solving this pair of equations simultaneously, we obtain the corrections $b = -0.17$ and $d = 0.33$, so that the corrected estimates are

$$\hat{\mu} = 1703 - 0.17 = 1702.83$$

$$\hat{\sigma} = 208 + 0.33 = 208.33$$

For the purposes of this illustration, a single cycle of iteration is considered sufficient, but the accuracy might be improved by repeating the above process using 1702.83 and 208.33 as new approximations.

The asymptotic variance-covariance matrix now follows approximately as

$$\begin{bmatrix} 0.00572 & -0.00268 \\ -0.00268 & 0.00717 \end{bmatrix}^{-1} = \begin{bmatrix} 212 & 79 \\ 79 & 169 \end{bmatrix}$$

Accordingly, $V(\hat{\mu}) \doteq 212$, $V(\hat{\sigma}) \doteq 169$, and $\mathrm{Cov}(\hat{\mu}, \hat{\sigma}) \doteq 79$.

Example 2. Life Test of Biological Specimens in a Stress Environment In this example, a total of 316 specimens were placed under observation. Life spans were recorded in days and the resulting data were grouped as shown in the accompanying table. Ten specimens were withdrawn (censored) after 36.5 days and ten more were withdrawn after 44.5 days. The p_i^* and the F_i^* were calculated using equations (4.15) and (4.16). These and the corresponding probit values are also shown in the frequency table.

Data for this example are summarized as follows: $N = 316$, $n = \sum_1^{14} f_i = 296$, $\sum_1^{14} r_i = 20$, $\bar{x} = 39.2703$, $s^2 = 20.16344$.

Life Distribution of Certain Biological Specimens in a Stress Environment

i	Boundaries T_i	Midpoints x_i	f_i	r_i	p_i^*	F_i^*	Probits y_i^*
0	24.5					.000000	
1	26.5	25.5	1	0	.003165	.003165	2.270
2	28.5	27.5	1	0	.003165	.006330	2.507
3	30.5	29.5	4	0	.012658	.018988	2.925
4	32.5	31.5	18	0	.056962	.075950	3.567
5	34.5	33.5	18	0	.056962	.132912	3.887
6	36.5	35.5	37	10	.117089	.250001	4.326
7	38.5	37.5	45	0	.148678	.398679	4.743
8	40.5	39.5	57	0	.188326	.587005	5.220
9	42.5	41.5	39	0	.128855	.715860	5.571
10	44.5	43.5	43	10	.142071	.857931	6.071
11	46.5	45.5	20	0	.086104	.944035	6.590
12	48.5	47.5	9	0	.038747	.982782	7.116
13	50.5	49.5	1	0	.004305	.987087	7.229
14	52.5	51.5	3	0	.012916	1.00000	
TOTALS			296	20			

The thirteen points (T_i, y_i^*), $(i = 1, 2, \ldots, 13)$ were plotted using rectangular coordinates and the probit regression line was sketched by eye as shown in Figure 5..First approximations to $\hat{\mu}$ and $\hat{\sigma}$ follow from the sketched line as $\mu_0 = 39.8$ and $\sigma_0 = 4.55$. One cycle of iteration gave $\mu_1 = 39.6$ and $\sigma_1 = 4.60$ as new approximations which were then used as the starting point for a second cycle of iteration. For this second cycle, equations (4.7) become

$$-14.591b + 0.801d = 0.261$$

$$0.801b - 28.067d = -0.33.$$

On solving this pair of equations simultaneously, we obtain as new corrections $b = -0.017$ and $d = 0.011$. The final estimates of μ and σ accordingly become

$$\hat{\mu} = 39.6 - 0.017 = 39.583$$

$$\hat{\sigma} = 4.60 + 0.011 = 4.611$$

The asymptotic variance-covariance matrix is now approximated as

$$\begin{bmatrix} 14.591 & -0.801 \\ -0.801 & 28.067 \end{bmatrix}^{-1} = \begin{bmatrix} 0.069 & 0.002 \\ 0.002 & 0.036 \end{bmatrix}$$

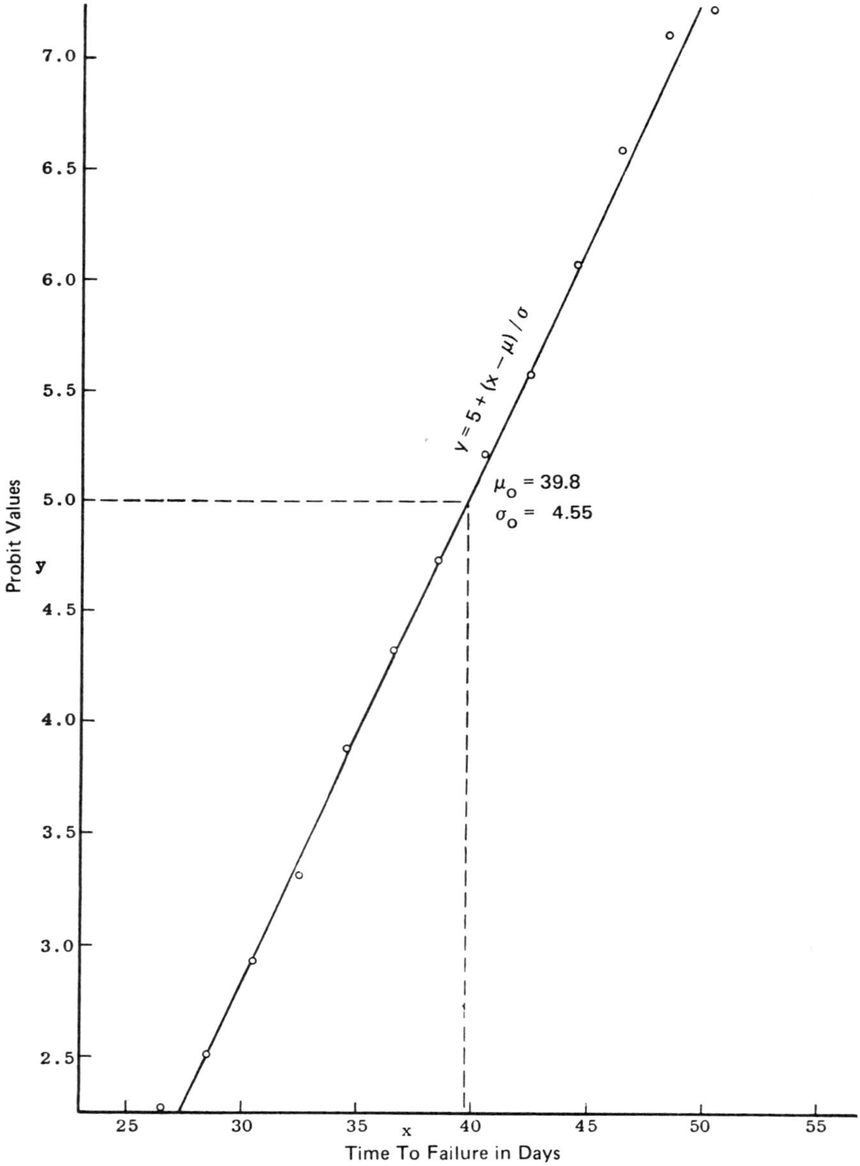

Figure 5 Probit Regression Line, example 2, life test of certain biological specimens.

and accordingly $V(\hat{\mu}) \doteq 0.069$, $V(\hat{\sigma}) \doteq 0.036$, $\text{Cov}(\hat{\mu}, \hat{\sigma}) \doteq 0.002$.

These two examples illustrate estimate calculations of normal distribution parameters (μ, σ^2) from progressively censored samples. In order to estimate lognormal parameters, (μ, σ^2, γ), we select a trial value γ, make the normalizing transformation $y = \ln(x - \gamma)$ and calculate conditional estimates $\hat{\mu}(\gamma)$ and $\hat{\sigma}^2(\gamma)$ by following the procedures illustrated in these two examples. The first two equations of (4.3) will therefore be satisfied. For *local maximum likelihood estimates* the third equation of (4.3) must also be satisfied. Accordingly, we repeat the calculations for $\hat{\mu}(\gamma)$ and $\hat{\sigma}^2(\gamma)$ using additional values of γ until we find values γ_i and γ_j in a sufficiently narrow interval such that $Q(\gamma_i) \lesseqgtr 0 \lesseqgtr Q(\gamma_j)$, where $Q(\gamma)$ designates the left side of the third equation of (4.3). Final estimates then follow by linear interpolation between γ_i and γ_j.

Modified estimates can be calculated by following a similar procedure. We calculate $\hat{\mu}(\gamma)$ and $\hat{\sigma}^2(\gamma)$ just as we did for the LMLE. However, the modified estimates must satisfy the inequalities either of (4.11) or (4.12) rather than the third equation of (4.3). After finding γ_i and γ_j in a sufficiently narrow interval, we interpolate linearly for the required estimates.

8. ESTIMATION FROM GROUPED DATA

When data from truncated or censored samples are grouped, estimating procedures differ from those employed with ungrouped data only with respect to summations involved in the estimating equations. For grouped data it is assumed that all observations in a given class are concentrated at the midpoint of that class. Unfortunately, when class intervals are large, grouping errors can lead to erroneous estimates. The effect of grouping errors, however, can be minimized by resorting to the probit technique as outlined in connection with estimation from progressively censored samples. This procedure leads to estimates $\hat{\mu}(\gamma)$ and $\hat{\sigma}(\gamma)$. The third estimating equation would then be either (4.11) or (4.12) for *modified estimates* or the third equation of (4.3) (taking grouping into account) for *local maximum likelihood estimates.*

As discussed in Chapter 4 in connection with estimation from grouped complete samples, it might be necessary to choose a somewhat arbitrary value between boundaries of the smallest class as x_1. As noted in discussing grouped data estimation for complete samples, the Chi-square and the Kolmorogov-Smirnov goodness of fit tests are available for determining whether or not agreement between expected and observed frequencies is satisfactory.

ACKNOWLEDGMENTS

Appreciation is expressed to the Editors and Publishers of *Biometrika* for permission to reproduce the chart designated here as Figure 4 from Cohen (1957). Appreciation is also expressed to the Editors and Publishers of *Technometrics* for permission to reproduce charts designated here as Figures 1, 2, 3, 5 and tables designated as Tables 1, 2, 4, plus portions of the text including illustrative examples from Cohen (1961, 1963, 1969).

REFERENCES

Bliss, C.I. and Stevens, W.L. (1937). The truncated normal distribution by Stevens is an appendix to the calculations of the time mortality curve by Bliss, *Ann. Appl. Biol. 24*, 815–852.

Cohen, A.C. (1950). Estimating the mean and variance of normal populations from singly truncated and doubly truncated samples, *Ann. Math. Statist.*, *21*, 557–569.

Cohen, A. C. (1957). On the solution of estimating equations for truncated and censored samples from normal populations, *Biometrika*, *44*, 225–236.

Cohen, A. C. (1959). Simplified estimators for the normal distribution when samples are singly censored or truncated, *Technometrics*, *1*, 217–237.

Cohen, A. C. (1961). Tables for maximum likelihood estimates: singly truncated and singly censored samples, *Technometrics*, *3*, 535–541.

Cohen, A. C, (1963). Progressively censored samples in life testing, *Technometrics*, *5*, 327–339.

Cohen, A. C. (1965). Maximum likelihood estimation in the Weibull distribution based on complete and censored samples, *Technometrics*, *5*, 579–588.

Cohen, A. C. (1975). Multi-censored sampling in the three-parameter Weibull distribution, *Technometrics*, *17*, 347–351.

Cohen, A. C. (1976). Progressively censored sampling in the three-parameter lognormal distribution, *Technometrics*, *18*, 99–103.

Cooley, C. G. and Cohen, A. C. (1970). Tables of maximum likelihood estimating functions for singly truncated and singly censored samples from the normal distribution. *NASA Contractor Report NASA CR-61330*. NASA George C. Marshall Space Flight Center, Marshall Space Flight Center, Alabama 35802.

Fisher, R. A. (1931). Properties and applications of *Hh* functions. *Introduction to Math. Tables*, *Vol 1*, British A.A.S., xxvi–xxxv.

Gajjar, A. V. and Khatri, C. G. (1969). Progressively censored samples from log-normal and logistic distributions, *Technometrics 11*, 793–803.

Galton, Sir Francis (1897). An examination into the registered speeds of American trotting horses with remarks on their value as hereditary data, *Proc. Roy. Soc. of London*, *62*, 310–314.

Giesbrecht, F. and Kempthorne, O. (1976). Maximum likelihood estimation in the three-parameter lognormal distribuion, *J. Roy. Statist. Soc.*, *Ser. B*, *38* 257–264.

Gupta, A. K. (1952). Estimation of the mean and standard deviation of a normal population from a censored sample, *Biometrika*, *39*, 260–273.

Hald, A. (1949). Maximum likelihood estimation of the parameters of a normal distribution which is truncated at a known point, *Skandinavisk Aktuarietidskrift*, *32*, 119–134.

Halperin, M. (1952). Estimation in the truncated normal distribution, *J. Amer. Statist. Assoc.*, *47*, 457–465.

Harter, H. L. (1961). Expected values of normal order statistics, *Biometrika*, *48*, 151–166.

Harter, H. L. (1969). Order statistics and their use in testing and estimation, Vol. 2. Table C1, 426–456, Aerospace Res. Lab., Wright-Patterson A.F. Base, Ohio.

Harter, H. L. and Moore, A. H. (1966). Local-maximum-likelihood estimation of the parameters of three-parameter lognormal populations from complete and censored samples, *J. Amer. Statist. Assoc.*, *61*, 842–851.

Herd, G. R. (1956). Estimation of the parameters of a population from a multi-censored sample, Ph.D. Dissertation, Iowa State College.

Herd, G. R. (1957). Estimation of reliability functions, *Proc. Third Nat. Symp. on Reliability and Quality Control*, 113–122, The Institute of Radio Engineers.

Herd, G. R. (1960). Estimation of reliability from incomplete data, *Proc. Sixth Nat. Symp. on Reliability and Quality Control*, 202–217, The Institute of Radio Engineers.

Hoel, P. G. (1954). *Introduction to Mathematical Statistics*, second edition, John Wiley and Sons, New York.

Ipsen, J. (1949). A practical method of estimating the mean and standard deviation of truncated normal distributions, *Human Biology*, *21*, 1–16.

Lemon, G. H. (1975). Maximum likelihood estimation for the three parameter Weibull distribution based on censored samples, *Technometrics*, *17*, 247–254.

Norgaard, N. (1975). Estimation of parameters in continuous distributions from restricted samples, Ph.D. Dissertation, University of Georgia, Athens, Georgia.

Pearson, Karl and Lee, A. (1908). On the generalized probable error in multiple normal correlation. *Biometrika*, *6*, 59–69.

Ringer, L. J. and Sprinkle, E. E. (1972). Estimation of the parameters of the Weibull distribution from multi-censored samples. *IEEE Transactions on Reliability*, *R-21*, 46.

Roberts, H. R. (1962a). Some results in life testing based on hypercensored samples from an exponential distribution, Ph.D. Dissertation, The George Washington University, Washington, D.C.

Roberts, H. R. (1962b). Life test experiments with hypercensored samples. *Proc. Eighteenth Annual Quality Control Conf.*, Rochester Soc. for Quality Control.

Sampford, M. R. (1952). The estimation of response-time distributions, Part II, *Biometrics*, *9*, 307–369.

Schmee, J., Gladstein, D., and Nelson, W. (1985). Confidence limits of a normal distribution from singly censored samples using maximum likelihood, *Technometrics*, *27*, 119–128.

Schmee, J. and Nelson W. B. (1977). Estimates and approximate confidence limits for (log) normal life distributions from singly censored samples by maximum likelihood. General Electric Company Technical Report 76CRD250, Schenectady, N.Y. (45 pages).

Stevens, W. L. (1937). The truncated normal distribution. (Appendix to paper by C. I. Bliss on: The calculation of the time mortality curve.) *Ann. Appl. Biol.*, *24*, 815–852.

Tiku, M. L. (1968). Estimating the parameters of log-normal distribution from censored samples, *J. Amer. Statist. Assoc.*, *63*, 134–140.

6

Bayesian Estimation

WILLIAM J. PADGETT Department of Statistics, University of South Carolina, Columbia, South Carolina

1. INTRODUCTION

In this chapter, the problem of Bayesian estimation of the parameters of the univariate lognormal distribution is considered. Since the lognormal distribution is frequently used as a lifetime model in reliability studies (see Mann, Schafer, and Singpurwalla, 1974, for many references), a great deal of attention is given here to Bayesian estimation of the reliability function of the $\Lambda(\mu, \sigma^2)$ distribution. In particular, Gupta (1962), Dumonceaux and Antle (1973), and Kotz (1973), among others, have studied various aspects of the lognormal distribution as a lifetime model.

The Bayesian approach to estimation in life testing is motivated by the notion that prior information concerning an unknown life parameter exists, generally in the form of life-test data on prototypes or data on a similar item in an allied product line. Often such knowledge can be translated into a prior distribution. Drake (1966) discussed the appealing aspects of a Bayesian approach to reliability problems, and various lifetime models have been examined in a Bayesian framework. For example, see Bhattacharya (1967), Canavos (1973), Padgett and Johnson (1983), Padgett and Tsokos

(1977), Padgett and Wei (1977), Soland (1969), Tsokos (1972), and Zellner (1971).

In the remainder of this chapter, Bayesian estimation of the parameters and reliability function with respect to both proper and noninformative prior distributions will be presented. In several of the cases considered, Type II censored samples will be assumed. In Section 2 Bayes estimates for several different proper prior distributions on the parameters will be given, including conjugate priors. Noninformative (or vague) priors will be discussed in Section 3. The problem of Bayesian lower bounds on the reliability function will be presented in Section 4, and in Section 5, some empirical Bayes estimation procedures will be outlined.

2. BAYES ESTIMATES FROM PROPER PRIOR DISTRIBUTIONS

Three cases will be considered in this section. First, it will be assumed that σ^2 is known. Then the case that μ is known will be discussed, and finally, the more realistic case that both parameters are unknown will be presented.

2.1 Prior Distributions for μ (σ^2 Known)

Suppose that the parameter μ is a value of a random variable M and consider a random sample of n items whose lifetimes are distributed as $\Lambda(\mu, \sigma^2)$. The n items are placed on a life test which is terminated after a predetermined number $r \leq n$ of the items fail. The ordered observed lifetimes $\mathbf{x}_r = (x_1, \ldots, x_r)$ are used to estimate μ, the reliability function $R(t) = \Phi((\mu - \ln t)/\sigma)$, $t > 0$, where Φ denotes the $N(0,1)$ distribution function, and other parameters. The observed lifetimes $\mathbf{x}_r$ constitute a Type II censored sample which for $r = n$ becomes a simple random sample from $\Lambda(\mu, \sigma^2)$.

The probability of observing r failures at times $x_1, \ldots, x_r$ and having $(n-r)$ items survive at least time x_r is given by the likelihood of the sample:

$$L(\mathbf{x}_r \mid \mu) = \frac{n!}{(n-r)!} \left[\prod_{i=1}^{r} \left(x_i \sigma \sqrt{2\pi} \right)^{-1} \right]$$

$$\cdot \exp\left[-\frac{1}{2\sigma^2} \sum_{i=1}^{r} (\ln x_i - \mu)^2 \right] \left[\Phi\left(\frac{\mu - \ln x_r}{\sigma} \right) \right]^{n-r}$$

2.1.1 *Uniform Prior Distributions.* Assume that M has a uniform prior distribution with density function

$$g(\mu) = \begin{cases} \dfrac{1}{\beta - \alpha} & \alpha \le \mu \le \beta \qquad (\alpha < \beta) \\ 0 & \text{otherwise} \end{cases}$$

Then the posterior density of M, given the sample $\mathbf{x}_r$, is

$$h(\mu \mid \mathbf{x}_r) = \left[\Phi\left(\frac{\mu - \ln x_r}{\sigma}\right)\right]^{n-r} \exp\left[-\frac{1}{2\sigma^2}(r\mu^2 - 2\mu T_r)\right] \Big/ I_1 \qquad (2.1)$$

$\alpha \le \mu \le \beta$, where $T_r = \sum_{i=1}^{r} \ln x_i$ and

$$I_1 = \int_\alpha^\beta \left[\Phi\left(\frac{\mu - \ln x_r}{\sigma}\right)\right]^{n-r} \exp\left[-\frac{1}{2\sigma^2}(r\mu^2 - 2\mu T_r)\right] d\mu$$

Thus, with respect to a squared-error loss function, the Bayes estimate of mean life, or mean time before failure (MTBF), θ, is the posterior expectation of $\theta = e^{M+\sigma^2/2}$, which is

$$\theta_{B,r} = \exp\left[\frac{T_r}{r} + \frac{\sigma^2}{2}\left(1 + \frac{1}{r}\right)\right] \cdot \frac{I_2}{I_3} \qquad (2.2)$$

where

$$I_2 = \int_\alpha^\beta \left[\Phi\left(\frac{\mu - \ln x_r}{\sigma}\right)\right]^{n-r} \exp\left[-\frac{r}{2\sigma^2}\left(\mu - \frac{T_r + \sigma^2}{r}\right)^2\right] d\mu$$

and

$$I_3 = \int_\alpha^\beta \left[\Phi\left(\frac{\mu - \ln x_r}{\sigma}\right)\right]^{n-r} \exp\left[-\frac{r}{2\sigma^2}\left(\mu - \frac{T_r}{r}\right)^2\right] d\mu$$

Similarly, the Bayes estimate of the reliability function at t with respect to a squared-error loss function is given by

$$R_B(t;r) = \int_\alpha^\beta \Phi\left(\frac{\mu - \ln t}{\sigma}\right)\left[\Phi\left(\frac{\mu - \ln x_r}{\sigma}\right)\right]^{n-r} \times \exp\left[-\frac{r}{2\sigma^2}\left(\mu - \frac{T_r + \sigma^2}{r}\right)^2\right] d\mu \times I_3^{-1} \tag{2.3}$$

For complete (simple random) samples, the Bayes estimates are somewhat simpler in form. Equation (2.2) reduces to

$$\theta_{B,n} = \exp\left[\frac{T_n}{n} + \frac{\sigma^2}{2}\left(1 + \frac{1}{n}\right)\right]\left[\frac{\Phi\left(\frac{n\beta - T_n - \sigma^2}{\sqrt{n}\sigma}\right) - \Phi\left(\frac{n\alpha - T_n - \sigma^2}{\sqrt{n}\sigma}\right)}{\Phi\left(\frac{n\beta - T_n}{\sqrt{n}\sigma}\right) - \Phi\left(\frac{n\alpha - T_n}{\sqrt{n}\sigma}\right)}\right]$$

and equation (2.3) becomes

$$R_B(t;n) = \frac{\int_{\frac{\alpha - \ln t}{\sigma}}^{\frac{\beta - \ln t}{\sigma}} \Phi(\mu)\sqrt{\frac{n}{2\pi}} \exp\left[-\frac{n}{2}\left(\mu - \frac{1}{\sigma}\left(\frac{T_n}{n} - \ln t\right)\right)^2\right] d\mu}{\Phi\left(\frac{n\beta - T_n}{\sqrt{n}\sigma}\right) - \Phi\left(\frac{n\alpha - T_n}{\sqrt{n}\sigma}\right)} \qquad t > 0$$

Also, for complete samples the Bayes estimate of μ with respect to squared-error loss is given by

$$\mu_{B,n} = \frac{T_n}{n} + \frac{\sigma\left\{\exp\left[-\frac{1}{2}\left(\frac{n\alpha - T_n}{\sqrt{n}\sigma}\right)^2\right] - \exp\left[-\frac{1}{2}\left(\frac{n\beta - T_n}{\sqrt{n}\sigma}\right)^2\right]\right\}}{\sqrt{2\pi n}\left[\Phi\left(\frac{n\beta - T_n}{\sqrt{n}\sigma}\right) - \Phi\left(\frac{n\alpha - T_n}{\sqrt{n}\sigma}\right)\right]}$$

Note that as $\alpha \to -\infty$ and $\beta \to \infty$, $\mu_{B,n}$ converges to $\hat{\mu} = T_n/n$ which is the maximum likelihood estimate of μ. Hence, the maximum likelihood estimate of μ is a "limit of Bayes rules" (Ferguson, 1967).

2.1.2 *Normal Prior Distributions.* For complete samples, the natural conjugate family for the lognormal distribution when σ^2 is known is the normal family. Hence, assume that M has a normal prior distribution with mean λ and variance ξ^2 and first consider Type II censored samples.

The posterior distribution of M, given the data $\mathbf{x}_r = (x_1, \ldots, x_r)$, is

$$h(\mu \mid \mathbf{x}_r) = \left[\Phi\left(\frac{\mu - \ln x_r}{\sigma}\right)\right]^{n-r} \times \exp\left[-\frac{r + \sigma^2/\xi^2}{2\sigma^2}\left(\mu - \frac{T_r + \lambda\sigma^2/\xi^2}{r + \sigma^2/\xi^2}\right)^2\right] I_4^{-1} \qquad -\infty < \mu < \infty \tag{2.4}$$

where

$$I_4 = \int_{-\infty}^{\infty} \left[\Phi\left(\frac{\mu - \ln x_r}{\sigma}\right)\right]^{n-r} \exp\left[-\frac{r + \sigma^2/\xi^2}{2\sigma^2}\left(\mu - \frac{T_r + \lambda\sigma^2/\xi^2}{r + \sigma^2/\xi^2}\right)^2\right] d\mu$$

Hence, for squared-error loss the Bayes estimate of mean life is

$$\theta_{B,r} = \exp\left[\frac{T_r + \lambda\sigma^2/\xi^2}{r + \sigma^2/\xi^2} + \frac{\sigma^2}{2}\left(1 + \frac{1}{r + \sigma^2/\xi^2}\right)\right] \cdot \frac{I_5}{I_4} \tag{2.5}$$

where

$$I_5 = \int_{-\infty}^{\infty} \left[\Phi\left(\frac{\mu - \ln x_r}{\sigma}\right)\right]^{n-r} \times \exp\left[-\frac{r + \sigma^2/\xi^2}{2\sigma^2}\left(\mu - \frac{T_r + \lambda\sigma^2/\xi^2 + \sigma^2}{r + \sigma^2/\xi^2}\right)^2\right] d\mu$$

Also, the Bayes estimate of the reliability function is

$$R_B(t; r) = \int_{-\infty}^{\infty} \Phi\left(\frac{\mu - \ln t}{\sigma}\right)\left[\Phi\left(\frac{\mu - \ln x_r}{\sigma}\right)\right]^{n-r} \times \exp\left[-\frac{r + \sigma^2/\xi^2}{2\sigma^2}\left(\mu - \frac{T_r + \lambda\sigma^2/\xi^2}{r + \sigma^2/\xi^2}\right)^2\right] d\mu \cdot I_4^{-1} \qquad t > 0 \tag{2.6}$$

For complete samples ($r = n$), the posterior distribution of M, given the data $\mathbf{x}_n = (x_1, \ldots, x_n)$, is the normal distribution with mean $(T_n + \lambda\sigma^2/\xi^2)/(n + \sigma^2/\xi^2)$ and variance $\sigma^2/(n + \sigma^2/\xi^2)$ when the prior is normal with mean λ and variance ξ^2. Hence, for $r = n$ the Bayes estimates (2.5) and (2.6) reduce to

$$\theta_{B,n} = \exp\left[\frac{T_n + \lambda\sigma^2/\xi^2}{n + \sigma^2/\xi^2} + \frac{\sigma^2}{2}\left(1 + \frac{1}{n + \sigma^2/\xi^2}\right)\right]$$

and

$$R_B(t;n) = \sqrt{\frac{n + \sigma^2/\xi^2}{2\pi\sigma^2}} \int_{-\infty}^{\infty} \Phi\left(\frac{\mu - \ln t}{\sigma}\right) \times \exp\left[-\frac{n + \sigma^2/\xi^2}{2\sigma^2}\left(\mu - \frac{T_n + \lambda\sigma^2/\xi^2}{n + \sigma^2/\xi^2}\right)^2\right] d\mu \qquad t > 0 \tag{2.7}$$

By Ellison's (1964, p. 93) result, after a change of variable of integration to $y = (\mu - \ln t)/\sigma$, (2.7) reduces to

$$R_B(t;n) = \Phi\left\{\left[\frac{T_n + \lambda\sigma^2/\xi^2}{n + \sigma^2/\xi^2} - \ln t\right] \times \sigma^{-1}\left[1 + \frac{1}{n + \sigma^2/\xi^2}\right]^{-1/2}\right\} \qquad t > 0 \tag{2.8}$$

Then it is easily shown that, as $\xi^2 \to \infty$, (2.8) becomes

$$\Phi\left[\frac{T_n/n - \ln t}{\sigma(1 + 1/n)^{1/2}}\right]$$

which is approximately equal to the maximum likelihood estimate of the reliability function for large sample size. So for large values of the prior variance ξ^2 and large sample size, the Bayes estimate of reliability with respect to the $N(\lambda, \xi^2)$ prior is approximately the maximum likelihood estimate. In addition, for squared-error loss the Bayes estimate of μ is simply

$$\mu_{B,n} = \frac{T_n + \frac{\lambda\sigma^2}{\xi^2}}{n + \frac{\sigma^2}{\xi^2}}$$

Again, note that as the prior variance $\xi^2 \to \infty$, $\mu_{B,n}$ converges to the maximum likelihood estimate $\hat{\mu}$, so that $\hat{\mu}$ is again a "limit of Bayes rules." Padgett (1978) gave a comparison of Bayes and maximum likelihood estimators for the lognormal distribution.

2.2 Prior Distributions for σ^2 (μ Known)

Assume that the parameter σ^2 is a value of a random variable Σ^2 and that the parameter μ is known. Two distributions for Σ^2 will be considered, the inverted-gamma distribution (the conjugate prior) with probability density function

$$g(\sigma^2;\lambda,\nu) = [\lambda\Gamma(\nu)]^{-1}(\lambda/\sigma^2)^{\nu+1}\exp(-\lambda/\sigma^2) \qquad \sigma^2 > 0 \qquad (\lambda,\nu > 0) \tag{2.9}$$

and the "general uniform" distribution (Tsokos, 1972) with density

$$h(\sigma^2;\alpha,\beta,a) = \frac{(a-1)(\alpha\beta)^{a-1}}{(\sigma^2)^a(\beta^{a-1}-\alpha^{a-1})} \qquad 0 < \alpha \le \sigma^2 \le \beta < \infty \qquad a \ge 0 \qquad a \ne 1 \tag{2.10}$$

Note that for $a = 0$, h is the uniform density on [0,1]. These two families provide a wide choice of appropriate priors for Σ^2.

Only complete samples ($r = n$), $\mathbf{x}_n = (x_1,\ldots,x_n)$, will be considered in this section.

2.2.1 *Inverted-Gamma Prior.* For the prior density $g(\sigma^2;\lambda,\nu)$, the posterior distribution is also an inverted-gamma distribution with parameters $\lambda^* = \sum_{i=1}^n(\ln x_i - \mu)^2/2+\lambda$ and $\nu^* = \nu+n/2$. Hence, for squared-error loss, the Bayes estimate of σ^2 is the posterior mean $\lambda^*/(\nu^*-1)$, which is

$$\sigma^2_{B,n} = \frac{\frac{1}{2}\sum_{i=1}^n(\ln x_i-\mu)^2+\lambda}{\nu+\frac{n}{2}-1}$$

2.2.2 *General Uniform Prior.* For the case that Σ^2 has the general uniform prior $h(\sigma^2;\alpha,\beta,a)$, the posterior distribution of Σ^2 is quite complicated,

$$h(\sigma^2 \mid \mathbf{x}) = \left(\frac{S_n}{2}\right)^{n/2+a-1} \frac{\exp(-S_n/2\sigma^2)}{(\sigma^2)^{n/2+a}\Gamma(n/2+a-1)} \div P\left[\frac{S_n}{\beta} \le \chi^2_\eta \le \frac{S_n}{\alpha}\right] \qquad 0 < \alpha \le \sigma^2 \le \beta$$

where $S_n = \sum_{i=1}^n(\ln x_i-\mu)^2$, $\eta = n+2a-2$, and χ^2_η denotes a chi-square random variable with η degrees of freedom. The Bayes estimates of σ^2, mean time before failure, and $R(t)$ may be obtained from $h(\sigma^2 \mid \mathbf{x})$ by numerical integration.

The distributions (2.9) and (2.10) will be used again in Section 4 in obtaining Bayesian lower bounds on the reliability (Padgett and Johnson, 1983).

2.3 Normal-Gamma Prior Distributions

Complete samples, $\mathbf{x}_n = (x_1, \ldots, x_n)$, from the $\Lambda(\mu, \sigma^2)$ distribution are assumed in this section.

A conjugate family for the $\Lambda(\mu, \sigma^2)$ when both parameters are unknown is the normal-gamma distribution. That is, it is assumed that (μ, σ^{-2}) is a value of a random vector (M, Σ^{-2}) so that the conditional prior distribution of M, given $\Sigma^2 = \sigma^2$, is normal with mean λ and variance $\tau\sigma^2$, where $-\infty < \lambda < \infty$ and $\tau > 0$, and the marginal prior distribution of Σ^{-2} is a gamma distribution $G(\alpha, \beta)$ with density $g(v) = \beta^\alpha/\Gamma(\alpha) v^{\alpha-1}\exp(-\beta v)$, $(v \geq 0)$, where λ, τ, α, and β are known constants (DeGroot, 1970, p. 170). Hence, the joint prior distribution of (M, Σ) is

$$\xi(\mu, \sigma) \propto \sigma^{1-\alpha}\exp[-\beta\sigma^2 - (\mu - \lambda)^2/(2\tau\sigma^2)]$$

The posterior distribution is also of the normal-gamma type. That is, the conditional posterior distribution of M, given $\Sigma^2 = \sigma^2$, is normal with mean $\lambda^* = (\tau^{-1}\lambda + n\bar{y})/(\tau^{-1} + n)$ and variance $\tau^*\sigma^2$, where $\bar{y} = n^{-1}\sum_{i=1}^n \ln x_i$ and $\tau^* = (\tau^{-1} + n)^{-1}$, and the marginal posterior distribution of Σ^{-2} is $G(\alpha^*, \beta^*)$, where $\alpha^* = \alpha + n/2$ and $\beta^* = \beta + 1/2\sum_{i=1}^n(\ln x_i - \bar{y})^2 + [\tau^{-1}n(\bar{y} - \lambda)^2]/[2(\tau^{-1} + n)]$. It can also be shown that the marginal posterior distribution of M is a t-distribution with location parameter λ^*, $2\alpha^*$ degrees of freedom, and precision $\alpha^*(\tau^{-1} + n)/\beta^*$. Therefore, the Bayes estimates of μ and σ^2 for squared-error loss are the means of the posterior marginal distributions,

$$\mu_B = \frac{\tau^{-1}\lambda + n\bar{y}}{\tau^{-1} + n} \quad \text{and} \quad \sigma_B^2 = \frac{\beta^*}{\alpha^* - 1}$$

since the marginal posterior of Σ^2 is inverted-gamma with parameters β^* and α^* (see (2.9)).

Padgett and Wei (1977) gave the Bayes estimator of the reliability function $R(t)$ with respect to the normal-gamma prior for (M, Σ^{-2}) and squared-error loss. The following lemma is a generalization of a result of Ellison (1964) and was applied to obtain the Bayes estimator.

Lemma 1 Let V be a random variable which has the gamma distribution with density function $g(v) = \beta^\alpha/\Gamma(\alpha)v^{\alpha-1}\exp(-\beta v)$, $(v > 0)$. Then

$$E_V[\Phi(cV^{1/2})] = P[T_{2\alpha} < c(\alpha/\beta)^{1/2}]$$

where Φ is the standard normal distribution function, c is any real number, and $T_{2\alpha}$ is a random variable having the t-distribution with 2α degrees of freedom (not necessarily an integer).

Let (M, Σ^{-2}) have the normal-gamma prior distribution. Applying Lemma 1 to the posterior marginal distribution of Σ^{-2}, the Bayes estimator of $R(t)$ for squared-error loss is given by

$$R_B(t) = 1 - P\left[T_{2\alpha^*} < \left(\frac{\alpha^*}{\beta^*}\right)^{1/2} \frac{(\ln t - \lambda^*)}{(1+\tau^*)^{1/2}}\right] \tag{2.11}$$

where α^*, β^*, λ^*, and τ^* are defined as before.

It was noted by Padgett and Wei (1977) that as $n \to \infty$, $\lambda^* \xrightarrow{\text{a.s.}} \mu_0$, $\beta^*/\alpha^* \xrightarrow{\text{a.s}} \sigma_0^2$, where μ_0 and σ_0^2 denote the true values of μ and σ^2, and $T_{2\alpha^*}$ has the $N(0,1)$ distribution as a limiting distribution. Thus, the estimator (2.11) is strongly consistent for $R(t)$. Also, as $\alpha \to 0$, $\beta \to 0$, and $\tau^{-1} \to 0$, it is easily seen that $\alpha^* \to n/2$, $\beta^* \to \sum_{i=1}^n (\ln x_i - \bar{x})^2/2$, and $\lambda^* \to \bar{x}$, with $\bar{x} = n^{-1}\sum_{i=1}^n \ln x_i$ as before. Hence, as α, β, $\tau^{-1} \to 0$, (2.11) converges to

$$1 - P\left[T_n < \frac{(\ln t - \bar{x})}{(1+1/n)^{1/2}} \left(\frac{n}{\sum_{i=1}^n (\ln x_i - \bar{x})^2}\right)^{1/2}\right] \tag{2.12}$$

3. NONINFORMATIVE PRIOR DISTRIBUTIONS

As mentioned by Bhattacharya (1967), in some situations the experimenter does not have any prior knowledge, or may have only vague prior knowledge, about the unknown parameters. In this case a noninformative prior distribution is used in a Bayesian analysis.

When σ^2 is known, Martz and Waller (1982, p. 438) present some results of Zellner (1971) for the noninformative prior distribution on μ given by (Box and Tiao, 1973)

$$g(\mu) \propto b \qquad b = \text{constant}$$

Then the posterior distribution of $\theta = \exp(\mu + \sigma^2/2)$, the mean of the $\Lambda(\mu, \sigma^2)$ distribution, is $\Lambda(\bar{x} + \sigma^2/2, \sigma^2/n)$ with $\bar{x} = n^{-1}\sum_{i=1}^{n} \ln x_i$ as before. Hence, the Bayes estimator of θ under squared-error loss is

$$\theta_B = \exp\left[\bar{x} + \frac{\sigma^2}{2}\left(1 + \frac{1}{n}\right)\right]$$

In the case of *relative* squared-error loss, $L = [(\hat{\theta} - \theta)/\theta]^2$, Zellner (1971) shows that the Bayes estimator of θ is the minimum mean-squared error estimator given by

$$\tilde{\theta} = \exp\left[\bar{x} + \frac{\sigma^2}{2}\left(1 - \frac{3}{n}\right)\right] \tag{3.1}$$

Bayesian probability intervals for θ are also given by Martz and Waller (1982).

Padgett and Tsokos (1977) consider a prior "quasi-density" for μ when σ^2 is known. This "quasi-density" is

$$g(\mu) \propto \exp(c\mu) \qquad -\infty < \mu < \infty \qquad (-\infty < c < \infty) \tag{3.2}$$

If a "Bayes" estimate can be formally obtained using (3.2) as a prior, it is referred to as a "generalized Bayes estimate" (Ferguson, 1967). Note that (3.2) includes the noninformative prior as a special case when $c = 0$.

For complete samples, the generalized Bayes estimate for μ with respect to the prior quasi-density (3.2) and a squared-error loss function is given by

$$\mu_{GB} = \frac{\int_{-\infty}^{\infty} \sqrt{\frac{n}{2\pi\sigma^2}}\,\mu \exp\left[-\frac{n}{2\sigma^2}\left(\mu^2 - 2\mu\frac{T_n}{n}\right) + c\mu\right] d\mu}{\exp\left[\frac{1}{2\sigma^2}\left(\frac{T_n + c\sigma^2}{n}\right)^2\right]}$$

$$= \frac{T_n + c\sigma^2}{n}$$

where $T_n = \sum_{i=1}^{n} \ln x_i$ as in Section 6.2. Hence, $\mu_{GB} = \frac{T_n}{n} = \hat{\mu}$ if and only if the prior quasi-density is the noninformative prior. Similarly, the generalized Bayes estimate of mean life θ with respect to the prior quasi-density (3.2) and squared-error loss is

$$\theta_{GB} = \frac{\int_{-\infty}^{\infty} \sqrt{\frac{n}{2\pi\sigma^2}} \exp\left[-\frac{n}{2\sigma^2}\left(\mu^2 - 2\mu\frac{T_n}{n}\right) + \mu(1+c) + \frac{\sigma^2}{2}\right] d\mu}{\exp\left[\frac{1}{2\sigma^2}\left(\frac{T_n + c\sigma^2}{n}\right)^2\right]}$$

$$= \exp\left[\frac{T_n}{n} + \frac{\sigma^2}{2}\left(1 + \frac{1+2c}{n}\right)\right] \quad (3.3)$$

Therefore, $\theta_{GB} = \exp\left[\frac{T_n}{n} + \frac{\sigma^2}{2}(1 - \frac{1}{n})\right]$, which is the minimum variance unbiased estimate of mean life, if and only if $c = -1$. Also, when $c = -2$, (3.3) equals (3.1), the minimal mean-squared error estimator of θ. Corresponding generalized Bayes estimates of the reliability function with respect to the prior quasi-densities may also be found without difficulty.

If both μ and σ^2 are unknown, the noninformative prior, or Jeffreys' invariant prior, for (M, Σ), given by $g(\mu, \sigma) \propto \sigma^{-1}$, $-\infty < \mu < \infty$, $\sigma > 0$, can be assumed (Padgett and Wei, 1977). In this case, the conditional posterior distribution of M, given $\Sigma = \sigma$, is $N(\bar{x}, \sigma^2/n)$, $\bar{x} = \sum_{i=1}^{n} \ln x_i/n$, and the marginal posterior distribution of s^2/Σ^2 is chi-square with $n-1$ degrees of freedom, where $s^2 = \sum_{i=1}^{n}(\ln x_i - \bar{x})^2$. Again, applying Lemma 1 of Section 2.3 gives the Bayes estimator of $R(t)$ as

$$R_J(t) = 1 - P\left[T_{n-1} < \frac{(\ln t - \bar{x})(n-1)^{1/2}}{(1 + 1/n)^{1/2} s}\right]$$

Notice the slight difference between (2.12) and $R_J(t)$, although they are asymptotically equivalent.

A Bayesian estimate of the mean θ with respect to squared-error loss and the joint noninformative prior $g(\mu, \sigma)$ above does not exist (see Zellner, 1971, or Martz and Waller, 1982, p. 441).

4. BAYESIAN LOWER BOUNDS ON THE RELIABILITY FUNCTION

Padgett and Wei (1978) obtained approximate Bayesian lower bounds on $R(t)$ with respect to the normal-gamma prior and the noninformative prior $g(\mu, \sigma) \propto \sigma^{-1}$. Padgett and Johnson (1983) gave Bayesian lower bounds on the reliability for several families of priors on σ^2 (μ known) and for the normal-gamma prior on (μ, σ^2).

Again, throughout this section, complete samples of n observations $\bar{x}_n = (x_1, \ldots, x_n)$ from the $\Lambda(\mu, \sigma^2)$ distribution are assumed. First, the prior distributions for σ^2 (μ known) given by (2.9) and (2.10) are considered.

Then the normal-gamma prior for (μ, σ^2) and the noninformative prior are used.

For the case that Σ^2 has an inverted gamma prior distribution given by (2.9), the posterior distribution is also inverted gamma with parameters $\lambda^* = \sum_{i=1}^n (\ln x_i - \mu)^2/2 + \lambda$ and $\nu^* = \nu + n/2$ as in Section 2.2. Then the random variable $Z = 2\lambda^*/\Sigma^2$, given $\mathbf{x}_n$, has the chi-square distribution with $2\nu^*$ degrees of freedom (not necessarily an integer). Hence, for a given $0 < \delta < 1$,

$$P\left[\frac{2\lambda^*}{\Sigma^2} \leq \chi^2_{1-\delta,2\nu^*} \mid \mathbf{x}\right] = 1 - \delta \tag{4.1}$$

and

$$P\left[\chi^2_{\delta,2\nu^*} \leq \frac{2\lambda^*}{\Sigma^2} \mid \mathbf{x}\right] = 1 - \delta \tag{4.2}$$

where $\chi^2_{1-\xi,2\nu^*}$ denotes the upper ξ percent point of the chi-squared distribution with $2\nu^*$ degrees of freedom. Therefore, since $R(t) = \Phi[(\mu - \ln t)/\sigma]$, $t > 0$, from (4.1) and (4.2), respectively, for $0 < \delta < 1$,

$$1 - \delta = P\left\{\Phi\left[\left(\frac{\chi^2_{1-\delta,2\nu^*}}{2\lambda^*}\right)^{1/2} (\mu - \ln t)\right] \leq R(t) \mid \mathbf{x}\right\} \qquad \text{if } \mu \geq \ln t$$

and

$$1 - \delta = P\left\{\Phi\left[\left(\frac{\chi^2_{\delta,2\nu^*}}{2\lambda^*}\right)^{1/2} (\mu - \ln t)\right] \leq R(t) \mid \mathbf{x}\right\} \qquad \text{if } \mu < \ln t$$

yielding a $(1 - \delta)$-level Bayesian lower bound for the reliability function $R(t)$ given by

$$\begin{array}{ll} \Phi\left[\left(\dfrac{\chi^2_{1-\delta,2\nu^*}}{2\lambda^*}\right)^{1/2} (\mu - \ln t)\right] & \text{if } \mu \geq \ln t \\ \Phi\left[\left(\dfrac{\chi^2_{\delta,2\nu^*}}{2\lambda^*}\right)^{1/2} (\mu - \ln t)\right] & \text{if } \mu < \ln t \end{array} \tag{4.3}$$

For the case that Σ^2 has the general uniform prior (2.10), the posterior density is given by $h(\sigma^2 \mid \mathbf{x})$ in Section 2.2. Integrating $h(\sigma^2 \mid \mathbf{x})$ from zero to y, after a change of variables, yields

$$P\left[\Sigma^2 \leq y \mid \bar{x}\right] = \frac{1}{p(\alpha,\beta)} \int_{S_n/y}^{S_n/\alpha} \frac{z^{n/2+a}\exp(-z/2)}{\Gamma(n/2+a-1)2^{n/2+a-1}}\,dz = \frac{p(\alpha,y)}{p(\alpha,\beta)} \quad (4.4)$$

where $S_n = \sum_{i=1}^{n}(\ln x_i - \mu)^2$ and

$$p(\alpha,\beta) = P\left[\frac{S_n}{\beta} \leq \chi_\eta^2 \leq \frac{S_n}{\alpha}\right] = P\left[\chi_\eta^2 \leq \frac{S_n}{\alpha}\right] - P\left[\chi_\eta^2 \leq \frac{S_n}{\beta}\right]$$

Therefore, given a value δ between zero and one, y can be chosen so that from (4.4),

$$P\left[\chi_\eta^2 \leq \frac{S_n}{y}\right] = P\left[\chi_\eta^2 \leq \frac{S_n}{\alpha}\right]\delta + P\left[\chi_\eta^2 \leq \frac{S_n}{\beta}\right](1-\delta) \quad (4.5)$$

Now use (4.5) to obtain a $(1-\delta)$-level Bayesian lower bound for $R(t)$ as follows: If $\mu \geq \ln t$, then choose a value z that satisfies

$$P\left[\chi_\eta^2 \leq z\right] = \delta P\left[\chi_\eta^2 \leq \frac{S_n}{\alpha}\right] + (1-\delta)P\left[\chi_\eta^2 \leq \frac{S_n}{\beta}\right]$$

and set $y = S_n/z$. Then for the value of y,

$$1-\delta = P\left[\Sigma^2 \leq y \mid \mathbf{x}\right] = P\left[\Phi\left(\frac{\mu - \ln t}{y^{1/2}}\right) \leq R(t) \mid \mathbf{x}\right]$$

Similarly, if $\mu < \ln t$, then choose a value z such that

$$P\left[\chi_\eta^2 \leq z\right] = (1-\delta)P\left[\chi_\eta^2 \leq \frac{S_n}{\alpha}\right] + \delta P\left[\chi_\eta^2 \leq \frac{S_n}{\beta}\right]$$

and

$$1-\delta = P\left[\Phi\left(\frac{\mu - \ln t}{y^{1/2}}\right) \leq R(t) \mid \mathbf{x}\right]$$

Therefore, a $(1-\delta)$-level Bayesian lower bound for the reliability function is given by $\Phi[(\mu - \ln t)/y^{1/2}]$, where y is chosen as above, according to whether $\mu \geq \ln t$ or $\mu < \ln t$.

Next, suppose that the prior distribution of (μ, σ^2) is the normal-gamma distribution described in Section 2.3. Then the posterior distribution is also normal-gamma as described in that section. The following lemma generalizes Lemma 1 of Section 2.3 (Padgett and Wei, 1978) and is used to obtain the Bayesian lower bounds for $R(t)$ in the present case.

Lemma 2 Let V be a random variable with gamma distribution $g(v)$ as before. Then for any constants c and $d \neq 0$

$$E_V\left[\Phi(c + dV^{1/2})\right] = P\left[T(-c; 2\alpha) \leq d(\alpha/\beta)^{1/2}\right]$$

where $T(-c; 2\alpha)$ is a noncentral-t random variable with 2α degrees of freedom (not necessarily an integer) and noncentrality parameter $-c$.

Now, to obtain a Bayesian lower bound for $R(t)$, consider the value c chosen so that, for a given δ, $0 < \delta < 1$,

$$1 - \delta = P[c \leq (M - \ln t)/\Sigma \mid \mathbf{x}] = \int_0^\infty \int_{\sigma c + \ln t}^\infty f(\mu, \sigma^2 \mid \mathbf{x})\, d\mu\, d\sigma^2 \qquad (4.6)$$

where $f(\mu, \sigma^2 \mid \mathbf{x})$ is the posterior density of (M, Σ^2), given the data. The integral in (4.6) is

$$\int_0^\infty g(y; \alpha^*, \beta^*) \left[\int_{c_*(y)}^\infty (2\pi)^{-1/2} \exp(-z^2/2)\, dz\right] dy \qquad (4.7)$$

where $g(y; \alpha^*, \beta^*)$ is a gamma density with the parameters α^*, β^* given before and $c_*(y) = [c - (\lambda^* - \ln t)y^{1/2}]/(\tau^*)^{1/2}$, where $\tau^* = 1/(n + \tau^{-1})$. Letting Y denote a random variable with density $g(y; \alpha^*, \beta^*)$, the right side of (4.7) is equal to

$$1 - E\left[\frac{\Phi(c - (\lambda^* - \ln t)Y^{1/2})}{(\tau^*)^{1/2}}\right]$$

By use of Lemma 2, equation (4.6) becomes

$$1-\delta = 1 - P\left[T\left(\frac{-c}{(\tau^*)^{1/2}};2\alpha^*\right) \leq (\ln t - \lambda^*)\left(\frac{\alpha^*}{\beta^*\tau^*}\right)^{1/2}\right] \tag{4.8}$$

where $T(-c/(\tau^*)^{1/2};2\alpha^*)$ denotes a noncentral-t random variable. Therefore, $\Phi(c)$ gives a $(1-\delta)$-level Bayesian lower bound for $R(t)$ for each time $t > 0$. The value of c can be determined easily for a given sample of lifetimes $\mathbf{x} = (x_1, \ldots, x_n)$, and values of $\alpha^*, \beta^*, \lambda^*, \tau^*$, and t from (4.8), by using the technique described on page 59 of Pearson and Hartley's (1972) tables.

If there is *vague* prior information concerning the behavior of (μ, σ^2), Jeffreys' prior may be utilized to obtain Bayesian lower bounds for reliability in a manner similar to that for normal-gamma priors.

5. SOME EMPIRICAL BAYES ESTIMATORS OF THE RELIABILITY FUNCTION

In this section the following kind of situation involving several testing "stages" will be considered. Let ψ denote an unknown parameter of the $\Lambda(\mu, \sigma^2)$ lifetime distribution, e.g. the mean time to failure, mean log-time to failure, or $R(t_0)$, the reliability function at time t_0. At the first testing stage, a random sample of n_1 failure times is observed from the lognormal failure time distribution with parameter ψ_1. An estimate of ψ_1 is obtained from this sample. At the second testing stage, another random sample of size n_2 is observed from the lognormal distribution with parameter ψ_2, which may be different from ψ_1 due to randomly changing environmental conditions between stages, such as imperfect control procedures. The value of ψ_2 is estimated from this second sample in the same manner as before. This situation continues until the jth (or present) testing stage at which time a random sample of size n_j is observed from the lognormal distribution with parameter ψ_j, and ψ_j is estimated. Hence, the ψ's vary randomly from one testing stage to the next and may be assumed to be chosen by "nature" from a probability density function $g(\psi)$ which is typically unknown. This procedure, where data are obtained in stages with estimation of the unknown parameter obtained at each stage based on current and past data without knowledge of the "prior distribution" $g(\psi)$, is referred to as an *empirical Bayes* (EB) procedure.

The object of this section is to present estimates of the unknown reliability function at each time t for the lognormal failure time distribution using empirical Bayes procedures. Empirical Bayes estimation was first introduced by Robbins (1955) and has frequently been used to estimate various parameters in failure models (see Mann, Schafer, and Singpurwalla, 1974, for some references). In particular, Padgett and Robinson (1978) obtained EB estimates of $R(t)$ for the lognormal distribution.

5.1 Estimation Based on the Empirical CDF

In this section the unknown prior distribution will be estimated at the current testing stage by the empirical *cumulative distribution function* (CDF) obtained from the estimates of the parameters in the previous testing stages. It is assumed that both parameters μ and σ^2 are unknown and that the prior density $g(\mu, \sigma^2)$ is to be estimated at each testing stage.

Let $\mathbf{x}_j = (x_{1j}, x_{2j}, \ldots, x_{n_j j})$ be a random sample of size n_j from the $\Lambda(\mu_j, \sigma_j^2)$ distribution at testing stage j, where the (μ_j, σ_j^2) are possibly different at each stage $j = 1, 2, 3, \ldots$. and are independently generated at each stage from $g(\mu, \sigma^2)$. Then

$$\hat{\mu}_j = n_j^{-1} \sum_{i=1}^{n_j} \ln x_{ij} \quad \text{and} \quad \hat{\sigma}_j^2 = n_j^{-1} \sum_{i=1}^{n_j} (\ln x_{ij} - \hat{\mu}_j)^2$$

are the maximum likelihood estimates of μ_j and σ_j^2 for the jth sample.

The Bayes estimator of $R(t)$ based on $\mathbf{x}_j$ is

$$\hat{R}_B(t) = \frac{\int_{-\infty}^{\infty} \int_0^{\infty} \Phi\left(\frac{\mu - \ln t}{\sigma}\right) f\left(\hat{\mu}_j, \hat{\sigma}_j^2 \mid \mu, \sigma^2\right) g\left(\mu, \sigma^2\right) d\sigma^2 d\mu}{\int_{-\infty}^{\infty} \int_0^{\infty} f\left(\hat{\mu}_j, \hat{\sigma}_j^2 \mid \mu, \sigma^2\right) g\left(\mu, \sigma^2\right) d\sigma^2 d\mu} \tag{5.1}$$

where $f(\hat{\mu}_j, \hat{\sigma}_j^2 \mid \mu, \sigma^2)$ is the conditional joint pdf of the sufficient statistics from the jth sample, given μ and σ^2, and $g(\mu, \sigma^2)$ is the prior density of (μ, σ^2). The unknown density $g(\mu, \sigma^2)$ can be estimated by the bivariate empirical CDF, i.e., point masses at the estimated values $(\hat{\mu}_1, \hat{\sigma}_1^2), \ldots, (\hat{\mu}_j, \hat{\sigma}_j^2)$. Then since $\hat{\mu}_j$ and $\hat{\sigma}_j^2$ are independent, and the (conditional) distribution of $\hat{\mu}_j$ (given μ, σ^2) is $N(\mu, \sigma^2/n_j)$ and the (conditional) distribution of $n_j \hat{\sigma}_j^2/\sigma^2$ is chi-square with $n_j - 1$ degrees of freedom, an EB estimator of $R(t)$ is obtained from (5.1) as

$$\hat{R}(t; j) = \frac{\sum_{i=1}^{j} \Phi\left(\frac{\hat{\mu}_i - \ln t}{\hat{\sigma}_i}\right) \hat{\sigma}_i^{-n_j} \exp\left\{-\frac{n_j}{2\hat{\sigma}_i^2}\left[(\hat{\mu}_j - \hat{\mu}_i)^2 + \hat{\sigma}_j^2\right]\right\}}{\sum_{i=1}^{j} \hat{\sigma}_i^{-n_j} \exp\left\{-\frac{n_j}{2\hat{\sigma}_i^2}\left[(\hat{\mu}_j - \hat{\mu}_i)^2 + \hat{\sigma}_j^2\right]\right\}} \qquad t > 0 \tag{5.2}$$

This is simply a weighted average of past and present estimators. Padgett and Robinson (1978) showed that it performs fairly well based on some simulation results.

5.2 A Smooth Empirical Bayes Estimator

Using the techniques of Bennett and Martz (1972), Padgett and Robinson (1978) developed a "smooth" EB estimator for $R(t)$ in the lognormal distribution. Assume first that σ^2 is known and fixed.

For the jth sample, $\hat{\mu}_j$ is the maximum likelihood estimator of μ_j, as before. The kernel estimator with the normal density as the kernel function can be used to estimate the prior density at stage j,

$$g_j(\mu) = \frac{1}{jh(j)\sqrt{2\pi}} \sum_{i=1}^{j} \exp\left[-(\mu - \hat{\mu}_i)^2/2h^2(j)\right] \qquad j = 1, 2, 3, \ldots$$

where $h(j)$ is a suitably chosen function of j. Then the empirical Bayes estimator of the reliability function is given by

$$\begin{aligned} \tilde{R}_j(t;\sigma^2) &= E[R(t) \mid \hat{\mu}_j] \\ &= \frac{\int_{-\infty}^{\infty} R(t;\mu,\sigma^2) f(\hat{\mu}_j \mid \mu) g_j(\mu)\, d\mu}{\int_{-\infty}^{\infty} f(\hat{\mu}_j \mid \mu) g_j(\mu)\, d\mu} \end{aligned} \tag{5.3}$$

Evaluation of the integral in the denominator of (5.3) is straightforward and can be shown to be

$$C_j \left[\frac{\sigma^2 h^2(j)}{\sigma^2 + h^2(j) n_j}\right]^{1/2} \sqrt{2\pi} \sum_{i=1}^{j} \exp\left[-\frac{n_j(\hat{\mu}_j - \hat{\mu}_i)^2}{2(\sigma^2 + h^2(j) n_j)}\right]$$

where $C_j = \sqrt{n_j}/2\pi j h^2(j)\sigma$, $j = 1, 2, 3, \ldots$. After some simplification, the numerator of (5.3) becomes

$$C_j \sqrt{2\pi} \left(\frac{\sigma^2 h^2(j)}{\sigma^2 + h^2(j) n_j}\right)^{1/2} \sum_{i=1}^{j} \exp\left[-\frac{n_j(\hat{\mu}_j - \hat{\mu}_i)^2}{2(\sigma^2 + h^2(j) n_j)}\right] \times E_M\left[\Phi\left(\frac{M - \ln t}{\sigma}\right)\right] \tag{5.4}$$

where M is a random variable which has a normal distribution with mean $\hat{\mu}_{ij}$ and variance $\sigma^2 h^2(j)/[\sigma^2 + h^2(j)n_j]$ and

$$\hat{\mu}_{ij} = \frac{n_j \hat{\mu}_j h^2(j) + \sigma^2 \hat{\mu}_i}{n_j h^2(j) + \sigma^2}$$

By an application of Ellison's (1964, p. 93) result, the expectation in (5.4) becomes

$$\Phi\left[\frac{\hat{\mu}_{ij} - \ln t}{\sigma\sqrt{1 + h^2(j)/(\sigma^2 + h^2(j)n_j)}}\right]$$

Therefore, the empirical Bayes estimator of $R(t)$ for the case that σ^2 is fixed and known is

$$\tilde{R}_j(t;\sigma^2) = \frac{\sum_{i=1}^{j} \Phi\left[\frac{\hat{\mu}_{ij} - \ln t}{\sigma\sqrt{1+h^2(j)/(\sigma^2+h^2(j)n_j)}}\right] \exp\left[-\frac{n_j(\hat{\mu}_j - \hat{\mu}_i)^2}{2(\sigma^2 + h^2(j)n_j)}\right]}{\sum_{i=1}^{j} \exp\left[-\frac{n_j(\hat{\mu}_j - \hat{\mu}_i)^2}{2(\sigma^2 + h^2(j)n_j)}\right]}$$

$$j = 1, 2, \ldots \qquad t > 0 \qquad (5.5)$$

The function $h(j)$ may be chosen to be either

$$h^2(j) = j^{-2/5}\,\mathrm{var}(\hat{\mu}_i) = j^{-2/5}\frac{\sigma^2}{n_i} \qquad i = 1, 2, \ldots, j$$

or

$$h^2(j) = j^{-2/5}\bar{\sigma}_j^2 \qquad \text{where } \bar{\sigma}_j^2 = \frac{1}{j}\sum_{i=1}^{j}\frac{\sigma^2}{n_i}$$

If σ^2 is unknown, the smoothing technique yields a complicated expression which does not seem to perform any better than the procedure of substituting the current estimate $\hat{\sigma}_j$ for σ in the expression (5.5). If σ^2 can be assumed to be approximately the same from one testing stage to another, then substituting the *average* of all past estimates $\hat{\sigma}_i^2$, $i = 1, 2, \ldots, j$, $\bar{\sigma}_j^2 = \frac{1}{j}\sum_{i=1}^{j} \hat{\sigma}_i^2$ into (5.5) is suggested.

5.3 Type II Censored Samples

For Type II censored samples at each stage, that is, the first r_j failure times are recorded out of the n_j items on test at stage j, EB estimates of $R(t)$ are more complicated than in the case of complete samples. However, they can still be numerically calculated.

Denote the first r_j ordered failure times by $x_{1j} \leq \ldots \leq x_{r_j j}$ and $y_{ij} = \ln x_{ij}$. Let $\mathbf{y}_{r_j}$ denote the censored sample of log-times.

For the case that both μ and σ^2 are unknown, the EB estimate of $R(t)$ at the jth testing stage based on the empirical CDF was given by Padgett and Robinson (1978) as

$$\hat{R}(t;j) = \frac{\sum_{i=1}^{j} \Phi\left(\frac{\hat{\mu}_i - \ln t}{\hat{\sigma}_i}\right) \hat{\sigma}_i^{-r_j} \exp(-S_{ij}/2\hat{\sigma}_i^2) \left[\Phi\left(\frac{\hat{\mu}_i - y_{r_j j}}{\hat{\sigma}_i}\right)\right]^{r_j}}{\sum_{i=1}^{j} \hat{\sigma}_i^{-r_j} \exp(-S_{ij}/2\hat{\sigma}_i^2) \left[\Phi\left(\frac{\hat{\mu}_i - y_{r_j j}}{\hat{\sigma}_i}\right)\right]^{r_j}}$$

where $S_{ij} = \sum_{k=1}^{r_j} (y_{kj} - \hat{\mu}_i)^2$ and $\hat{\mu}_j$ and $\hat{\sigma}_j^2$ are solutions to the two nonlinear equations

$$\mu_j = y_{r_j j} + \frac{\sigma_j \phi_j}{\Phi_j}$$

and

$$\sigma_j^2 = r_j^{-1} \sum_{i=1}^{r_j} (y_{ij} - \mu_j)^2 - \frac{r_j \sigma_j \phi_j}{\Phi_j} (\mu_j - y_{r_j j})$$

with $\phi_j = \phi((\mu_j - y_{r_j j})/\sigma_j)$ and $\Phi_j = \Phi((\mu_j - y_{r_j j})/\sigma_j)$, where ϕ denotes the standard normal density function. The solution of these two equations may be obtained using the methods of Harter and Moore (1966).

The smooth EB estimate of $R(t)$, assuming σ^2 is known, is given by

$$\hat{R}_s(t;j) = \frac{\sum_{i=1}^{j} \exp(-b_{ij}) E_{W_i} \left[\Phi\left(\frac{W_i - \ln t}{\sigma}\right) \Phi^{r_j}\left(\frac{W_i - y_{r_j j}}{\sigma}\right)\right]}{\sum_{i=1}^{j} \exp(-b_{ij}) E_{W_i} \left[\Phi^{r_j}\left(\frac{W_i - y_{r_j j}}{\sigma}\right)\right]}$$

where E_{W_i} denotes expected value with respect to W_i which has $N(\hat{\mu}_{ij}, 1/a_j)$ distribution with

$$a_j = \frac{\sigma^2 + r_j h^2(j)}{2\sigma^2 h^2(j)} \qquad \hat{\mu}_{ij} = \frac{r_j \bar{y}_{r_j} h^2(j) + \sigma^2 \hat{\mu}_i}{\sigma^2 + r_j h^2(j)}$$

$$b_{ij} = \frac{\sigma^2 \hat{\mu}_i^2 r_j - r_j^2 \bar{y}_{r_j}^2 h^2(j) - 2\sigma^2 \hat{\mu}_i r_j \bar{y}_{r_j}}{2\sigma^2[\sigma^2 + r_j h^2(j)]}$$

and $\bar{y}_{r_j} = r_j^{-1} \sum_{k=1}^{r_j} y_{k_j}$. The function $h(j)$ may be chosen as in Section 5.2 and the expectations in $\hat{R}_s(t;j)$ may be obtained by numerical integration. If σ^2 is unknown, the estimates of σ^2 may be substituted into $\hat{R}_s$ in a manner similar to that discussed in Section 5.2.

Simulation results of Padgett and Robinson (1978) indicated that in general all of the EB estimators exhibited smaller estimated mean squared errors than the corresponding maximum likelihood estimators. The smooth EB estimators were almost always the best in this sense.

REFERENCES

Bennett, G. K. and Martz, H. F., Jr. (1972). A continuous empirical Bayes smoothing technique, *Biometrika, 59*, 361–368.

Bhattacharya, S. K. (1967). Bayesian approach to life testing and reliability estimation, *J. Amer. Statist. Assoc., 62*, 48–62.

Box, G. E. P., and Tiao, G. C. (1973). *Bayesian Inference in Statistical Analysis*, Addison-Wesley, Reading, MA.

Canavos, G. C. (1973). An empirical approach for the Poisson life distribution, *IEEE Trans. Reliability, R-22*, 91–96.

DeGroot, H. (1970). *Optimal Statistical Decisions*, McGraw-Hill, New York.

Drake, A. W. (1966). "Bayes statistics for the reliability engineer," Proc. 1966 Annual Sympos. Reliability, pp. 315–320.

Dumonceaux, R. and Antle, C. E. (1973). Discrimination between the log-normal and the Weibull distributions, *Technometrics, 15*, 923–926.

Ellison, B. E. (1964). Two theorems for inferences about the normal distribution with applications in acceptance sampling, *J. Amer. Stat. Assoc., 59*, 89–95.

Ferguson, T. S. (1967). *Mathematical Statistics: A Decision Theoretic Approach*, Academic Press, New York.

Gupta, S. S. (1962). Life test sampling plans for normal and lognormal distributions, *Technometrics, 4*, 151–175.

Harter, H. L., and Moore, A. H. (1966). Iterative maximum-likelihood estimation of parameters of normal populations from singly and doubly censored samples, *Biometrika, 53*, 205–213.

Kotz, S. (1973). Normality vs. lognormality with applications, *Communications in Statistics, 1*, 113–132.

Mann, N. R., Schafer, R. E., and Singpurwalla, N. D. (1974). *Methods for Statistical Analysis of Reliability and Life Data*, John Wiley and Sons, New York.

Martz, H. F., and Waller, R. A. (1982). *Bayesian Reliability Analysis*, John Wiley and Sons, New York.

Padgett, W. J. (1978). Comparison of Bayes and maximum likelihood estimators in the lognormal model for censored samples, *Metron, 30-VI*, 79–98.

Padgett, W. J., and Johnson, M. P. (1983). Some Bayesian lower bounds on reliability in the lognormal distribution, *Canadian J. Statist., 11*, 137–147.

Padgett, W. J., and Robinson, J. A. (1978). Empirical Bayes estimators of reliability for lognormal failure model, *IEEE Trans. Reliability, R-27*, 332–336.

Padgett, W, J., and Tsokos, C. P. (1977). Bayes estimation of reliability for the lognormal failure model, *The Theory and Applications of Reliability, Vol. II* (C. P. Tsokos and I. Shimi, eds.), Academic Press, New York, p. 133–161.

Padgett, W. J., and Wei, L. J. (1977). Bayes estimation of reliability for the two-parameter lognormal distribution, *Commun. Statist.-Theor. Meth., A6*, 443–447.

Padgett, W. J. and Wei, L. J. (1978). Bayesian lower bounds on reliability for the lognormal model, *IEEE Trans. Reliability, R-27*, 161–165.

Pearson, E. S., and Hartley, H. O. (1972). *Biometrika Tables for Statisticians. Volume II*, Cambridge Univ. Press.

Robbins, H. (1955). An empirical Bayes approach to statistics, *Proc. 3rd Berkeley Sympos. Probability and Statist., Vol. 1*, pp. 157–163.

Soland, R. M. (1969). Bayesian analysis of the Weibull process with unknown scale and shape parameters, *IEEE Trans. Reliability, R-18*, 181–184.

Tsokos, C. P. (1972). Bayesian approach to reliability using the Weibull distribution with unknown parameters and its computer simulation. *Rep. Stat. Appl. Res., Japanese Union of Scientists and Engineers, 19*, 123–134.

Zellner, A. (1971). Bayesian and non-Bayesian analysis of the log-normal distribution and log-normal regression, *J. Amer. Stat. Assoc., 66*, 327–330.

7

Poisson-Lognormal Distributions

S. A. SHABAN Faculty of Commerce, Economics and Political Science, Department of Insurance and Statistics, Kuwait University, State of Kuwait

1. DEFINITION AND PROPERTIES

The Poisson-lognormal distribution—hereafter referred to as P-LN—is defined as the mixture of Poisson distributions with parameter λ, where λ follows a lognormal distribution with parameters μ and σ^2: λ is $\Lambda(\mu, \sigma^2)$. Thus the probability mass function for the P-LN is given by

$$P(X = r) = P_r(\mu, \sigma) = \frac{1}{\sigma\sqrt{2\pi}}\frac{1}{r!}$$

$$\int_0^\infty e^{-\lambda}\lambda^{r-1}\exp\left\{-\frac{(\ln\lambda - \mu)^2}{2\sigma^2}\right\} d\lambda \qquad r = 0, 1, 2, \ldots \quad (1.1)$$

This integral cannot be expressed in a simpler form. Also, an explicit expression for the generating function is not known. The kth factorial moment is

$$\mu_{[k]} = \alpha^k \omega^{k^2/2} \qquad k = 1, 2, \ldots, \quad (1.2)$$

where

$$\alpha = e^{\mu}$$
$$\omega = e^{\sigma^2}$$

The first four moments about the origin are

$$\mu_1'(X) = \alpha\sqrt{\omega} \tag{1.3}$$
$$\mu_2'(X) = \alpha\sqrt{\omega}\left[\alpha\omega\sqrt{\omega} + 1\right] \tag{1.4}$$
$$\mu_3'(X) = \alpha\sqrt{\omega}\left[\alpha^2\omega^4 + 3\alpha\omega\sqrt{\omega} + 1\right] \tag{1.5}$$
$$\mu_4'(X) = \alpha\sqrt{\omega}\left[\alpha^3\omega^7\sqrt{\omega} + 6\alpha^2\omega^4 + 7\alpha\omega\sqrt{\omega} + 1\right] \tag{1.6}$$

Central moments are

$$\mu_2(X) = \alpha\sqrt{\omega}\left[\alpha\sqrt{\omega}(\omega - 1) + 1\right] \tag{1.7}$$
$$\mu_3(X) = \alpha\sqrt{\omega}\left[\alpha^2\omega(\omega - 1)^2(\omega + 2) + 3\alpha\sqrt{\omega}(\omega - 1) + 1\right] \tag{1.8}$$
$$\begin{aligned}\mu_4(X) = \alpha\sqrt{\omega}\big[&\alpha^3\omega\sqrt{\omega}(\omega - 1)^2(\omega^4 + 2\omega^3 + 3\omega^2 - 3)\\ &+ 6\alpha^2\omega(\omega - 1)(\omega^2 + \omega - 1) + \alpha\sqrt{\omega}(7\omega - 4) + 1\big]\end{aligned} \tag{1.9}$$

The first two moment-ratios are

$$\begin{aligned}\beta_1 = &\left[\alpha^2\omega(\omega - 1)^2(\omega + 2) + 3\alpha\sqrt{\omega}(\omega - 1) + 1\right]^2\\ &\div \alpha\sqrt{\omega}\left[1 + \alpha\sqrt{\omega}(\omega - 1)\right]^3\end{aligned} \tag{1.10}$$
$$\begin{aligned}\beta_2 = 3 + \big[&\alpha^3\omega\sqrt{\omega}(\omega - 1)^3(\omega^3 + 3\omega^2 + 6\omega + 6) + 6\alpha^2\omega(\omega - 1)^2\\ &\times (\omega + 2) + 7\alpha\sqrt{\omega}(\omega - 1) + 1\big] \div \alpha\sqrt{\omega}\left[\alpha\sqrt{\omega}(\omega - 1) + 1\right]^2\end{aligned} \tag{1.11}$$

Hence the distribution is positively skewed. We note that

$$\begin{aligned}\frac{\beta_2 - 3}{\beta_1} = &\left[\alpha\sqrt{\omega}(\omega - 1) + 1\right]\big[\alpha^3\omega\sqrt{\omega}(\omega - 1)^3(\omega^3 + 3\omega^2 + 6\omega + 6)\\ &+ 6\alpha^2\omega(\omega - 1)^2(\omega + 2) + 7\alpha\sqrt{\omega}(\omega - 1) + 1\big]\\ &\div \left[\alpha^2\omega(\omega - 1)^2(\omega + 2) + 3\alpha\sqrt{\omega}(\omega - 1) + 1\right]^2\end{aligned} \tag{1.12}$$

Values of the ratio (1.12) for various values of ω—fixing $\alpha = 1$—i.e., $\mu = 0$, are shown in the following table:

σ	0.1	0.5	1	2	3	4	5
$(\beta_2 - 3)/\beta_1$	1.020	1.488	2.931	53.825	8102.1	8.886(6)*	7.2(10)*

* $7.2(10) = 7.2 \times 10^{10}$

The ratio is an increasing function of $\omega = e^{\sigma^2}$. This behavior is similar to that of the lognormal distributions.

Since λ follows $\Lambda(\mu, \sigma^2)$, which is the probability density function of a positive, unimodal and absolutely continuous variable, then according to the theorem due to Holgate (1970), the non-negative, integer-valued random variable defined by (1.1) is said to have a unimodal lattice distribution. (Lattice distributions (univariate) are a class of discrete distributions in which the intervals between values of the random variable for which there are non-zero probabilities are integral multiples of one quantity (Johnson and Kotz 1970, p. 31)).

2. HISTORICAL REMARKS AND GENESIS

The Poisson-lognormal distribution was first suggested as a model for commonness of species by Preston (1948). The analysis leading him to suggest this distribution was as follows: Suppose that the number of each species caught in a trap is assumed to be a single sample from a Poisson distribution with mean λ. Now λ itself will vary among species, and if the variation in λ were lognormal, the number of species, represented by $r = 1, 2, \ldots$ individuals, would conform to a truncated P-LN. Preston, however, termed this distribution "a truncated lognormal."

Anscombe (1950) investigated the sampling properties of eight distributions used in sampling butterflies, and in the collection of moths by means of a light-trap. These distributions include the P-LN, negative binomial, Thomas distribution, Fisher *Hh* distribution, Polya-Aeppli distribution and Neyman distributions of types *A*, *B* and *C*. Ordering these distributions according to their skewness and kurtosis, the P-LN is the most skewed and leptokurtic distribution. Anscombe pointed out that the P-LN distribution suffers from the disadvantage that its probability function involves an untabulated integral. Holgate (1969) noted that a clear distinction between the P-LN and the lognormal has not always been maintained. An example may be found in the work of Bliss (1966), who explained the advantage of the P-LN, then proceeded to fit the lognormal distribution to five sets of data from moth trap experiments. However, it was argued by Bliss that after trapping has been in operation for a long time, the distribution of the empirical abundance in the collection will be close to the abundance distribution of the population. This implies that the sequence of normalized P-LN distributions converges to the lognormal. Similar findings were reported in Cassie (1962), who proposed the P-LN distribution as a model for plankton ecology. Cassie (1962) indicated that for most plankton populations, samples with $\mu = 100$ would not depart significantly from a lognormal distribution.

3. TABLES

Grundy (1951) published tables for the P-LN of $(1 - P_0)$ and P_1 for the range of values of $\mu \log_{10} e = -2(0.25)3.0$ and $\sigma^2 = 2(1)16$, to four decimal places. The tables of Brown and Holgate (1971), however, are more detailed and present the probabilities P_j for $j = 0(1)\infty$ and $m = 0.1(0.1)0.9$, $V = m + 0.1(0.1)1.0$, in Table 1 and $m = 1(1)9$, $V = m + 1(1)10$ in Table 2 to five decimal places, where m and V are the P-LN's mean and variance respectively, namely

$$m = \alpha\sqrt{\omega}$$
$$V = m^2(\omega - 1) + m$$

4. APPROXIMATIONS

There is a considerable variety of approximations to the individual probabilities. Bulmer (1974) has obtained the following approximation for large r:

$$P_r \cong \frac{1}{\sigma\sqrt{2\pi}}\frac{1}{r}\exp\left\{-\frac{1}{2\sigma^2}(\log r - \mu)^2\right\} \times \left[1 + \frac{1}{2r\sigma^2}\left(\frac{(\log r - \mu)^2}{\sigma^2} + \log r - \mu - 1\right)\right] \tag{4.1}$$

This approximation—as stated by Bulmer—has a relative error less than 10^{-3} when $r \geq 10$, for values of μ and σ^2 likely to be encountered in practice.

Holgate (1969) has reported that, if μ is held constant and $\sigma^2 \to 0$, the dominant term in the expansion of Laplace transform of the P-LN distribution gives Poisson probabilities, and if the next term is included, then

$$P_r \simeq \exp(-e^{\mu})e^{r\mu}\left[1 + \sigma^2\left\{(r - e^{\mu})^2 - e^{\mu}\right\}\right]/r! \tag{4.2}$$

This is a probability distribution if $\sigma^2 \leq e^{-\mu}$. A crude bound for these probabilities is

$$P_1 \geq \tfrac{1}{2}\left[e^{\mu+1/2\sigma^2}\right]\exp\left\{-e^{\mu+\sigma^2}\right\} \tag{4.3}$$

$$P_r \leq \frac{1}{r!}e^{r\mu+r\sigma^2-r^2\sigma^2/2} \qquad r > 1 \tag{4.4}$$

An approximation of similar form can be suggested using a general result developed by Paul and Plackett (1978) for approximating the probabilities of the compound Poisson distribution. This formula is

$$P_r = e^{-\theta}\frac{\theta^r}{r!} + \tfrac{1}{2}\nu\nabla^2\left(e^{-\theta}\frac{\theta^r}{r!}\right)$$

where ∇^2 stands for the second backward difference operator (operates on r) and θ and ν are the mean and variance respectively of the compounder. Applying this formula to the P-LN we obtain

$$P_r = e^{-m}\frac{m^r}{r!} + \frac{1}{2}\nu\nabla^2\left(e^{-m}\frac{m^r}{r!}\right) \tag{4.5}$$

where

$$\begin{aligned} m &= e^{\mu+\sigma^2/2} \\ &= \alpha\sqrt{\omega} \end{aligned}$$

and

$$\nu = m^2(\omega - 1)$$

Note that (4.5) does not define a probability mass function. The second term may be considered as a correction to the Poisson density similar to the correction given by Holgate (1969) in equation (4.2).

This suggests that the Poisson density may also be considered as an approximation to the P-LN.

To study the performance of the above approximations, a numerical investigation was carried out. As a measure of the degree of fit we define the quantity

$$\varepsilon^2 = \sum_i (p_i - p_i^*)^2/p_i \tag{4.6}$$

where p_i and p_i^* are the probabilities of the P-LN and the approximating distribution respectively; small values of ε^2 indicate a better fit (Johnson and Kotz (1970) p. 221). The exact probabilities of P–LN were obtained from the Brown and Holgate (1971) tables. In calculating ε^2, we included all values of $P_i > 0$. The study covers the range of values of (m, V) considered in Brown and Holgate tables. The following summarizes the results of this study.

(1) For certain values of (m, V), these approximations—except the Poisson—do not define proper probabilities (negative values or probabilities exceeding one occur). To be specific, we give below the (m, V) combinations that produce improper probabilities for the different approximations under consideration.

Bulmer's approximation (4.1):

(a) $m = 0.3, 0.5, 0.6, 0.7, 0.9, 3, 4, 5, 6, 7, 8, 9$	and	$V = m + 1$
(b) $m = 2, 3, 4, 5, 6, 7, 8$	and	$V = m + 2$
(c) $m = 5, 6, 7$	and	$V = m + 3$

Holgate's approximation (4.2):

(a) $m = 2$	and	$V = 8, 9, 10$
(b) $m = 3$	and	$V = 9, 10$
(c) $m = 4$	and	$V = 10$

Paul and Plackett's (4.5):

(a) $m = 0.1, 0.2, 0.3, 0.4$	and	$V = m + 0.5(.1)1.0$
(b) $m = 0.1, 0.2, 0.3$	and	$V = m + 0.4$
(c) $m = 0.1, 0.2$	and	$V = m + 0.3$
(d) $m = 0.1$	and	$V = m + 0.2$
(e) $m = 1, 2$	and	$V = m + 5(1)10$
(f) $m = 2, 3$	and	$V = 4$
(g) $m = 3$	and	$V = 10$

(2) The Poisson approximation is more efficient (possesses smaller ε^2) than Holgate's approximation for all values of (m, V) considered.

(3) In the range of values of (m, V) for which it is a proper probability, Paul and Plackett's approximation is more efficient than the Poisson approximation, except in the following cases:

(a) $m = 0.1, 0.2$	and	$V = m + 0.1$
(b) $m = 0.3$	and	$V = m + 0.2(0.1)0.3$
(c) $m = 0.4$	and	$V = m + 0.3(0.1)0.4$
(d) $m = 0.5$	and	$V = m + 0.4(0.1)0.5$
(e) $m = 1$	and	$V = m + 1(1)2$
(f) $m = 2$	and	$V = m + 3(1)4$
(g) $m = 3$	and	$V = m + 5(1)6$

(4) Bulmer's approximation is valid only for P_r, $r > 0$. To study its efficiency, we considered the P-LN and Poisson density with P_0 omitted, and ε^2 is then calculated. The study showed that Bulmer's approximation is more efficient than Poisson in the following cases:

(a) $m = 0.1$ and $V = m + 0.1(0.1)1.0$
(b) $m = 0.2$ and $V = m + 0.1(0.1)1.0$
(c) $m = 0.3$ and $V = m + 0.3(0.1)1.0$
(d) $m = 1$ and $V = m + 1(1)10$
(e) $m = 2$ and $V = m + 6(1)10$
(f) $m = 3$ and $V = m + 5(1)10$

5. ESTIMATION OF PARAMETERS

5.1 Moment Estimators

The simplest method of estimating the parameters μ and σ^2 is to equate the sample mean and variance to the corresponding population values. From equations (1.3) and (1.4) it follows that the estimators of μ and σ^2 based on the method of moments are

$$\hat{\sigma}_m^2 = \log\left(\frac{m_2' - m_1'}{m_1'^2}\right) \tag{5.1}$$

and

$$\hat{\mu}_m = \log \frac{m_1'^2}{\sqrt{m_2' - m_1'}} \tag{5.2}$$

where

$$m_1' = \bar{x} \qquad m_2' = \frac{\sum x_i^2}{n}$$

The statistic $\sqrt{n}(\hat{\mu}_m - \mu, \hat{\sigma}_m^2 - \sigma^2)'$ converges to a bivariate normal with mean $\mathbf{0}$ and covariance matrix $\sum_{\mathrm{ME}}$ where

$$\sum_{\mathrm{ME}} = G\Lambda G' \tag{5.3}$$

$$\Lambda = \begin{bmatrix} \mu_2 - \mu_1^2 & \mu_3 - \mu_2\mu_1 \\ \mu_3 - \mu_2\mu_1 & \mu_4 - \mu_2^2 \end{bmatrix}$$

and

$$\mu_i = E(x^i) \qquad i = 1, 2, 3, 4$$

also

$$G = \left[\frac{\partial g_i(\mu_1, \mu_2)}{\partial \mu_j} \right] \qquad i, j = 1, 2$$

where

$$g_1(\mu_1, \mu_2) = \mu = \log(\mu_1^2 / \sqrt{\mu_2 - \mu_1})$$
$$g_2(\mu_1, \mu_2) = \sigma^2 = \log\left((\mu_2 - \mu_1)/\mu_1^2\right)$$

Calculating the elements of G and Λ, we find that

$$\begin{aligned} V(\hat{\mu}_m) &= (ae + bf)a + (af + bg)b \\ \mathrm{Cov}(\hat{\mu}_m, \hat{\sigma}_m^2) &= (ae + bf)c + (af + bg)d \\ V(\hat{\sigma}_m^2) &= (ce + df)c + (cf + dg)d \end{aligned}$$

where

$$\begin{aligned} a &= (1 + 4\alpha\omega\sqrt{\omega}) \div (2\alpha^2\omega^2) \\ b &= -(2\alpha^2\omega^2)^{-1} \\ c &= -(1 + 2\alpha\omega\sqrt{\omega}) \div (\alpha^2\omega^2) \\ d &= (\alpha^2\omega^2)^{-1} \\ e &= \alpha\sqrt{\omega}(1 + \alpha\sqrt{\omega}(\omega - 1)) \\ f &= \alpha\sqrt{\omega}(\alpha^2\omega^2(\omega^2 - 1) + \alpha\sqrt{\omega}(3\omega - 1) + 1) \end{aligned}$$

and

$$\begin{aligned} g = \alpha\sqrt{\omega}(&\alpha^3\omega^3\sqrt{\omega}(\omega^4 - 1) \\ &+ 2\alpha^2\omega^2(3\omega^2 - 1) + \alpha\sqrt{\omega}(7\omega - 1) + 1) \end{aligned}$$

5.2 Maximum Likelihood Estimators

The maximum likelihood estimation in the case of truncated P-LN was discussed by Bulmer (1974). The likelihood function for the non-truncated case is

$$L = \sum n_r \log P_r \tag{5.4}$$

and the likelihood equations are

$$\sum_{r=0} \frac{n_r}{P_r} [rP_r - (r+1)P_{r+1}] = 0$$

$$\sum_{r=0} \frac{n_r}{P_r} \left[r^2 P_r - (r+1)(2r+1)P_{r+1} + (r+1)(r+2)P_{r+2}\right] = 0$$

If we write $\theta_1 = \mu$, $\theta_2 = \sigma^2$, then the variance-covariance matrix of the maximum likelihood estimates for large sample size is

$$\sum_{\text{MLE}} = \left[- \left(\frac{\partial^2 L}{\partial\theta_i \partial\theta_j} \right)_{\hat{\theta}=\theta} \right]^{-1} \tag{5.5}$$

where

$$\frac{\partial^2 L}{\partial\theta_i^2} = \sum \frac{n_r}{P_r} \left[\frac{\partial^2 P_r}{\partial\theta_i^2} - \frac{1}{P_r} \left[\frac{\partial P_r}{\partial\theta_i} \right]^2 \right] \qquad i = 1, 2$$

$$\frac{\partial^2 L}{\partial\theta_1 \partial\theta_2} = \sum \frac{n_r}{P_r} \left[\frac{P_r}{\partial\theta_1 \partial\theta_2} - \frac{1}{P_r} \left[\frac{\partial P_r}{\partial\theta_1} \frac{\partial P_r}{\partial\theta_2} \right] \right]$$

To evaluate the above expressions we need the second partial derivatives, $\partial^2 P_r/\partial\theta_1^2$, $\partial^2 P_r/\partial\theta_2^2$, $\partial^2 P_r/\partial\theta_1\partial\theta_2$. Their explicit forms are

$$\frac{\partial^2 P_r}{\partial\theta_1^2} = r^2 P_r - (r+1)(2r+1)P_{r+1} + (r+1)(r+2)P_{r+2}$$

$$\frac{\partial^2 P_r}{\partial \theta_2^2} = \tfrac{1}{4}\left[P_r - (r+1)(2r+1)(2r(r+1)+1)P_{r+1} + (r+1)(r+2)\{2r(r+1)+(2r+1)(2r+3)+1\}P_{r+2} - 4(r+1)^2(r+2)(r+3)P_{r+3} + (r+1)(r+2)(r+3)(r+4)P_{r+4}\right]$$

$$\frac{\partial^2 P_r}{\partial \theta_1 \partial \theta_2} = \tfrac{1}{2}\left\{r^3 P_r - (r+1)(3r(r+1)+1)P_{r+1} + 3(r+1)^2(r+2)P_{r+2} - (r+1)(r+2)(r+3)P_{r+3}\right\}$$

Obtaining the maximum likelihood estimates and their asymptotic variance-covariance matrix requires numerical integration routines together with optimizing techniques. Reid (1981) reported that the relative efficiency of the moment estimates relative to the maximum likelihood as measured by the quantity $\det(\sum_{\mathrm{MLE}})/\det(\sum_{\mathrm{ME}})$ ranges from 1% to 39% for the particular applications given in his paper.

6. RELATED DISTRIBUTIONS

6.1 Truncated and Censored P-LN

If P_j denote the probabilities of the untruncated distribution, then the probabilities for the zero truncated distribution are

$$p_j = \frac{P_j}{1 - P_0} \qquad j = 1, 2, \ldots$$

The truncated distribution is widely used in species abundance studies, since n_0, the number of species which are not represented at all in the collection, is usually unknown.

The censored distribution arises when the frequencies $n_1, n_2, \ldots, n_k$ are known for $k \le K$, and only the total number of species with more than K is known.

Probability values and parameter estimates for both truncated and censored distributions can be obtained by suitable modifications of the results in previous sections.

6.2 Poisson-Lognormal Convolution (P-LNC) and Negative Binomial Lognormal Convolution (NB-LNC)

Uhler and Bradley (1970) proposed two models to describe a phenomenon in petroleum exploration. These two models used the sum of lognormal random variables as a compounder of Poisson and negative binomial distributions. The assumptions under which these models were derived are as follows:

(1) The spatial occurence of petroleum reservoirs can be represented by a Poisson process.
(2) The sizes of the individual reservoirs are lognormally distributed.
(3) The total of petroleum reserves is distributed as a sum of lognormal random variables where the number in the sum is determined by a Poisson process.

More precisely, let $\{X_k\}$ be a sequence of mutually independent and identically distributed lognormal variables. We are interested in sums

$$Z(N) = X_1 + X_2 + \cdots + X_N \tag{6.1}$$

where the number N of terms is a random variable independent of the x_j. Let

$$g(n) = P(N = n) \tag{6.2}$$

be the distribution of N. For the distribution of $Z(N)$ we get from the fundamental formula for conditional probabilities

$$f(z) = \sum_{n=0}^{\infty} f(z \mid n)g(n) \tag{6.3}$$

Since the distribution of the sum of lognormal variates cannot be expressed in explicit form, so also $f(z)$ cannot. However, the moments of any order of $f(z)$ can be obtained when $g(n)$ is specified. Accordingly

(1) If $g(n)$ is Poisson with parameter λ, we may term the resulting distribution a Poisson-Lognormal convolution (P-LNC). The first four moments of this distribution are

$$\mu_1'(Z) = \lambda\alpha\sqrt{\omega} \tag{6.4}$$

$$\mu_2'(Z) = \lambda\alpha^2\omega(\omega + \lambda) \tag{6.5}$$

$$\mu_3'(Z) = \lambda\alpha^3\omega\sqrt{\omega}(\omega^3 + 3\lambda\omega + \lambda^2) \tag{6.6}$$

$$\mu_4'(Z) = \lambda\alpha^4\omega^2(\omega^6 + 4\lambda\omega^3 + 3\lambda\omega^2 + 6\lambda^2\omega + \lambda^3) \tag{6.7}$$

Hence the central moments are

$$\mu_2(Z) = \lambda\alpha^2\omega^2 \tag{6.8}$$

$$\mu_3(Z) = \lambda\alpha^3\omega^4\sqrt{\omega} \tag{6.9}$$

$$\mu_4(Z) = \lambda\alpha^4\omega^4(\omega^4 + 3\lambda) \tag{6.10}$$

The moment-ratios are

$$\beta_1 = \frac{\omega^3}{\lambda} \tag{6.11}$$

$$\beta_2 = 3 + \frac{\omega^4}{\lambda} \tag{6.12}$$

Hence the distribution is positively skew. Further, as $\lambda \to \infty$, (β_1, β_2) approaches the normal point (0, 3). We note that β_1, β_2 do not depend on the lognormal parameter μ, and

$$\frac{\beta_2 - 3}{\beta_1} = \omega \tag{6.13}$$

which is an increasing function of σ^2 similar to the behavior reported in section 1 for the P-LN.

(2) If N is Poisson with parameter λ, and if we assume that λ itself varies—due to the tendency for petroleum reservoirs to cluster—and in addition it follows a gamma distribution with parameters a and γ, then

$$\begin{aligned} g(n) = P(N = n) &= \int_0^\infty e^{-\lambda}\frac{\lambda^n}{n!}\frac{a^\gamma}{\Gamma(\gamma)}\lambda^{\gamma-1}e^{-a\lambda}\,d\lambda \\ &= \frac{\Gamma(n+\gamma)}{\Gamma(n+1)\Gamma(\gamma)}\left(\frac{1}{1+a}\right)^n\left(\frac{a}{1+a}\right)^\gamma \qquad n = 0, 1, 2, \ldots \end{aligned} \tag{6.14}$$

which is negative binomial with $p = 1/(1+a)$, $q = 1 - p = a/(1+a)$. Therefore we may term the resulting distribution a negative bionomial-lognormal convolution (NB–LNC). If we let $Q = p/q$, then the first four moments of this distribution are

$$\mu_1'(Z) = \gamma Q\alpha\sqrt{\omega} \tag{6.15}$$

$$\mu_2'(Z) = \gamma Q\alpha^2\omega\left[\omega + Q(\gamma+1)\right] \tag{6.16}$$

$$\mu_3'(Z) = \gamma Q\alpha^3\omega\sqrt{\omega}\left[\omega^3 + 3(\gamma+1)Q\omega + (\gamma+1)(\gamma+2)Q^2\right] \tag{6.17}$$

$$\begin{aligned} \mu_4'(Z) = \gamma Q\alpha^4\omega^2 \big[&\omega^6 + 4(\gamma+1)Q\omega^3 + 3(\gamma+1)Q\omega^2 \\ &+ 6(\gamma+1)(\gamma+2)Q^2\omega + (\gamma+1))(\gamma+2)(\gamma+3)Q^3\big] \end{aligned} \tag{6.18}$$

Hence the central moments are

$$\mu_2(Z) = \gamma Q\alpha^2\omega\,[\omega + Q] \tag{6.19}$$

$$\mu_3(Z) = \gamma Q\alpha^3\omega\sqrt{\omega}\,\left[\omega^3 + 3Q\omega + 2Q^2\right] \tag{6.20}$$

$$\mu_4(Z) = \gamma Q\alpha^4\omega^2\left[\omega^6 + 4Q\omega^3 + 3(\gamma+1)Q\omega^2 + 6(\gamma+2)Q^2\omega + 3(\gamma+2)Q^3\right] \tag{6.21}$$

The moment-ratios β_1 and β_2 are

$$\beta_1 = \frac{(\omega^3 + 3\omega Q + 2Q^2)^2}{\gamma Q(\omega + Q)^3} \tag{6.22}$$

$$\beta_2 = 3 + \frac{\omega^6 + 4Q\omega^3 + 3Q\omega^2 + 12Q^2\omega + 6Q^3}{\gamma Q(\omega + Q)^2} \tag{6.23}$$

The distribution is positively skew. Neither β_1 nor β_2 depends on the lognormal parameter μ.

6.3 Poisson-Inverse Gaussian Distribution (P-IG)

The probability mass function of the P-IG distribution is given by

$$P_x(\varsigma, \phi) = \frac{1}{x!}\sqrt{\frac{2\varsigma\phi}{\pi}}\exp(\phi)\left[\frac{\varsigma\phi}{2(1+\varsigma\phi/2)}\right]^{x-1/2} \cdot K_{x-1/2}\left[\sqrt{2\varsigma\phi(1+\phi/2\varsigma)}\right] \qquad x = 0, 1, 2, \ldots \tag{6.24}$$

where $K_\gamma(Z)$ represents a modified Bessel of the third kind of order γ and ς and ϕ are the inverse Gaussian parameters (Shaban, 1981). It is expected that this distribution can be used in cases where the Poisson-Lognormal (P-LN) is applicable. This is conjectured because of the similarity between the lognormal and the inverse Gaussian distribution. Besides the P-IG is more tractable than the P-LN. A numerical study was performed to evaluate the performance of the P-IG as an approximation to the P-LN.

The mean and variance of P-IG, which are equal to ς and $\varsigma\phi^{-1}(\varsigma + \phi)$, respectively, were matched with the values of m and V, the mean and variance of P-LN in the Brown and Holgate tables. P-IG probabilities were calculated using these values. As a measure of the degree of fit we use the quantity ε^2, as defined in equation (4.6). The values of ε^2 were compared with the values obtained using the other approximations given in Section 4. The following is a summary of the results:

(1) The P-IG is more efficient than the Poisson, Paul and Plackett (4.5), and Holgate (4.2) approximations for all values of (m, V) considered in the Brown and Holgate tables.

(2) Since Bulmer's approximation (equation (4.1)) is not defined for $r = 0$, we compare ε^2 of P-IG with P_0 omitted with that of the Bulmer approximation. Bulmer's approximation is found to be more efficient than P-IG in the following cases:

(a) $m = 0.1$ and $V = m + 0.2(0.1)1.0$,
(b) $m = 0.2$ and $V = m + 0.3(0.1)1.0$,
(c) $m = 0.3$ and $V = 1.0$.

As an example of our comparison we give the following two tables.

In Table 1, the mode of the P-LN is at $r = 0$, while in Table 2, it is at $r = 5$. It is of interest to note the mode of P-IG occurs at the same points as for P-LN.

Table 1 Comparison of P-LN with P-IG, Poisson, Paul and Plackett, Holgate, and Bulmer Approximations ($m = 0.1$, $V = 0.2$)

	$P(X = r)$					
r	P-LN (1.1)	P-IG (6.24)	Poisson	Paul and Plackett (4.5)	Holgate (4.2)	Bulmer (4.1)
0	.92387	.92941	.90484	.95008	.90226	N/d†
1	.06211	.05366	.09048	.00453	.09313	.05172
2	.00950	.01049	.00452	.04094	.00451	.00689
3	.00254	.00353	.00015	.00423	.00010	.00194
4	.00094	.00147	–	.00022	–	.00076
5	.00043	.00069	–	.00001	–	.00036
6	.00022	.00034	–	–	–	.00019
7	.00013	.00018	–	–	–	.00011
8	.00008	.00010	–	–	–	.00007
9	.00005	.00005	–	–	–	.00005
10	.00003	.00003	–	–	–	.00003
11	.00002	.00002	–	–	–	.00002
12	.00002	.00001	–	–	–	.00002
13	.00001	.00001	–	–	–	.00001
14	.00001	–*	–	–	–	.00001
15	.00001	–	–	–	–	.00001
16	.00001	–	–	–	–	.00001
ε^2		.00226	.02014	.16087	.02289	.06220

* A dash, –, means probability is zero to five decimal places.

† N/d means the approximation is not defined for $r = 0$.

Table 2 Comparison of P-LN with P-IG, Poisson, Paul and Plackett, Holgate, and Bulmer Approximations ($m = 6$, $V = 10$)

	$P(X = r)$					
r	P-LN	P-IG	Poisson	Paul and Plackett	Holgate	Bulmer
0	.00870	.00867	.00248	.00744	.01286	N/d†
1	.03402	.03406	.01487	.03470	.05221	.00009
2	.07159	.07176	.04462	.07932	.10034	.07105
3	.10791	.10809	.08924	.11898	.12065	.26690
4	.13084	.13086	.13385	.13385	.10355	.13558
5	.13589	.13573	.16062	.12493	.07570	.01798
6	.12573	.12547	.16062	.10708	.06536	.05490
7	.10643	.10619	.13768	.09179	.07520	.10570
8	.08401	.08387	.10326	.08031	.08862	.11180
9	.06271	.06269	.06884	.06884	.09054	.08901
10	.04476	.04483	.04130	.05507	.07816	.06055
11	.03080	.03092	.02253	.04005	.05784	.03746
12	.02058	.02070	.01126	.02628	.03741	.02184
13	.01343	.01352	.00520	.01560	.02149	.01226
14	.00859	.00865	.00223	.00842	.01112	.00671
15	.00541	.00544	.00089	.00416	.00524	.00362
16	.00336	.00338	.00033	.00189	.00227	.00194
17	.00207	.00207	.00012	.00080	.00091	.00103
18	.00126	.00125	.00004	.00031	.00034	.00055
19	.00077	.00075	.00001	.00012	.00012	.00029
20	.00046	.00045	–*	.00004	.00004	.00016
21	.00028	.00027	–	.00001	.00001	.00008
22	.00017	.00016	–	–	–	.00005
23	.00010	.00009	–	–	–	.00003
24	.00006	.00005	–	–	–	.00001
25	.00004	.00003	–	–	–	.00001
26	.00002	.00002	–	–	–	–
27	.00001	.00001	–	–	–	–
28	.00001	.00001	–	–	–	–
ε^2		.000018	.08492	.01975	.17916	.44126

* A dash, –, means probability is zero to five decimal places.

† N/d means the approximation is not defined for $r = 0$.

REFERENCES

Anscombe, F.J. (1950). Sampling theory of the negative binomial and logarithmic series distributions, *Biometrika*, *37*, *358–382*.

Bliss, C.I. (1966). An analysis of some insect trap records. *Sankhyā*, *A28*, 123–136.

Brown, S. and Holgate, P. (1971). Table of the Poisson-log normal distribution. *Sankhyā*, *B33*, 235–258.

Bulmer, M.G. (1974). On fitting the Poisson lognormal distribution to species-abundance data, *Biometrics*, *30*, 101–110.

Cassie, R.M. (1962). Frequency distribution models in the ecology of plankton and other organisms, *Journal of Animal Ecology*, *31*, 65–92.

Grundy, P.M. (1951). The expected frequencies in a sample of an animal population in which the abundances of species are log-normally distributed, Part I. *Biometrika*, *38*, 427–434.

Holgate, P. (1969). Species frequency distributions, *Biometrika*, *56*, 651–660.

Holgate, P. (1970). The modality of some compound Poisson distributions, *Biometrika*, *57*, 666–667.

Johnson, N.L. and Kotz, S. (1970). *Distributions in Statistics: Discrete Distributions*, John Wiley and Sons, N.Y.

Paul, S.R. and Plackett, R.L. (1978). Inference sensitivity for Poisson mixtures, *Biometrika*, *65*, 591–602.

Preston, F.W. (1948). The commonness, and rarity, of species, *Ecology*, *29*, 254–283.

Reid, D.D. (1981). The Poisson lognormal distribution and its use as a model of plankton aggregation, *Statist. Distr. Sci. Work*, *6*, 303–316.

Shaban, S.A. (1981). Computation of the Poisson-inverse Gaussian distribution, *Commun. Statist. Theor. Meth.*, *A10*, 1389–1399.

Uhler, R.S. and Bradley, P.G. (1970). A stochastic model for determining the economic prospects of petroleum exploration over large regions, *J. Amer. Statist. Assoc.*, *65*, 623–630.

8

The Lognormal as Event-Time Distribution

RAYMOND J. LAWRENCE Department of Marketing, University of Lancaster, Bailrigg, Lancaster, England

1. INTRODUCTION

The time to an event is distinguished from other stochastic processes occurring in time, often expressed as differential equations, by use of the term "point processes." The terminology was developed by Wold and explained by Cox and Miller (1965, p. 338):

> Roughly speaking, we consider a process as a point process when interest is concentrated on the individual occurrences of the events themselves, events only being distinguished by their position in time. Thus, although the linear birth-death process is characterized by point events, namely the births and deaths, we would not normally think of it as a point process, since we are usually interested in the number of individuals alive at a particular time rather than directly in the instants at which births and deaths occur.

The distinction is therefore a matter of emphasis or interest. Cox and Isham (1980, p. 3) pointed out that it was not a sharp one, and that any stochastic process in continuous time in which the sample paths are step functions is associated with a point process.

In applications, the lognormal is used with point process data. Interest is on the timing of events and the probability of one or more events happening by a given time. Independent variables are often introduced by their effect on the mean time to an event. Time is the metric of processes, and the lognormal lends itself to explanation of the underlying process in terms of Kapteyn's law—random or manifold factors producing change multiplicatively. Many authors refer explicitly to this interpretation.

The case would be different if time was not simply a metric but a *causative* factor. In everyday language a man can be said to die of old age, implying that age caused his death, although a doctor's certificate would assign some more immediate cause, such as heart failure. Laurent (1975) took up an idea first introduced in biology by Teissier in 1934, that mortality can be due to "pure ageing," without lethal accidents or specific causes of death. His theory used the concept of life potential, subject to a failure rate through time. Conceptually, the individual starts at birth or at some threshold time with a stockpile of life units, which are gradually depleted. The assumptions led to a model based on exponentials with a shape like a normal curve but slightly right-skewed. Normally, however, investigators treat time simply as the metric of underlying processes.

2. EVENTS IN MEDICAL HISTORIES

2.1 The Incubation Period of Disease

The incubation or latent period of an infectious disease is the time from the initial infection to the onset of symptoms in the host. It is the period during which the parasitic organism is developing.

Sartwell (1950) studied seventeen sets of data on incubation periods collected from identified sources. More than one series was available for measles, typhoid fever, chickenpox and serum hepatitis. Plotting the data in grouped time intervals on lognormal probability paper, he concluded that the frequency curve for most infectious diseases resembled a lognormal curve, equally for diseases with very short and very long incubation periods. A small minority, including a typhoid epidemic, proved exceptions. The typical σ was about 0.4. Hill (1963) later remarked on the little variation in σ even between quite different diseases, but criticized the value of the third parameter representing the latency period in Sartwell's fitting of $X' = X - \tau$ since the date of infection was often uncertain. He derived the likelihood of τ and applied it to 18th century data on smallpox symptoms, where the infection began at a known inoculation date.

In a major study, Kondo (1977) found 86 epidemics in the literature, including 11 of those used by Sartwell, 32 Japanese cases, and 13 reported to the Center for Disease Control in Atlanta, Georgia. Multiple series were available for food poisoning, bacillary dysentery, typhoid and scarlet fevers, serum hepatitis, measles, and the common cold. Kondo selected cases where the date of infection was known and the population exposed was homogeneous (in one nearly ideal example, 5,000 young soldiers at Camp Polk, Louisiana, were all inoculated on the same day from the same batch of faulty vaccine, 20% of them developing serum hepatitis). Classified by goodness of fit, 61 of the 86 examples, or about 70%, could be accepted as lognormal, either noncentral or central with $\tau = 0$, at the 5% level of significance or better. The remaining cases were rejected. Kondo concluded that the lognormal distribution in general represented the incubation time. He quoted the mutually independent causes acting multiplicatively in sequence explanation, and ran simulations to show the results of assuming two types of genes combined in varying proportions as causal agents, leading to lognormal distributions of incubation times.

2.2 Time to Recovery

After the onset of a disease, two main events occur: the patient either recovers or he dies. There is evidence that the lognormal can represent the elapsing time period in both cases.

The earliest work seems to be the study of the duration of periods of sickness undertaken in the context of industrial insurance claims. Amoroso fitted seven data series by the lognormal in 1934 (for references see Chapter 9, section 2.4). Oldham (1965) studied average days off work among 300 patients treated for chronic bronchitis, and found the distribution quite closely lognormal. He was more concerned to analyse the effects of alternative treatments than to give details of fit. He gave a reference from the medical literature to a lognormal distribution of physicians' consultation times.

Eaton and Whitmore (1977) analysed records of length of stay in hospital for 2,311 schizophrenic patients taken from the Maryland Psychiatric Case Register over seven years. The data base was sufficiently large to classify hospitalization into 32 time intervals. The mixed exponential, Type XI, Weibull, gamma, lognormal and inverse Gaussian (IG) distributions were fitted to the data using simple parameter estimates taken from two or three sample quantiles. The lognormal and IG produced the best fits by a Kolmogorov type of test statistic, the former being marginally better and supported by three other papers quoted by the authors. They preferred the IG distribution, however, because of its interpretation as Brownian motion with drift towards an absorbing barrier. The schizophrenic patient exemplified a 'drift towards health.' They found it difficult to conceptualize the individual-level process in the case of the lognormal, although they had quoted the many factors affecting length of stay: seriousness of illness, frequency of case review by physicians, availability of bedspace, and so on.

The Eaton and Whitmore study was strongly criticized by Von Korff (1979) for the inefficiency of the parameter estimation methods used. He refitted the mixed exponential using maximum likelihood estimators and obtained a fit comparable to those given by the other two distributions, and better by the chi-squared test. The tails in particular were better fitted, although it is to be noted that the parameters of the other distributions were not revised.

In a smaller study, Balintfy (1962) fitted the lognormal to the time spent in the therapeutic, post-operative phase by 165 surgical patients at the Johns Hopkins hospital. Heterogeneity among patients in terms of seriousness of operation, age, state of health and so on was catered for by suggesting a lognormal distribution of mean times of recovery, with each patient having a lognormal distribution of discharge time given his mean. The resulting lognormal had a variance which was the sum of the mean time and individual variances (σ_0^2 and σ^2 in Aitchison and Brown (1957), p. 110). However, the important requirement that σ^2 should be constant for all patients was not examined.

2.3 Time to Death

Time to death rather than recovery was first studied statistically by Boag (1949). He collected data on the survival time of cancer patients in four hospitals, with several hundred cases in each of eight sets, and found that the data appeared to agree well with the lognormal specification. The exponential distribution was tested and found to provide a better fit in one case, cancer of the breast, but there were reasons to expect that the discrepancy would disappear with a longer data base. Since a proportion, c, of cancer patients is permanently cured while the remaining $(1 - c)$ are liable to die of cancer if they do not die previously of other causes, Boag gave the maximum likelihood estimators of μ, σ^2, and c for the lognormal case.

Feinleib (1960) was able to quote seven references from specialist medical journals where the distribution of duration of survival in several diseases rather closely approximated the lognormal. His own data covered survival times of 234 male patients with chronic lymphocytic leukemia. Months-to-death were recorded from the date of diagnosis of the disease, but Feinleib found that survival times required a three-parameter lognormal. His estimate of τ in the $X - \tau$ distribution was $\tau = 4$ months. He was careful to point out that the statistical estimate of τ did not necessarily indicate the time between the onset of the disease and its diagnosis. However, if a biological interpretation was desired, τ might be described as the apparent origin in time at which the cumulative factors which lead to death began to act.

As survival data generally include patients still alive at the end of the observation period, the three-parameter distribution is also right censored. Feinleib gave two useful estimators. If p is the proportion of survivors after the last observation period ending at t, the mean time-to-death t^* to ascribe to those survivors (in moment estimation for example) is by integration

$$t^* = \frac{\exp(-t^2/2)}{p\sqrt{2\pi}}$$

A shortcut estimate of τ is given by Gaddum's note that if z_1, z_2 and z_3 are three values from a lognormal distribution which are equidistant on the normal deviate scale then $z_2 - \tau$ is the geometric mean of $z_1 - \tau$ and $z_3 - \tau$. An approximation of τ is therefore

$$\tau^* = \frac{z_1 z_3 - z_2^2}{z_1 + z_3 - 2z_2}$$

Using lognormal probability paper, z_2 can be taken as the median of the sample distribution and z_1 and z_3 can be taken from percentiles equidistant from z_2.

Death from cancer was also studied by Whittemore and Altschuler (1976), who gave six references to discussion in medical journals of the lognormal as survival time distribution. Their own work was on the smoking histories of over 34,000 male doctors in Britain in 1951, with changes in smoking habits recorded in 1956 and 1966. Figures for years to death by lung cancer were fitted by the Weibull and lognormal distributions, using hazard functions to estimate parameters. The distributions were found to approximate the facts equally well.

Farewell and Prentice (1979) used data on survival times of 137 lung cancer patients previously analysed by exponential and Weibull models, finding that the cases needed to be divided into two groups. For the 97 patients who received no prior therapy, the test to discriminate among Weibull, gamma and lognormal models, which was the main feature of Farewell and Prentice's article, marginally indicated a lognormal rather than a Weibull distribution. Bennett (1983) later applied the log-logistic to the same data for 97 patients, noting that the distribution was very similar in shape to the lognormal. He advocated a distribution with a nonmonotonic hazard function, as the course of the disease was such that mortality reaches a peak after a finite period and then slowly declines. The Weibull did not provide this but the log-logistic did so. It may be noted that the lognormal also has a hazard function of the desired shape.

3. LOGNORMAL EVENT-TIMES IN BUSINESS AND ECONOMICS

3.1 Time in a Job: Renewal Processes

Labor turnover rates have been studied as a measure of work force stability. Lane and Andrew (1955) developed a full stochastic model for employment within an organization, allowing for entrants, leavers, and a length of service distribution. Figures were compiled from labor department records of completed length of service in four branches of the United Steel Companies Limited in the UK. The numbers of employees in the data sets ranged from 2,400 to over 7,000, with durations of employment from one month to twenty years or more. All four sets were well fitted by the lognormal, σ varying from 2.15 to 2.62 but μ showing a much wider range, -0.46 to 1.35, measured in years. The latter statistic was suggested for use as a comparative measure of labor stability. The lognormal also fitted satisfactorily two sets of completed service data covering a two-year period in the Glacier Metal Company which had previously been fitted by Pearson Type XI curves (Silcock, 1954) and were later fitted by two further models (Bartholomew, 1959, Clowes, 1972). Lane and Andrew broke employees into subgroups, fitting lognormals separately to those living in their own homes and in hostels, single and married workers, and younger against older entrants, bringing out stability differences between the groups. In the discussion following the presentation of the paper, Mr. J. Aitchison suggested that the displaced lognormal $X - \tau$, with τ equal to half a month, would improve the fit. He thought that Kapteyn's law might hold: the chance of an employee with x years service remaining in service for another y years depended only on the ratio y/x.

Bartholomew (1967) took Kapteyn's law as implying that the length of time which an employee spends in his current job would be a random multiple of the time he spent in his previous job. In his book, Bartholomew developed a renewal model for an organization and was perhaps the first to consider the lognormal as event-time distribution in a renewal process, although it had been mentioned as a possible candidate by Cox (1962).

Renewal theory began with the study of components such as light bulbs which are replaced immediately on failure, but developed on more general lines. If the time-to-failure of a component is a random variable, X, with probability density function $f(x)$, the sum of n such independent lifetimes is

$$S_n = X_1 + X_2 + \cdots + X_n$$

so that S_n is the total failure time of n successive components, or the convolution sum of the X_is. A post in a large organization can meet renewal theory conditions: each incumbent of a post has a probability $f(x)$ of leaving or being transferred, to be replaced at once by another person with the same leaving probability, and so on. A basic concept of renewal theory is the renewal density, which represents the underlying rate of the process. In a job context, it is simultaneously the rate of recruitment and the rate of leaving or wastage, since by definition each event involves both the gain of a new person and the loss of the previous one. If $h(t)$ is the probability of a gain or loss in the small interval $(t, t + dt)$, the integral equation of renewal theory is

$$h(t) = f_1(t) + \int_0^t h(t - x) f(x)\, dx$$

The right-hand side represents the probability of the first renewal in $(t, t + dt)$ plus the probability of a later renewal in that interval. The case $f_1(x) = f(x)$ defines an ordinary renewal process, with the time to the first event distributed identically to the others—in particular, starting at $x = 0$. In these circumstances it is known that the solution for $h(t)$ is $h(t) = 1/\alpha$ as t goes to infinity, where α is the mean of $f(x)$.

The convolution indicated by the integral can be handled by Laplace transforms in many cases, but the lognormal does not have one. Bartholomew (1963, 1967) found an approximation for the integral in the form

$$\frac{F^2(t)}{\int_0^t [1 - F(x)]\, dx}$$

In the case of the lognormal, the denominator integrates to (using $z = (\log t - \mu)/\sigma$)

$$t\,[1 - \Phi(z)] + \alpha\Phi(z - \sigma)$$

Bartholomew demonstrated that, with the lognormal, the renewal density approaches its equilibrium value very slowly. With $\mu = 0$, giving the median time in a job as one year, and $\sigma = 2$, as found in practice, the density takes nearly twenty years to reach half its value after one year and is still about twice its equilibrium value. The mixed exponential, in contrast, reaches equilibrium in a matter of a few years. Bartholomew used this distribution, with its simple Laplace transform, when fitting his model to actual company data.

3.2 Durations of Strikes

Another aspect of work force activity is its propensity to continue a strike. Lawrence (1984) fitted the lognormal to many sets of data on the duration of strikes in the U.K. taken from government statistics. Figures for strikes in 1965 were available for eight separate industries. Surprisingly, the variance of strike durations could be taken as the same in all cases ($\sigma^2 = 1.95$). Also, $\mu = 0.6$ applied to all industries except Transport. Left-truncated distributions were fitted as the number of stoppages lasting one day or less were not fully recorded. This allowed a test of the lognormal against the Inverse Gaussian (IG) distribution, which had been used by Lancaster (1972) to fit the same data. When extrapolated to the left of the truncation point, the lognormal predicted more than the under-recorded numbers; the IG failed to do so. The difference between the two distributions was mainly near the origin for the parameter values in question.

Lawrence found that the lognormal fitted U.K. strike durations for every year in the period 1975–82, with some doubt about 1978 (the "Winter of Discontent"). The estimates of μ and σ^2 varied from year to year but appeared to be negatively correlated: when one increased, the other fell. The modal strike duration $\exp(\mu - \sigma^2)$ therefore fluctuated much more widely than the mean $\exp(\mu+\sigma^2/2)$, in which the contrary movements were offsetting. American strike data for 1961, analysed by Horvath (1968) as displaced Weibull distributed, could be taken as lognormal up to a duration of 140 days. Thereafter the lognormal overpredicted. It was suggested that U.S. records of strikes *settled* failed to include strikes which did not end in settlement, as when they petered out or plants shut down and reopened elsewhere after a long stoppage.

The intensity function (or hazard function) of a lognormal distribution reaches its maximum at a time which is the solution for t of

$$\frac{d\Lambda(t)/dt}{1-\Lambda(t)} = \frac{\log t + \sigma^2 - \mu}{\sigma^2 t}$$

With the parameters for 1965 British strikes, the maximum occurred at $t = 0.4$ days. It was pointed out that when grouped data are used, an apparently monotonically declining intensity could be misleading as the maximum could be lost within the first data period (Morrison and Schmittlein (1980) made this error).

Lawrence also looked at the mean residual lifetime in the lognormal case to answer the question "Given a strike has already lasted t days, how much longer is it expected to continue?" He gave graphs showing the path of expected lifetime curves and calculated points at which the expected future duration was less than elapsed time as under two months in Britain but perhaps $2\frac{1}{2}$ years for American strikes. The small dip immediately after the origin in the lognormal mean residual lifetime curve, pointed out by Watson and Wells (1961), was illustrated and discussed.

3.3 Time to Purchase

The lognormal has been applied to times to first purchase in a population of potential buyers and to interpurchase times for consumer goods kept permanently in stock.

Bain (1964) examined the growth path of television ownership in the U.K. He referred to previous work in which the logistic, representing a "simple epidemic" diffusion process, had been fitted to ownership data but found the symmetry of the curve inappropriate. The Gompertz, with its maximum rate of growth (point of inflexion) always proportionate to a parameter, was thought too inflexible for general use. In a cross-section analysis, the lognormal fitted survey research data on the growth of TV ownership both in total and for subgroups categorized by socio-demographics such as social class and household size. The estimates of σ were the same for all subclasses except regional groups, but the exception could be due to differences in time-origin. Given constant σ, the mean adoption time was taken in a time-series analysis as influenced by variables representing the introduction of commercial television, credit restrictions, prices of sets, and even the coronation of Elizabeth II in 1953. The effects of the independent variables were multiplicative. Overall, an R^2 of 0.78 was obtained.

Bonus (1973) added three further references to the lognormal growth of TV ownership: in the US (an MIT dissertation by W.F. Massy in 1960), in Germany and in Finland (originals in German and Swedish). Bonus himself had data for 33 cross-sections of ownership of consumer durables over the period 1956-67, taken from nationally representative samples of West German households. For each cross-section, a lognormal distribution was fitted to household income. The standard deviation turned out to be roughly constant at $\sigma = 0.5$, while means increased over time after cost-of-living adjustment. Bonus was interested in establishing the conditions whereby sometimes the lognormal and sometimes the logistic curve fitted the growth of ownership. With F_t for the cumulative proportion of owners by time t, he used quasi-Engel curve analysis to propose the model

$$F_t = F_t^* \int_0^\infty \Lambda(\alpha \mid \mu_{1t}, \sigma_1^2)\, d\Lambda(\alpha \mid \mu_{2t}, \sigma_2^2)$$

The distribution at the right was the income distribution at time t, with the time subscript omitted from σ_2^2 as it was found to be roughly constant. The distribution to its left represented the probability of ownership as a function of income, also with time subscript omitted as the data permitted treating the parameter as a constant, though varying from one good to another. F_t^* was the proportion of potential owners in the population, with asymptote at a ceiling ownership level. By the convolution property of the lognormal integral,

$$\frac{F_t}{F_t^*} = \Phi(y_t \mid 0, 1)$$

with

$$y_t = \frac{-\mu_{1t} + \mu_{2t}}{\sqrt{\sigma_1^2 + \sigma_2^2}}$$

With $F_t^* = 1$, as with most goods, ownership growth would appear to be logistically distributed in view of the very close correspondence between the normal and logistic curves. With $\mu_{1t} = \mu_1$, a constant, growth would be fully induced by the increase in median income μ_{2t}. No "diffusion process" at all need be assumed. Bonus called this Type I growth and found that it applied to cameras and projectors. Income had increased exponentially in West Germany over the period as a linear μ_{2t} required. Type II growth occurred when μ_{1t} declined faster than μ_{2t} increased, interpreted as a shift downward in the median income at which the goods became purchased, perhaps due to declining prices or a revision of plans among potential buyers to allow for actual ownership at a lower income. The growth curve would again appear like the logistic but more steeply rising, as in the case of refrigerators, automobiles and vacuum cleaners. Type III growth applied to TV sets and washing machines, where F_t^* could not be set to 1 but instead had to be considered time-dependent: more and more households became susceptible to the innovation, and market potential increased, possibly due to learning effects or genuine diffusion. The growth curve would be closer to the log-logistic and hence to the lognormal. Bonus had used the log-logistic or logit model in lieu of the lognormal for parameter estimation purposes, with a considerable gain in mathematical tractability.

In interesting contrast, Lawrence worked with fast-moving consumer goods and with not-new products, i.e., familiar products with quality, flavor or packaging improvements which spread progressively but did not "create a usage situation for the first time" (Lawrence 1985). His model used four assumptions:

(1) A population of continuous stockists of a product class, conforming to renewal theory requirements by repurchasing a product as soon as the old product was finished, to keep a stock on hand for use as required.
(2) A lognormal distribution $\Lambda(\alpha \mid \mu_0, \sigma_0^2)$ of α, the mean product lifetime and repurchase period, within the stockist population.
(3) A lognormal distribution of time-to-purchase for each household with given α, $\Lambda(t \mid \mu, \sigma^2)$.
(4) A constant individual purchase time variance, σ^2.

Renewal theory gives a simple form to the expected number of events in time t from a random start (as opposed to a start at time $t = 0$ or taking the origin at the occurrence of a previous event, as in the ordinary renewal process referred to above). It is t/α, where α is the mean time between events. If α is $\Lambda(\mu_0, \sigma_0^2)$ then $1/\alpha$ is $\Lambda(-\mu_0, \sigma_0^2)$ and the distribution of the expected number of events in time t is $\Lambda(t/\alpha \mid \log t - \mu_0, \sigma_0^2)$. This was the distribution which Lawrence (1980) found applicable to his toothpaste buying data, including subgroups of the population, as discussed in Chapter 9, Section 3.2.

The probability of a purchase occurring by time t, measured from a random start at $t = 0$, is

$$\int_0^\infty \Lambda\left(t \mid \log\alpha - \frac{\sigma^2}{2}, \sigma^2\right) d\Lambda(\alpha \mid \mu_0, \sigma_0^2) = \Lambda\left(t \mid \mu_0 - \frac{\sigma^2}{2}, \sigma^2 + \sigma_0^2\right)$$

Aggregated over a panel of N purchasers, this lognormal needs multiplying by Nt/α_0, with $\alpha_0 = \exp(\mu_0 - \sigma_0^2/2)$, to give the proportion of total purchases bought by time t. Lawrence (1982) was able to show that the theoretical derivation held for over 18,000 toothpaste purchasing trips made in one year. He did not estimate parameters directly but used his previous estimates of μ_0 and σ_0^2. He was in the fortunate position of being able to examine σ^2, the missing individual purchase time variances, since purchases recur and are not single events such as death which allow no repeated observations of the same individual and so no σ^2 estimate. Grouping buyers by frequency of purchasing trips, he was able to show the distribution of σ^2 in 2,700 households. The distributions became more sharply peaked and narrower around a central value with increasing numbers of trips, as would be expected. But they conformed to the theoretical chi-squared distribution of sample variances; and the central value could be taken as the same in all cases: $\sigma^2 = 0.43$. The variances for all households could be taken as one constant, irrespective of purchase frequency.

The same article introduced a formula of perhaps wider application. With a consumer panel, the early and repeat buyers tend to be heavy buyers. The mean inter-purchase time as calculated from short period records is therefore biassed in their favor and unrepresentative of the panel as a whole. Using a Bayesian method, the general mean α_0 can be estimated from the observed mean by t, m, using

$$E(\alpha_0 \mid m) = am^b \text{ where } a = \exp \frac{\sigma^2(\sigma_0^2 + \mu_0)}{(\sigma^2 + \sigma_0^2)}$$

$$b = \frac{\sigma_0^2}{\sigma^2 + \sigma_0^2}$$

Only in the extreme case of $\sigma^2 = 0$, i.e., all buyers buy at exactly regular intervals, is it true that $E(\alpha_0 \mid m) = m$.

Renewal theory gives the only form of $f_1(t)$, the density of time to the first event, which is consistent with a renewal process of constant intensity $1/\alpha$. The density is $[1 - F(t)]/\alpha$, where $F(t)$ is the distribution function of the interevent time distribution and α is its mean (see Cox, 1962). Lawrence (1979) gave the lognormal version of the density integrated over the whole range of α as

$$\frac{1}{\alpha}\int_0^\infty [1 - \Lambda(t \mid \mu, \sigma^2)]\, d\Lambda(\alpha \mid \mu_0, \sigma_0^2) = \frac{1 - \Lambda(t \mid \mu_0 - \sigma_0^2 - \sigma^2/2, \sigma^2 + \sigma_0^2)}{\alpha'}$$

with $\alpha' = \exp(\mu_0 - \sigma_0^2/2)$, and noted that mean and variance estimators would not be independent because they contained elements in common. Integration of the density yields the expected time to first purchase in the population as

$$\left(\frac{t}{\alpha'}\right)\left[1 - \Lambda\left(t \mid \mu_0 - \sigma_0^2 - \frac{\sigma^2}{2}, \sigma^2 + \sigma_0^2\right)\right] + \Lambda\left(t \mid \mu_0 + \frac{\sigma^2}{2}, \sigma^2 + \sigma_0^2\right)$$

Taken from a time origin random in relation to any individual's repurchase cycle, such as a particular date or the start of a panel data period, this expression represents the expected penetration or proportion of people first buying by time t. Lawrence (1985) fitted the model to recorded first purchases at a new dairy products store and (1979) related the curve to the negative exponential. By simulation he found that a negative exponential would almost always closely approximate the lognormally-derived penetration curve, but only when fitted retrospectively. If progressively estimated through time as buying data became available, negative exponentials would fit but with changing parameter values, since they varied from the lognormal penetration path.

4. OTHER EVENT-TIMES IN HUMAN AFFAIRS

4.1 Time to Marriage and Divorce

A few investigators have looked at the distribution of time to marriage. Wicksell (1917) first suggested the three-parameter lognormal and developed much of the theory when studying the distribution of the ages of men and women married in Sweden during the years 1901–10. The threshold parameter was needed since the distribution of time-to-marriage does not start with birth at $t = 0$ but in Western societies requires a delay until puberty or a legal minimum age. One practical difficulty was noted by Pretorius (1930), who observed that there was unquestionably a misstatement of ages by persons under 21, the chief motive being the avoidance of legal requirements. He was using data on the ages of bride and groom at marriage in Australia, 1907–14, and found that neither Pearson curves nor the lognormal fitted the figures satisfactorily. However, government statistics on marriage ages confuse first with second and later marriages, so that the data do not represent a single time distribution. Similarly the ages of mothers and fathers at the birth of a child, which Pretorius attempted to fit with lognormal and Pearson curves, cover more children than the first.

Preston (1981) looked at marriage ages, together with data of his own on the height of nests in forests and the life of tumblers in a restaurant, in putting forward an alternative distribution to the lognormal based on exponentials. He noted that marriage age in official statistics was not necessarily the age at first marriage, but assumed that in most cases it would be. A graph for Danish women illustrated a close fit for the lognormal, and it was described as fairly good for U.K. women in 1973. U.S. Census reports for 1966 related to first marriages, and again the fit was good. Preston took the zero points for the distributions as 15 years of age.

The end of marriage in divorce was studied by Agrafiotis (1981) using data for ten cohorts of U.S. couples married between 1949 and 1958. Their divorces were recorded for each of the succeeding nine years. The lognormal proved a good representation of all ten series. It was more accurate than two other models: the exhaustible Poisson used by another researcher to account for marriage times, and the mover-stayer model (Blumen et al., 1955).

4.2 Periods on Welfare Dependency and Recidivism

Amemiya and Boskin (1974) applied the lognormal to the duration of welfare dependency. A State of California survey reported households receiving welfare payments over the period 1965–70. The analysis followed 658 households which came on welfare in the first month of the study, and the total number of months spent on aid was treated as the dependent variable. Because welfare payments came to an end after five years, a right-truncated distribution was fitted.

The independent variables were socio-demographic details of households such as age, health and preschool child together with indexes computed to represent the expected market wage and the expected duration of unemployment for the household head, and non-wage income. These were taken as a K-component vector of unknown constants, x_t, entering into the arithmetic means and variances of time-on-welfare distributions with weights β, a K-component vector of unknown parameters. The log-transformed time-on-welfare variable then had

$$E(\log y_t) = \log \beta' x_t - \frac{\sigma^2}{2} \qquad \mathrm{Var}(\log y_t) = \sigma^2 = \log(1 + \eta^2)$$

Truncation was allowed for by setting an upper limit on the range of y_t. The authors set out the maximum likelihood estimators for σ^2 and the βs. Solution was by Newton-Raphson iteration and problems in choosing the starting values were discussed (see also Amemiya, 1973). Substantively, expected market wage rates were found to have the most important influence on the duration of time on aid, among the independent variables chosen by the researchers.

The methods used by Amemiya and Boskin were applied by Witte and Schmidt (1977) to the length of time between release from prison and reconviction, using a sample of 600 men released in North Carolina in 1969 or 1971. In the follow-up period which varied but averaged 37 months, three-quarters of the sample had returned to jail so right-truncated distributions were used. The lognormal was reported to fit the data fairly well, and rather better than the exponential, although some subgroups were less satisfactorily predicted. The study concentrated on the effects of explanatory variables such as age, race, addiction, and the number of previous convictions.

4.3 Lengths of Telephone Calls

It seems that the length of telephone calls may be lognormally distributed, with the interesting feature that two lognormals are needed to account for the data. Fowlkes (1979) used the mechanically-recorded lengths of nearly 65,000 calls at a switching office, divided into 31 one-minute segments and an over 30 minute group. The histogram was not bimodal, i.e., there were no two peaks to indicate a mixture. To illustrate wider applications in fields such as biology and reliability testing, figures for the lengths of two different strains of protozoon and test-to-failure of lasers were given. Fowlkes worked through the stages of the analysis. Plotted on lognormal paper, the cumulative distribution of the two populations combined shows an S-shaped curve, both ends of the S tailing off to straight lines as one of the two populations comes to dominate the data. The most effective plotting to reveal the presence of a mixture is to take the sample mean and variance to form the unit normal variates z_i corresponding to each of the n observations. These are taken as the x coordinates for plotting with y coordinates $\Phi(z) - p_i$, with $p_i = (i - 1/2)/n$. Fowlkes showed with simulated data how the resulting plot approximated a straight line on the null hypothesis of no mixture, i.e., one normal distribution. The shape of the plot when a mixture is present was illustrated and discussed. Plotting also helps to establish an initial estimate of p, the proportion of one population in the mixture, crucial because of local minima encountered in the maximum likelihood and least squares estimators. The data proportions can then be adjusted to p_i/p and $(p_i - p)/(1 - p)$ respectively for the two populations, i.e., a rough allowance for one set being truncated from above and the other from below, to allow first estimates of the two means and variances. With these starting values, a minimization routine for the likelihood functions can be used, found by simulation to have more bias but a smaller standard deviation than the least squares method. Empirically, the telephone calls were well fitted by two lognormals assuming 80% of short calls with median length 1.9 minutes and 20% of long calls of median 13.7 minutes. As Fowlkes pointed out, the analysis was valuable in planning a rate structure for calls which accurately reflected telephone use.

4.4 Problem-Solving Times

The lognormal has been considered for the distribution of time to solve a problem within a group of people.

Brée (1975) began by questioning a stochastic model put forward by Restle and Davis (1962), who suggested that a problem was solved in strictly sequential stages. If the probabilities that a subject solved a stage were identical for all stages and all individuals, a gamma distribution of problem solving times might be expected. Brée thought these assumptions highly unrealistic and found that improved parameter estimates worsened the fit obtained in the original study. His own data recorded the times taken by 73 students to solve the Missionaries and Cannibals problem, wherein a river must be crossed in a two-person boat. The gamma fitted the data well but with a parameter of 2, implying two stages in the problem solving process. The data included "thinking aloud" protocols as students worked their way through the problem, whose solution required 11 boat trips, and Brée was unable to find any way that enlightenment could dawn in just two stages. A negative exponential with a shift of origin fitted the data well, as did the lognormal. Brée discussed the multiple-factor theory of intelligence and the lognormal as a form of null hypothesis, in that many factors affecting solution time might interact, leading to the lognormal representing "some kind of chance process." However, a likelihood ratio test favored the exponential so that Brée accepted problem solving as a single-stage birth process operating after a delay equivalent to the fastest possible solution time, the same for all subjects.

REFERENCES

Agrafiotis, G.K. (1981). A lognormal model for divorce by marriage cohort, *Hum. Relations, 34*, 835–845.

Aitchison, J. and Brown, J.A.C. (1957). *The Lognormal Distribution*, Cambridge University Press, Cambridge.

Amemiya, T. (1973). Regression analysis when the variance of the dependent variable is proportional to the square of its expectation, *J. Amer. Statist. Assoc., 68*, 928–934.

Amemiya, T. and Boskin, M. (1974). Regression analysis when the dependent variable is truncated lognormal, with an application to the determinants of the duration of welfare dependency, *Internat. Econ. Rev., 15*, 485–496.

Bain, A.D. (1964). *The Growth of Television Ownership in the United Kingdom Since the War*, University of Cambridge Department of Applied Economics Monograph No. 12, Cambridge University Press, Cambridge.

Balintfy, J.L. (1962). Mathematical models and analysis of certain stochastic processes in general hospitals, Dissertation for the School of Engineering, The Johns Hopkins University, Baltimore.

Bartholomew, D.J. (1959). Note on the measurement and prediction of labour turnover, *J. Roy. Statist. Soc. A, 122*, 232–238.

Bartholomew, D.J. (1963). An approximate solution of the integral equation of renewal theory, *J. Roy. Statist. Soc. B, 25*, 432–441.

Bartholomew, D.J. (1967). *Stochastic Models for Social Processes*, John Wiley & Sons, London-New York-Sydney-Toronto.

Bennett, S. (1983). Log-logistic regression models for survival data, *Appl. Statist.*, *32*, 165–171.

Blumen, I., M. Kogan, and P. J. McCarthy (1955). *The Industrial Mobility of Labor as a Probability Process*, Cornell University Press, Ithaca, New York.

Boag, J.W. (1949). Maximum likelihood estimates of the proportion of patients cured by cancer therapy, *J. Roy. Statist. Soc. B*, *11*, 15–53.

Bonus, H. (1973). Quasi-Engel curves, diffusion, and the ownership of major consumer durables, *J. Political Econ.*, *81*, 655–677.

Brée, D.S. (1975). The distribution of problem-solving times: an examination of the stages model, *Brit. J. Math. and Statist. Psychol.*, *28*, 177–200.

Clowes, G.A. (1972). A dynamic model for the analysis of labour turnover, *J. Roy. Statist. Soc. A*, *135*, 242–256.

Cox, D.R. (1962). *Renewal Theory*, Methuen & Co. Ltd., London.

Cox, D.R. and Isham, V. (1980). *Point Processes*, Chapman and Hall Ltd., London-New York.

Cox, D.R. and Miller, H.D. (1965). *The Theory of Stochastic Processes*, Methuen & Co. Ltd., London.

Eaton, W.W. and Whitmore, G.A. (1977). Length of stay as a stochastic process: a general approach and application to hospitalization for schizophrenia, *J. Math. Sociol.*, *5*, 273–292.

Farewell, V.T. and Prentice, R.L. (1979). A study of distributional shape in life testing, *Technometrics*, *19*, 69–75.

Feinleib, M. (1960). A method for analyzing log-normally distributed survival data with incomplete follow-up, *J. Amer. Statist. Assoc.*, *55*, 534–545.

Fowlkes, E.B. (1979). Some methods for studying the mixture of two normal (lognormal) distributions, *J. Amer. Statist. Assoc.*, *74*, 561–575.

Hill, B.M. (1963). The three-parameter lognormal distribution and Bayesian analysis of a point-source epidemic, *J. Amer. Statist. Assoc.*, *58*, 72–84.

Horvath, W.J. (1968). Statistical model for the duration of wars and strikes, *Behav. Sci.*, *13*, 18–28.

Kondo, K. (1977). The lognormal distribution of the incubation time of exogenous diseases, *Jap. J. Human Genet.*, *21*, 217–237.

Lancaster, T. (1972). A stochastic model for the duration of a strike, *J. Roy. Statist. Soc. A*, *135*, 257–271.

Lane, K.F. and Andrew, J.E. (1955). A method of labour turnover analysis, *J. Roy. Statist. Soc. A*, *118*, 296–314.

Laurent, A.G, (1975). Failure and mortality from wear and ageing. The Teissier model, *Statistical Distributions in Scientific Work*, Vol 2 (Patil, G.P., Kotz, S. and Ord, J.K., eds), Reidel Publishing Co., Dordrecht-Boston, 301–320.

Lawrence, R.J. (1979). The penetration path, *European Res.*, *7*, 98–108.

Lawrence, R.J. (1980). The lognormal distribution of buying frequency rates, *J. Marketing Res.*, *17*, 212–220.

Lawrence, R.J. (1982). A lognormal theory of purchase incidence, *European Res.*, *10*, 154–163.

Lawrence, R.J. (1984). The lognormal distribution of the duration of strikes, *J. Roy. Statist. Soc., A, 147*, 464–483.

Lawrence, R.J. (1985). The first purchase: models of innovation, *Marketing Intell. and Planning, 3*, 57–72.

Morrison, D.G. and Schmittlein, D.C. (1980). Jobs, strikes, and wars: probability models for duration, *Org. Behav. and Human Performance, 25*, 224–251.

Oldham, P.D. (1965). Note: on estimating the arithmetic means of lognormally-distributed populations, *Biometrics, 21*, 235–239.

Preston, F.W. (1981). Pseudo-lognormal distributions, *Ecology, 62*, 355–364.

Pretorius, S.J. (1930). Skew bivariate frequency surfaces, examined in the light of numerical illustrations, *Biometrika, 22*, 109–223.

Restle, F. and Davis, J.H. (1962). Success and speed of problem solving by individuals and groups, *Psychol. Rev., 69*, 520–536.

Sartwell, P.E. (1950). The distribution of incubation periods of infectious disease, *Amer. J. Hygiene, 51*, 310–318.

Silcock, H. (1954). The phenomenon of labour turnover, *J. Roy. Statist. Soc. A, 117*, 429–440.

Von Korff, M. (1979). A statistical model of the duration of mental hospitalization: the mixed exponential distribution, *J. Math. Sociol., 6*, 169–175.

Watson, G.S. and Wells, W.T. (1961). On the possibility of improving the mean useful life of items by eliminating those with short lives, *Technometrics, 3*, 281–298.

Whittemore, A. and Altschuler, B. (1976). Lung cancer incidence in cigarette smokers: further analysis of Doll and Hill's data for British physicians, *Biometrics, 32*, 805–816.

Wicksell, S.D. (1917). On logarithmic correlation with an application to the distribution of ages at first marriage, *Medd. Lunds Astr. Obs.*, No. 84.

Witte, A.D. and Schmidt, P. (1977). An analysis of recidivism, using the truncated lognormal distribution, *Appl. Statist., 26*, 302–311.

9

Applications in Economics and Business

RAYMOND J. LAWRENCE Department of Marketing, University of Lancaster, Bailrigg, Lancaster, England

There are tantalizing indications that the lognormal represents numerous aspects of economic life.

Evidence from many sources can be seen as complementary when the distribution is taken as a unifying perspective. In economics, a classical distinction identifies firms and individuals as the units of analysis, to be treated at the macro and micro levels. When firms are taken in aggregate, the lognormal frequently fits their distribution by size measured as number of employees or asset values. Many of the measures represent yields—yields of goods produced, value added, earnings and so on, with Stock Exchange prices reflecting the value of the yields, past and expected in the future. At the micro level, the firm can find inventory, sales and demand lognormal over the goods it produces. In the case of individuals, income has been found lognormal in aggregate, at least for the main body of income earners. The amounts and prices of products they buy, their shareholdings, the amounts of their insurance claims, their expenditure on goods as a function of income, and many other aspects of economic life have been represented by the lognormal distribution.

Analysis frequently deals with change over time. The genesis of the lognormal via the law of proportionate effects (Chapter 1, 3.1) could then apply, where variables such as size or price are affected by multiple, independent external factors acting multiplicatively. Many authors explicitly refer to such a process. Through its operation in the past, present static distributions could be explained. However, there are cases such as the amounts of insurance claims where the concept of classification would be more appropriately applied.

If the lognormal relates to many business affairs, it is surprisingly underrepresented in management science, applied business statistics, or economic analysis textbooks. Often they simply quote the equation and give a sketch of one curve, but take it no further. The poor coverage reflects the preference given to other distributions in much theoretical work, sometimes for technical reasons: the availability of Laplace transforms or monotonic intensity functions, for example. But the test of theory is practical application. The emphasis here is therefore on findings related to data. The main issue is taken to be: to what extent is the lognormal distribution actually found to give a good account of phenomena in the real world of economic and business affairs?

The emphasis does not preclude many interesting theoretical developments. Frequently, researchers have had to extend theory to cope with the special problems they have encountered in applied work. For example, lognormals with shifting variance have been much studied in financial applications. The poor fit in the tails of the distribution has led economists to suggest models to account for the discrepancies. The best work on fitting two overlapping lognormals occurs in connection with the duration of telephone calls (Fowlkes, 1979). Convolutions of lognormal variables are important in insurance and actuarial science. Each discipline offers some developments which could also be relevant and useful in other areas. Lognormal applications taken as a focus reveal common interests in work otherwise linked only by references to the classical source book so frequently acknowledged, Aitchison and Brown (1957).

1. THE LOGNORMAL DISTRIBUTION IN ECONOMICS

The main applications of the lognormal in economics relate to the distribution of firms by size and to income earners by the amount they earn.

In both cases, the lognormal has been consistently used over many years, despite claims for alternative distributions. Interest has often focussed on obtaining some measure of concentration or inequality, to make comparisons over time or between countries and to relate the measure to other criteria of economic performance. Hart (1975) showed that most such measures, including the Hirschman, Herfindahl and entropy indexes, could be expressed using the lognormal variance, σ^2. Bourguignon (1979) proved that only one measure of inequality, equivalent to σ^2 in the lognormal case, met basic requirements and gave decomposability—the property that the inequality measure for a population can be expressed as the sum of inequality measures for subgroups within it.

The lognormal shape parameter is therefore important, even if the evidence for the distribution itself is not conclusive. In general it is the tails which give rise to doubt, especially in the case of income. The Pareto has been preferred to fit the higher levels of income although several authors have suggested additional factors which would favor the lognormal as the basic distribution.

1.1 The Size of Firms

In his classic development of the lognormal, Gibrat (1931) began with a theory: the law of proportionate effect. The third part of his book showed its application to French statistics for firms classified by number of employees, in thirty-two industries for each of three years. The actual and estimated numbers of firms were given in many cases, showing the closeness of the fitted lognormals. Gibrat suggested a form of the standard deviation as concentration index and gave comparative values for several countries and industries within them. He used the three-parameter distribution, recommending that the threshold parameter should be treated like the other two to improve the fit. He offered no interpretation when it varied from its natural value as the lowest figure recorded in the data.

Hart and Prais (1956) extended the analysis over time, using the market value (share selling price times the number of shares) for all companies with shares quoted on the London Stock Exchange in each of six years from 1885 to 1950. Companies were "born" or "died" as they first appeared or disappeared from the listed shares. The two-parameter lognormal was found a reasonable approximation to the distribution of company values in each year. To study the dynamics of change, transitions of firms from one size class to another were tabulated, taking a base 2 logarithmic scale for class size boundaries. Probabilities of growth or decline were normally distributed on this scale for small, medium and large firms, in accordance with Gibrat's law. Two factors offset the continually increasing variance of the model, treated from a theoretical standpoint by Kalecki (1954). One was a reversion-to-the-mean effect (the original sense of Galton's "regression"). The second was the continual appearance of new firms which tended to be smaller, with smaller size variance.

Simon and Bonini (1958) confirmed the normal distribution of size change on a logarithmic scale, using transition matrices for the 500 largest U.S. industrial corporations from 1954 to 1955 and 1954 to 1956. In addition to the law of proportionate effect, they assumed a relatively constant rate for the birth of new firms, leading to the Yule and Pareto distributions as derived by Simon (1955) and later developed with weaker assumptions (Ijiri and Simon, 1964). Adelman (1958) used the asset values of 100 U.S. steel companies to develop a transition matrix from which an equilibrium size distribution was derived. Steindl (1965) found that the lognormal gave a good fit to manufacturing plants, measured by number of employees, in the U.S. in 1939 and 1954 and in Austria in 1953. The fit to U.S. plants by the value of gross output was satisfactory for 1939, but less so in the upper tails in 1954 when some individual industries were analysed separately.

Many alternative distributions, including the Champernowne, lognormal, double exponential, and three types of Pareto, were compared by Quandt (1964). His data included firms with net worth over $1 million in 1958 and 1960 and *Fortune* sales figures for the 500 largest U.S. firms in 1955 and 1960. Although several distributions fitted the figures adequately, the Pareto of the second kind and the lognormal were the most satisfactory on the whole. The lognormal did less well on a test statistic for the randomness of successive differences between observed and predicted values.

Quandt (1966) repeated his analysis on the *Fortune* figures and added 1963 company asset values in each of thirty selected SIC code industries, noting that small firms tended to be underrepresented. Again sample sizes precluded clear discrimination between the distributions. The lognormal gave a good fit for twenty of the thirty industries. Quandt concluded (1966, p. 431) "In all likelihood the Law of Proportionate Effect operates together with such complicated conditions on birth and death as to be incapable of yielding a pure test of the law itself."

The concentration of British firms in individual manufacturing industries and in total was studied by Prais (1976), using Census of Production figures for firms grouped by number of employees since the beginning of the century. A second impression of his book, with an updating preface, appeared in 1981. Despite some discrepancies, Prais believed that "the lognormal distribution seems to capture the broad essentials of the matter" (p. 193). His σ estimates were obtained by calculating a standard deviate, z, representing the proportion of firms up to a given employment size and then calculating the first moment equivalent (proportion of all workers employed by those firms) with a theoretical value $z - \sigma$. Transition matrices showing movements of firms from one employment size group to another between successive periods and models of the growth process were developed. Hart (1982) referred to the graphic methods of parameter estimation used by Prais and gave reasons for preferring the variance of the logarithms of the first moment distribution to measure business concentration, without commitment to a particular theoretical distribution such as the lognormal. Using the same data source, he showed the results of several alternative concentration measures for the 1958–78 period, with allowance for the limitations of the Census figures.

The Law of Proportionate Effect has been questioned. Singh and Whittington (1975) looked at U.K. companies in twenty-one industries, recording the net asset values of firms with Stock Exchange quotations in both of two years six years apart. There was a mildly positive relation between their growth rates and the log of their initial size: larger firms tended to show higher average growth than smaller ones, although sparsity of data failed to yield statistical significance in most cases. The standard deviation of growth rates declined with size. When two consecutive periods were examined, a small degree of persistence in growth rates was found, as well as a tendency for company growth rates to regress to the mean growth rate of the industry. Births and deaths of firms were related to size and not independent of it, but led in aggregate to a decrease in the log variance. Chesher (1979) also found some serial correlation in the growth rates of a small sample of British firms, interpreted as "windfall" growth or decline which continued over the three-year data period.

Clarke (1979) carefully reviewed earlier evidence and gave results for the employment size distributions of 147 U.K. industries in 1968. Using the method of areas to estimate parameters, he found that only one quarter of plants passed three tests of lognormality (chi-squared, skewness and kurtosis) at the 1% level. The discrepancies were mainly in the righthand tail, with cases of too few and too many large firms in roughly equal numbers.

1.2 Income and Expenditure

The most widely accepted result on the distribution of income is paradoxical: the lognormal approximates incomes in the middle range but fails in the upper tail, where the Pareto distribution is more appropriate.

Pareto initiated the statistical study of income, using data from England, Prussia, Saxony and Italy to find a remarkable regularity known as Pareto's law. If N is the number of incomes exceeding a given income level X, the relationship is

$$N = AX^{-\alpha}$$

where A and α are constants. Pareto found α generally in the region of 1.5.

But there are problems with his generalization. The corresponding density function is monotonically declining, which does not check with the facts of income distribution. The Pareto can only represent the modal income and greater. Also, the variance only exists for α larger than 2.

The lognormal was first applied to income by Kapteyn in the early years of the century. Gibrat (1931) acknowledged his work and illustrated the law of proportionate effect with extensive income data from many countries and over many years. One of his tables showed the actual and fitted figures for forty-eight income groups covering nearly 15 million Prussians. Generalizing from all the evidence, Gibrat found little variation by country or by time period in the index of inequality of income and wealth, around a central value equivalent to a standard deviation of 1.6. Such a high value brought the mode close to the origin, accounting for Pareto's observations. However, the inequality of salaries was much less, around a standard deviation of 0.3. The lognormal was also found to apply to many income-related figures: wealth, rents, dividends, profits, the value of inheritances and share-holdings in a bank, for example.

Aitchison and Brown (1957) gave early references and their own findings for homogeneous groups of manual workers such as cowmen. The data for 1950 were well fitted by the lognormal, in some cases with the third parameter. However, the authors found that aggregated income data for the USA in the 1940s showed systematic divergences: the lowest and highest income classes were underpredicted, leading them to suggest that higher incomes would probably be more accurately graded by a Pareto-type curve. Similar findings were quoted from Quensel (1944) in an earlier paper (Aitchison and Brown, 1954).

Lydall (1968) confirmed the suggestion, based on income statistics from thirty countries ranging from Argentina to Yugoslavia. His treatment was mainly graphical and incorporated adjustments when the figures available did not correspond with his own standard, for example when they included part-time workers or aggregated male with female earnings. His general conclusions were: the central part of the distribution, from perhaps the tenth to the eightieth percentile, was close to lognormal; the upper tail often approximately followed the Pareto law, for at least the top 20 percent of the aggregate frequency.

In the same year Thatcher (1968) provided more detail about individual employment groups, working from a large British survey of over 170,000 employees. The lognormal was found to give a good approximation to the observed distribution of earnings of full-time adult manual workers, both male and female. The effects on the distribution were shown when part-timers and other groups were included in the figures. Thatcher (1976) later confirmed these findings for the years 1970 to 1972 from the same data base. The earnings of manual men, measured in the same week in successive years, were bivariate lognormal with regression to the mean. The situation was different for nonmanual men, however. The top ten percent of earners departed clearly from the lognormal and were approximately Pareto. The parameter α increased from about 3.4 in 1970 to about 4 in 1973. When figures for all men were combined, the two-parameter Champernowne (1953, 1973) distribution

$$F(x) = \left(1 - \frac{2}{\pi}\right) \tan^{-1} \left(\frac{m}{x}\right)^{\alpha}$$

could be fitted.

Fase (1971) used Dutch data from firms in banking and industry to show the lognormal distribution of income at a given age, with median income changing over the earnings lifespan. Cowell (1975) corrected an error in this paper and in a book (Cowell, 1977) confirmed the general relevance of the lognormal for the middle range of income and the Pareto for the upper tail, using U.K. pre-tax income figures for 1973–74. He pointed out the heterogeneity of the data which would not aggregate to a lognormal unless, among other things, the variances of the component distributions were identical. Harrison (1979) showed how variances in fact differed between occupational groups. Working with British statistics for seven separate categories such as managers, engineers and technologists, and transport industry employees in 1972, he found little to choose between the Pareto and lognormal for the upper levels of income in each case.

One of the most detailed studies was published by Airth (1985), based on 21 government surveys of adult male manual worker earnings in Britain during the period 1886 to 1980. Standardising the distributions by relating earnings to median earnings in each year, Airth found the lognormal standard deviation to fluctuate closely around 0.29 over the whole period, with sequences of higher or lower values hinting at a cycle of 12–13 years. Systematic departure from lognormality began only below the first or second and above the 96th percentiles.

Airth suggested an underlying lognormal paradigm for manual men's earnings extending over nearly a century. He illustrated departures from the paradigm distribution in each of the fourteen data years and argued that these were due to union bargaining, shifts in power between industries, and in general random perturbations typical of the real world. But in a free market economy, there would be forces leading to the re-establishment of familiar wage differentials represented by the underlying distribution. A matrix of empirical year-to-year transitions from one income class to another, with allowance for men entering and leaving the work force, would incorporate the random perturbations. Airth found that two such matrices, when transitioned to their equilibrium value, gave rise to a lognormal vector income distribution. He pointed out that the random disturbances would tend to displace the median upwards, giving a false estimate of the steady state vector. Also, working with more and smaller income groups, he found departures from the lognormal in both tails of the distribution. The data points fell between lognormal and log-logistic curves with the same mean and median, although there was no theoretical justification for such a hybrid distribution.

Airth's work throws light on a criticism made by Ransom and Cramer (1983), who pointed out that virtually all attempts to fit distributions to income data failed lamentably by the chi-squared test statistic for goodness of fit. They thought the test too strict, because of errors in the data. But perturbation in any one set of figures is to be expected, as Airth made clear. Their own work on U.S. family income data 1960–69 indicated that a normally distributed error term, added via convolution to standard distributions, improved substantially the fit of both gamma and lognormal, although not to standards passing a chi-squared test. Two distributions with three parameters, the beta and a generalization of the Pareto and Weibull suggested by Singh-Maddala (1976), offered no improvement.

Economists who accept the departure of the lognormal towards the Pareto at high income levels have offered several explanations of the anomaly. Rhodes (1944) referred to the distribution of talents determining an individual's income earning capacity. Simon (1957) and Lydall (1968) suggested a connection with graded pay in an organizational hierarchy. Grubb (1985) developed the idea of ability and power distributions, showing the applicability of his model, which involved a nonstandard integral, to very high income earners in eight developed countries.

The anomaly can be escaped by assuming alternative distributions. Many have been suggested in addition to those also relating to the size of firms. They include the sech-squared (Fisk, 1961), the gamma (Salem and Mount, 1974), the beta (McDonald and Ransom, 1979), stable distributions of a log-transformed variable (Mandelbrot, 1960, Van Dijk and Kloek, 1980), and a distribution derived from a beta-distributed Lorenz curve (Kakwani, 1980). In many cases the models were proposed because they described some data adequately, but theoretical justification was not offered and little attention was spared for the nature and compositon of the income data used. Hart (1983) questioned several of these studies on the grounds that adjustments were needed to allow for household size and composition. On an income per equivalent adult basis, data thought not to be lognormal proved to be so. Hart gave reasons for preferring the bivariate lognormal to fit British earnings of male manual workers in 1971 and 1978, showing from the transition matrix no change in inequality but considerable mobility.

Creedy (1985) pointed out the neglect of aggregation problems. Income does not remain the same over an individual's working life. A lognormal distribution of income within each age group will not normally lead to a lognormal total, as he clearly proved from extensive British household survey data for 1965–71. Among professional groups such as solicitors, doctors and engineers, average earnings at age t could be represented by a quadratic in t, leading to a $\Lambda(\mu_0 + \theta t - \delta t^2, \sigma_0^2 + \sigma_u^2 t)$ distribution of income at a given age. Parameters were estimated from published data. The deviations from linearity in the mean and from constancy in the variance, with weights δ and σ_u^2, although found to be small, limited the fit of the lognormal to aggregate figures for all age groups. In addition to many developments of the lognormal, Creedy showed errors in Pareto's original work and proved that a criticism of σ as a measure of inequality because it failed Dalton's Principle of Transfers could be dismissed as a curiosity.

The amount of money which people spend is closely related to the amount they receive, so that per capita expenditure would follow a similar form. Jain (1977), for example, used Indian national sample survey data for 1961–65 and discussed methods for fitting the three-parameter lognormal. The very small threshold expenditure varied between years and areas. He also noted some departure at the upper extreme end, leading him to suggest some other distribution such as the Pareto although the sparsity of observations precluded fitting it.

The Engel curve is the relationship between expenditure on one commodity and total expenditure. Iyenger (1960) used lognormal assumptions to estimate Engel elasticities from Indian survey data for 1955–56. Aitchison and Brown (1957) showed the application in the case of inferior goods, where a limit represented saturation expenditure on a commodity however large the income. They proved that jointly lognormal distributions of that limit and of mean income would lead to a lognormal form for the mean expenditure on the commodity in the population.

1.3 Perceptions of Income and Price

As well as a cash value, income can have a perceived value. Van Praag (1968) argued that individuals relate their own earnings to other levels which they can assess, particularly the best and worst income levels they can imagine. They can nominate incomes rated as "excellent," "barely sufficient," "very bad," and so on, thus defining their personal income welfare function. As evidence, samples of about 2,000 people in Belgium and Holland were surveyed in successive years from 1969. Van Praag (1978) gave references to resulting articles and summarised the main results. Individual welfare functions, as defined by the measurement technique, were closely approximated by the lognormal. The standard deviation varied between individuals classified by socio-demographic groups, but not widely from an average of 0.5. The mean was a function of family size, the individual's own net post-tax income, and reference group income. People perceived income as multiples or fractions of a reference group income, in accordance with the Weber-Fechner law.

Van Praag (1978) drew further conclusion using the lognormal for both actual income distributed over individuals and for perceived income. For example, a proportionate change in all incomes would be seen as reducing inequality by richer people but as increasing it by the poorer. In view of location of the mode of the lognormal, the individual welfare function was not concave at low income levels perceived as "bad." (This contradiction of accepted theory led one discussant of Van Praag's paper to suggest measurement error, or respondents not trying hard enough to imagine what their life would be like at income levels well below their own.) The same double lognormal assumptions have been used to derive an optimal personal income tax policy (Van Herwaarden and de Kam, 1983).

Price perceptions have also upset ideas important to economists. Gabor and Granger (1961, 1968) asked 3,000 housewives in Nottingham, England, the maximum and the minimum prices which they would be prepared to pay for various textiles and food products. Below a lower limit, women would doubt the quality of the goods and refuse to buy. The implied individual demand curves with negative slope, barely allowed by theory in the rare case of Giffen goods, contradicts the fundamental law of Demand (Boland, 1977). Taken over all respondents, the upper and lower price limits plotted close to straight lines on lognormal probability paper. French researchers had been the first to suggest the method and to quote similar results from consumer surveys, although not in the economics literature: Stoetzel et al. (1954) for radio sets, Adam (1958) for articles of clothing, gas lighters and refrigerators, and Fouilhé (1960) for packet soups and washing powders. In the latter study, correlations between the price limits for the two products were established and differences found between age and social classes.

In addition to the lognormal distribution of price limits in a population, Van Herwaarden and Kapteyn (1981) used the Van Praag method and sample base to establish a lognormal distribution of price perceptions at the individual level. People were asked to nominate prices which would suit them perfectly, moderately, not at all, etc. The means and variances of the curves were analysed by socio-demographic classes.

A similar line of analysis led Granger and Billson (1972) to test the hypothesis that the proportion of consumers buying brand X, given that they bought either X or Y, was a normal distribution function of the log price ratio, $\log p_x/p_y$, in the case of soft drinks. Evidence for the Weber-Fechner law was reviewed by Monroe (1973).

2. FINANCE AND RISK

If price is a sluggish variable at one extreme, changing little from a fair or just price used as a point of reference, at another it is highly volatile, changing by the minute. Heavily traded shares quoted on Stock Exchanges worldwide are the prime example. Financial analysts with large resources to invest ensure continual interest. "Of all economic time series, the history of security prices, both individual and aggregate, has probably been most widely and intensively studied." (Roberts, 1959)

The lognormality of prices and price changes has been investigated empirically but the evidence is relatively limited compared to the extensive use of the lognormal in financial theory, in turn leading to practical applications. The data base is huge, thanks to the demand for share price information. But parameters change in the real world. As a result, model-builders have produced many ideas to allow for a lognormal variance which varies over time.

The amount of an insurance claim can be regarded as the price of an accident, damage or loss. The lognormal has been frequently used to fit amounts of claims in various classes of insurance business. Considered jointly with the probability of n claims arising, convolutions of lognormal variables have relevance to the risk reserves which insurance companies need to maintain.

2.1 The Random Walk of Share Prices

It is generally accepted that share price movements are random. Bachelier (1900) was well before his time in developing a complete stochastic model for prices, arriving at the diffusion equation of physics which describes the movement of a particle in Brownian motion. Working (1934) noted that commodity and stock prices appeared to move randomly, with results that would look like an economic time series. Kendall (1953) carried out a major study using industrial share prices in fifteen British industries, recorded at weekly intervals for the years 1928–38, and Chicago wheat and New York spot cotton prices over even longer periods. Taking first differences to measure price changes, he found that serial correlations of the first ten orders were for most practical purposes negligible in all series except cotton prices, where the first order correlation for the period 1816–1950 was 0.31.

The exception was explained by Working (1960). Kendall's figures had been monthly averages of daily prices. The expected serial correlation due to averaging was given by

$$E(r) = \frac{m^2 - 1}{2(2m^2 + 1)}$$

where m was the number of time intervals in the average. Most of the observed correlation was therefore explained.

Roberts (1959) found that American data, both for indexes and for individual companies, were entirely consistent with those of Kendall. In the same year Osborne (1959) introduced the lognormal distribution, working inductively from the direct examination of prices and also deductively, using the Weber-Fechner law of proportionate effect. Prices could be taken as a stimulus, and investors were more interested in the proportionate changes in the value of stocks than in absolute dollar values. Empirically, the distributions of all common (not preferred) stock prices on the American and the New York Stock Exchanges at the end of one day's trading were approximately lognormal, although his plots on probability paper indicated discrepancies in the tails. The log of the ratio of price changes or log price relative $Y = \log(P(t+T)/P(t))$ was also roughly normal for $T = 1$ month. Osborne took a sample of stock prices for various values of T on a trading time scale, i.e., allowing for Stock Exchange opening hours, and used the semi-interquartile range to estimate the increasing variance over time, which he found proportional to T. In a later article, Osborne (1962) discussed modifications to Brownian motion on a logarithmic scale, due for example to the level of trading activity, variation between hours and days of trading, and price levels which provoked support or resistance, modelled as Brownian motion with partly reflecting barriers. Cootner (1962) found evidence for such barriers in short-run price changes, ascribed to the actions of professional Stock Exchange operators.

The random walk of stock prices implies that the past history of the series is unrelated to its future. Consequently, investment analysis and attempts to anticipate price shifts would be fruitless. In the 1960's, several researchers investigated the evidence for this alarming conclusion. Granger and Morgernstern (1963) used spectral analysis on the Standard and Poor, Dow Jones, and several other stock prices indexes taken over long time periods. The random walk model fitted the data very well, with modest indications of periodicity at eight months. Moore (1964) examined weekly price changes of thirty randomly selected stocks on the New York Stock Exchange and found an average serial correlation of −0.06. Fama (1965) used daily price changes of Dow Jones stocks 1957–62 and found serial correlations with lags of one to ten days to be too small and variable to serve predictive purposes. A runs test showed no significance in most cases.

In an effort to controvert the assertion of randomness, Alexander (1964) worked out "filter" trading rules for when to buy and sell stocks after troughs or peaks in prices, concluding from share price data that profits could be made on this basis. A more thorough test of the trading rules by Fama and Blume (1966) showed that, after allowing for dividend payments and brokerage commissions, the rules offered no advantage over the simple policy of buying stock and holding on to it. Even professionally-managed funds failed to beat the market average (Sharpe, 1966; Jensen, 1968).

The evidence that stock price changes follow a random walk has been used to prove the efficiency of the market. Current prices fully reflect all the information available; there are no exploitable patterns in the price series themselves (Mittra and Gassen, 1981, p. 597). But the efficiency argument does not apply to company earnings, to which stock prices rapidly adjust although expected earnings are better predictors of share prices than actual earnings (Elton, Gruber and Gultekin, 1981). Earnings might be thought dependent on longer-term factors such as company strength in research and development, management skill, or progress up a learning curve. Considerable surprise was produced in 1962 by an article "Higgledy piggledy growth" by Little, who later confirmed in a book (Little and Rayner, 1966) that changes in earnings of British firms followed a random walk. Historical rates of growth gave no clue to later growth. Confirmatory findings for American corporations are summarised by Lorie and Hamilton (1973, Chapter 9). Some limited predictive power was found by Jones and Litzenberger (1970) from quarterly earnings of companies which showed an unusual change in trend.

2.2 The Lognormal and Related Models

The random walk model requires only that successive price changes be independent and conform to *some* probability distribution. Kendall (1953) suggested that it was roughly normal but found the actual distributions leptokurtic, with more large price changes than would be expected (in fact he omitted some outliers from moment calculations). Sprenkle (1961) looked at 37 companies which issued warrants in two periods 1923–32 and 1953–59, chosen because price movements were lively. Their stock prices at one month intervals were mostly fitted adequately by the lognormal but a more skewed distribution was indicated. Mandelbrot (1963) questioned the assumption of normality and introduced the stable Paretian distributions, which Levy had proven to be the only limit distributions of sums of random variables. The hypothesis was applied to thirty Dow Jones stocks by Fama (1965), who found higher peaks and more extreme price changes at monthly intervals than the normal curve would allow. The "fatter tails" were adequately fitted by stable Paretians.

After a period of popularity, however, the assumption was questioned. Praetz (1972) and Blattberg and Gonedes (1974) found data were better fitted by the Student's t distribution. Hsu, Miller and Wichern (1974) reported that no stable Paretian fitted daily log price relatives for four large American companies, although monthly figures could be fitted. The stability condition was not fulfilled because the characteristic exponent a varied over time: twenty firms all showed a shift from an a of about 1.4 prewar to nearer 2 postwar (unstable behavior of Fama's a was shown from a large data base by Simkowitz and Beedles (1980)). Simulation showed that a stable Paretian would fit an artificial mixture of two normal variables with the same mean but different variances. There was no compelling reason to take a less than 2, the normal case, and the authors proposed a normal probability model for log price relatives with a nonstationary variance subject to step changes at irregular time points.

Similar findings were reported by Boness, Chen and Jatusipitak (1974). While the stable Paretian fitted log price relatives for 33 utility companies, with normality in a third of the cases, variances changed within the period. The companies were chosen for having issued additional debt. A comparison of pre- and post-issue period share price changes showed that variances differed significantly in half the cases. Hsu (1984) documented the variance shifts in Standard and Poor's index of 500 common stocks and in quoted prices for twenty randomly-selected firms, using blocks of one hundred daily quotations over the period 1926–80.

Several methods of incorporating a changing variance into a lognormal model have been suggested, so producing the fatter tails otherwise described by the stable Paretian hypothesis. Resek (1975) suggested that the successive values of log prices relatives were uncorrelated but connected with each other by a nonlinear relationship. He assumed a lognormal distribution within each time period with a fixed mean throughout but with serial correlation in the variances of successive distributions, so that prices would pass though periods of relatively high and relatively low variation. Box-Jenkins methods were applied to the daily price quotations of 32 stocks over a seven year period, to deduce an ARMA (1,1) model and to fit parameters which could then be related to the time path of the variance. Resek showed how this autoregressive process produced decreasing kurtosis as the period of observation lengthened. The stable symmetric family implied one extreme of the process while independence of observations implied the other. The data were best fitted by an intermediate value between the two extremes.

Some researchers have used the idea of information arrival to explain changing variance. Clark (1973) suggested that the price series for cotton futures evolved at different rates in periods of the same length. When there was no news, trading was slow. When old expectations changed it became brisk. A stochastic process $T(t)$, called the directing process, represented the rate of price evolution and transformed chronological time t into variables T serving as operational time. Taking $X(t)$ as price at time t, a new process $X(T(t))$ was formed. If the independent increments $\Delta X(t)$ were distributed $N(0, \sigma_2^2)$ and the independent increments $\Delta T(t) = v$ were $\Lambda(\mu, \sigma_1^2)$ the process would have lognormal-normal increments given by

$$f(x; \mu, \sigma_1^2, \sigma_2^2) = (2\pi\sigma_1^2\sigma_2^2)^{-1} \int_0^\infty v^{-3/2} \exp\left(\frac{-(\log v - \mu)^2}{2\sigma_1^2} - \frac{x^2}{2v\sigma_2^2}\right) dv$$

which could be approximated by numerical integration techniques. The introduction of $T(t)$ had the effect of making the price distribution more leptokurtic, with variance $\sigma_2^2 \exp(\mu + \sigma_1^2/2)$ and kurtosis $3\exp(\sigma_1^2)$. The kurtosis could therefore increase as much as desired while keeping an unchanged variance. Clark tested the $f(x)$ distribution on two samples of 1,000 daily cotton futures prices against stable distributions, taking many sets of parameters rather than attempting to estimate them, and found that the likelihood of a lognormal-normal having generated the samples was very significantly higher than the likelihood of any stable distribution. He commented that it was not clear theoretically why operational time should be lognormally distributed.

An interesting point was the connection with $V(t)$, the volume of trading on day t. Trading volume could be taken as an "imperfect clock" measuring the speed of evolution of the price change process. It would then be correlated with price change variance, and a curvilinear relationship was in fact found. In practice, large price changes could result either from the arrival of ambivalent information, provoking much trading, or from clear information viewed in the same way by all traders leading to little volume change. The connection between trading volume and price change variance was supported by Epps and Epps (1976), with a correction of Clark's formulation, and by Westerfield (1977) using the subordinated stochastic process model. Empirical evidence on the connection between trading volume and price changes had been provided earlier by Ying (1966).

A large-scale investigation to discriminate between Clark's subordinated stochastic process (SSP) and the stable Paretian (SP) models was undertaken by Upton and Shannon (1979), using monthly returns for each of fifty companies randomly selected from New York Stock Exchange firms over the twenty years 1956–75. Their approach was to examine returns data at time intervals of 1, 3, 6 and 12 months. The characteristic exponent a in the SP model would be the same theoretically in each case, but the SSP model would move toward the lognormal asymptotically as the interval lengthened, provided that the parameters were constant over the period. It was found that monthly returns were not lognormal in 29% of cases, but the longer interval figures improved markedly toward lognormality. Skewness and kurtosis measures also moved toward lognormal values as the time periods lengthened. When shares were grouped into portfolios of ten stocks each, the statistical tests were overwhelmingly in favor of lognormality even for monthly data: the increase in stationarity due to portfolio effects more than offset other factors.

A series of stochastic models allowing for jumps in share prices were proposed by Cox and Ross (1976), who referred to the lognormal diffusion process as "the workhorse of the option pricing literature." This would express the percentage of a stock price, S, in the next instant as

$$\frac{dS}{S} = \mu dt + \sigma dz$$

showing the drift term μdt and a normally distributed stochastic term σdz with mean 0 and variance $\sigma^2\, dt$, making the percentage change in stock value from t to $t + dt$ normal with mean μdt and variance $\sigma^2\, dt$.

Merton (1976) suggested an alternative which allowed for the random arrival of important news about a firm or possibly its industry, leading to active trading. The revision was

$$dS/S = (\alpha - \lambda k)dt + \sigma dz + (Y - 1)$$

where α was the mean of the stock return, λ was the mean number of Poisson information arrivals per unit time and k was the expected value of $(Y - 1)$, the random variable proportional change in the stock price if the Poisson event occurred. The final term $(Y - 1)$ represented an impulse function producing a finite jump in S to SY. If the distribution of Y was lognormal, the distribution of $S(t)/S$ would be lognormal with the variance parameter a Poisson random variable. By 1984, an article in a financial journal was able to claim "...the consensus today seems to be that the incidence of extreme rates of return is best modeled as arising from a jump (Poisson) process superimposed upon a lognormal diffusion process" (Marcus and Shaked 1984, p. 68).

In summary, evidence that the lognormal adequately describes the distribution of price relatives is available from Osborne (1959) and Cootner (1962). Relevant findings were reported by Kendall (1953), Sprenkle (1961), Moore (1964) and Hsu, Miller and Wichern (1974), as noted above. Two less readily available sources quoted in the financial literature are Lintner (1972) and Rosenberg (1973). More recently Tehranian and Helms (1982) showed that most random portfolios of twenty stocks 1961–76 yielded returns fitted by the lognormal, so that there was no need to use the more complex stochastic dominance model. In the case of Treasury bills, Dale (1981) found that the futures market showed lognormal behavior patterns with no signs of resistance or support levels. Marsh and Rosenfeld (1983) used three models to fit rates of return on Treasury bills with one week and one month to maturity, with the lognormal turning out to be the most likely. They noted the changing variance through time and the positive relation between interest rate changes and their levels.

Although most applied work has dealt with price movements, a set of prices can be considered as an ensemble at a given point in time. Osborne (1959) took the closing prices on one day for all stocks on two Exchanges, to find that they plotted as approximately lognormal. Further evidence that an ensemble may be lognormal comes from the Sears Roebuck catalogue, which in its heyday listed almost everything an ordinary consumer might wish to buy. Hermann and Montroll (1977) sampled every eighth page of sixteen catalogues issued between 1900 and 1976, noting the price of all items on the page. To a rather good approximation the logarithms of the prices were normally distributed in the range 2–99%. The σs were almost the same every year, deviating from the average $\sigma = 2.26$ with a standard deviation of only 0.17. The larger deviations were usually associated with an external event such as the great depression affecting the 1932–33 catalogue, when average prices also dropped exceptionally.

In another application, Marcus and Shaked (1964) assumed that the assets held by banks would be lognormally distributed. They found the assets of forty American banks and calculated the probability that earnings would be below payments to depositors, leading to insolvency. Most banks were extremely safe but there were a few exceptions, correlated with high variance of returns and some financial ratios such as loans to deposits. Shaked (1985) later applied the same method to the assets of life insurance companies.

2.3 Extensions of the Lognormal in Financial Applications

The lognormal frequently appears in theoretical developments leading to practical application in investment and finance.

The mean and variance of stock returns were first used analytically by Markowitz (1952, 1959). Past returns could be averaged to give an expected return, to be taken literally as an expectation of returns in the future. Investors aimed to maximize expected returns but they did not put all their money into the one stock with the highest value. Instead, they diversified their holdings to reduce risk. Markowitz equated riskiness with the variance of a return, measured by share price fluctuations, i.e., neglecting dividend or interest payments. A set of mean-variance efficient portfolios could therefore be defined, whose expected return could not be increased without also incurring greater variance and whose variance could not be reduced without surrendering some expected return. Given an investor's utility function for risk against return and the past history of Stock Exchange prices, the one efficient portfolio could be picked by computer program. By implication financial advice offered only inferior alternatives.

The work of Markowitz is acknowledged as the starting point of portfolio theory. It stirred the investment community and led to a considerable literature, with applications as far afield as the product portfolios of companies (Cardozo and Smith, 1983; Devinney, Stewart and Shocker, 1985). Elton and Gruber (1974) followed the work of Fama with the stable Paretian hypothesis by assuming a multivariate lognormal distribution for investment relatives, one plus the rate of return, and provided an algorithm for calculating efficiency frontiers. Ohlsen and Ziemba (1976) used the same assumptions to develop a nonlinear programming solution when investor's utility was a power or logarithmic function of wealth.

A problem with the Markowitz analysis was that share prices were not independent but tended to vary differentially with the swings of the market as a whole. The variance of a portfolio therefore needed to include terms for all the pairs of constituent share price covariances. Sharpe (1963) introduced a great simplification by relating the riskiness of a stock to the riskiness of the market, represented by a comprehensive stock price index such as Standard and Poor's. In his later Capital Asset Pricing Model or CAPM, Sharpe (1964) used β_i to relate the expected return on a particular stock, $E(R_i)$, to the market risk by the equation

$$E(R_i) = R_f + \beta_i(E(R_M) - R_f)$$

where $E(R_M)$ was the expected return for the market portfolio of risky assets and R_f was the rate of return on a riskfree asset such as a Treasury bill. Sharpe (1965) provided evidence that the linear relationship was broadly correct. There was a correlation of 0.84 between average annual rates of return and the standard deviations of those returns for 34 mutual funds 1954–63. His work was convincing both theoretically and practically. "In the early aftermath of the publication of Sharpe's work on the price of risk, a new industry was created to manufacture and distribute beta coefficients" (Lorie and Hamilton 1973, p. 211).

Troubles arose when attempts were made to check the model (for references see Fogler, 1978). β_is could be calculated in various ways and past values failed to predict later ones. Black, Jensen and Scholes (1972) showed that the β_is were not proportional to risk premiums: actual returns on high beta stocks were lower than predicted and returns on low beta stocks were higher. But the CAPM provides the optimum selection rule only if investors have quadratic utility functions or share prices are normally distributed. Merton (1971) pointed out that both assumptions were objectionable. If instead prices were lognormally distributed, Tobin's analysis held: In theory, investors had only to divide their money between two funds, one risky and one risk-free, irrespective of preferences (the form of the utility function), wealth distribution, or time horizon. The geometric Brownian motion assumption for assets in the risky portfolio allowed proof of strong theorems about the optimum consumption-portfolio rules, and was appropriate for an economy where expectations about future returns had settled down. Working on these lines, Bawa and Chakrin (1979) reformulated the CAPM for an equilibrium securities market in which all asset returns and the market return were lognormally distributed. The efficiency frontier for a portfolio was part of a hyperbola on these assumptions and the revised formula for the CAPM was, in the notation used above,

$$\frac{E(R_i) - R_f}{E(R_i)} = \frac{E(R_M) - R_f}{E(R_M)}\beta_i \qquad \beta_i = \frac{\exp(\rho\sigma_i\sigma_M) - 1}{\exp(\sigma_M^2) - 1}$$

with ρ as the correlation between R_i and R_M. Bawa and Chakrin suggested that some poor results obtained in attempts to verify the traditional CAPM might be due to incorrect betas not estimated on the basis of their formulation.

One assumption in the CAPM is that all investors have mean-variance efficient portfolios. However, both the mean and the variance have been criticized as inadequate measures. Since one investment can be switched into another, an individual's final wealth per dollar of initial wealth is the product of the returns that the money has earned in each investment. A wealth or growth maximizing policy therefore requires the geometric mean, and not the arithmetic mean, of successive holdings (Latané 1959, Latané and Tuttle 1967). Unfortunately a portfolio which has the largest geometric mean may not be mean-variance efficient (Hakansson, 1971). However, Elton and Gruber (1974) proved that if returns are lognormally distributed the portfolio which maximizes the geometric mean lies on the (arithmetic) mean-variance efficient frontier. Approximations and statistical analysis of geometric means are important for financial analysis (Young and Trent (1969), Hilliard and Clayton (1982), Hilliard (1983), Cheng and Karson (1985), Brennan and Schwartz (1985)), not least because Stock Exchange indexes such as the Dow Jones are calculated on an arithmetic basis.

The variance measure of riskiness was criticized by Bawa and Lindenberg (1977) who proposed instead a criterion based on the lower partial moments of an asset return distribution, i.e., moments taken up to the risk-free rate of return. Markowitz (1959) had originated the idea under the title of semi-variance, arguing that better portfolios could be picked if stocks were measured by their chance of incurring a loss than by calculating their variance, which covered both losses and gains. This was particularly important when the distribution of returns was skewed. Working on these lines, Bawa and Lindenberg reformulated the beta coefficients of the CAPM using colower partial moments. The logic of the model was attractive, deriving from stochastic dominance rules for portfolio selection which had attracted much attention. However, Price, Price and Nantell (1982) computed these coefficients using 300 securities in six periods 1927–68. Their own assumption was that the joint density of R_i and R_M, the return on an individual stock and the market return, was bivariate lognormal. Results were consistent with this hypothesis as opposed to the Bawa-Lindenberg formulation.

Another large body of theoretical work and practical applications followed from the Black-Scholes option pricing model, which offered the perfect hedge: The owner of a stock could protect himself from the market risk of price change by taking a long position in the stock and a short position in the stock option. Black and Scholes (1973) assumed that the price of an asset at the end of a period t, $R_i(t)$, would be distributed $\Lambda((\mu_i - \sigma_i^2/2)t, \sigma_i^2 t)$ with $\mu_i > \sigma_i^2/2$ if ultimate worthlessness was to be avoided. The hedge position was maintained by determining the number of options to be sold short against one share of stock long, depending on a partial derivative representing the change in an option price relative to the change in price of the underlying stock. The hedge, maintained by continuous adjustments as prices changed, made the asset return certain. In a perfect market, therefore, the return must be the same as the prevailing short term rate of interest, r. This condition allowed for the solution of a differential equation (the heat-transfer equation of physics) to give the value v of a call option as

$$v = x_i \Phi(d_1) - c\Phi(d_2)e^{-rt}$$

$$d_1 = \frac{\ln(x_i/c) + (r + \sigma_i^2/2)t}{\sigma_i \sqrt{t}}$$

$$d_2 = d_1 - \sigma_i \sqrt{t}$$

where x_i is the stock price, t is the time to the expiration of the option, and c is the option exercise or striking price. These variables were directly observable. The only unknown was the log variance σ_i^2 of the individual stock.

As an indicator of the interest aroused by Black-Scholes, their option pricing formula became available for programmable hand calculators to facilitate the search for option prices which differed from their theoretical values (Mittra and Gassen 1981, p. 770). Many theoretical articles were written retaining the lognormal but testing, relaxing or changing some simplifying assumptions of the model, e.g., Frankfurter, Stevenson and Young (1979), Merton (1980), Hansen and Singleton (1983), Stapleton and Subrahmanyam (1984). Rogalski (1978) and Beckers (1980) queried the formula from applications to data. Merton (1976) added random jumps to the Brownian motion of share prices when evaluating options, a line followed by Jarrow and Rudd (1982), who found improvements over the Black-Scholes model.

Lee, Rao and Auchmuty (1981) introduced into option pricing the Bawa-Chakrin revision of the CAPM, in which the stock and market returns R_i and R_M were correlated bivariate lognormal. The Black-Scholes valuation became the special case when the expected return on a hedged stock was always equal to the riskfree rate, R_f. When individual stock returns and their covariance with the market were considered, additional parameters and terms in the option valuation equation were needed. In favor of their model, the authors quoted evidence that Black-Scholes overpriced deep out-of-the-money options and underpriced deep in-the-money ones, which the new model corrected. However, Brown and Huang (1983) pointed out an error in assumption: if an asset has a lognormal return, then an option on it will not. Krausz and Harpaz (1984) found, using data on closing prices for a sample of nearly one thousand call options, that the model was subject to systematic mispricing and did not significantly outperform the Black-Scholes formulation.

2.4 Insurance and Risk

Insurance claims represent the prices of events causing accidents, damage or death. The randomness of a set of claims stems from the unpredictability of these events. Sophisticated analysis has been applied to claims amount distibutions. The upper tail is highly important: the possibility of very large claims threatens the solvency of an insurance company, with implications for the levels of premiums, risk reserves and reinsurance.

The first application of the lognormal was by L. Amoroso (1934), who used it to fit seven sets of data on sickness claims and sickness duration. A further eight sets were added later (E. Amoroso, 1942). Workmen's compensation costs were analysed as lognormal by Bühlmann and Hartmann (1956), and by Latscha (1956), workmen's permanent disability claims by Dropkin (1966), and health insurance claims by de Wit and Kastelijn (1977). Beard (1957) gave an example of the lognormal and log-Pearson type I distributions in an experience of fire claims in Denmark. Benckert (1962) fitted the lognormal to eight sets of claims in fire, accident and motor third party insurance. Ramel (1960) used lognormals for 100 fire insurance claim ratios (claims over sums insured) 1857–1956, and further claim ratios 1916–50, previously analyzed as Pearson type I, were shown to be potentially lognormal by Seal (1969, p. 6). Ferrara (1971) fitted the distribution to nine sets of fire claims data from different industries, and domestic fire insurance claims in Sweden 1956–69 were represented by lognormal and Pareto curves in Benckert and Jung (1974).

In the field of automobile insurance, the lognormal was used by Henry (1937), Bailey (1943), Harding (1968) for the original amount of a claim falling under excess of loss reinsurance, and Bickerstaff (1972) for the size of collision claims. O'Neill and Wells (1972) reported that the two-parameter lognormal would usually fit damage claims, especially when they were grouped into homogeneous classes such as claims for damage to specific car makes and types. Collision claims classified by age of driver required the three-parameter model. The authors found that grouping data by logarithmic increments rather than by equal arithmetic increments improved the fit and significantly increased the efficiency of the parameter estimates. They gave the information matrix for grouped data, for use in iterative corrections to a set of initial estimates.

In other areas of insurance, Finger (1976) used the lognormal for claim amounts in liability insurance. Ziai (1979) worked with substantial data on general insurance claims, finding a good fit with the lognormal and inverse Gaussian distributions truncated to eliminate the smallest claims. The gamma and three-parameter Weibull were unsuitable. Decision rules on how much to pay for reinsurance were suggested by Samson and Thomas (1983) on the assumption of lognormal claims. Marlin (1984) considered deductibles, the minimum damage below which claims were not payable, which effectively left-truncate a lognormal distribution of claims. He assumed that truncation points were themselves lognormally distributed and gave the maximum likelihood estimators for numerical solution by Newton's method. Some illustrative examples were given from property damage claims for large industrial risks, with a warning that data could contain heterogeneous risk classes.

The above examples of lognormal applications cannot be taken to prove the case for the distribution. Statistical tests of fit were not always used, and there is doubt about the accurancy of the number of small claims recorded in some cases. In a comprehensive review of the literature to his time, Seal (1969) referred to work using Pearson distribution types I to IV, as well as others. A particularly important contender is the Pareto for the upper tails. According to leading authorities, "experience has shown that the behaviour of the tail in practice is often between that of the Pareto and log-normal types" (Beard, Pentikäinen and Pesonen, 1984, p. 76). They described Pareto's with parameter $a < 2$ as "very dangerous" distributions because of the chance of very large claims, and quoted several intermediate forms developed in risk and insurance work. For example, the quasi-lognormal is given by

$$F(z) = 1 - b\left(\frac{z}{z_0}\right)^{-\alpha-\beta\ln(z/z_0)}$$

where z_0 is the point in the tail from which the function is fitted. $\beta = 0$ gives the Pareto case while curves closely approximate the lognormal for positive β values.

An important development by Scandinavian authors has become known as the collective theory of risk. It deals with the whole risk business of an insurance portfolio, treated as a stochastic process with ruin as a possible outcome. A basic model expresses claim liabilities, X, over a period t via the distribution function

$$F(x,t) = \sum_{n=0}^{\infty} p_n(t)B^{n*}(x)$$

where $p_n(t)$ is the probability of n claims and $B(x)$ is the probability that the amount of a claim is less than x. The last term is the n-fold convolution of $B(x)$.

Lundberg (1940) first developed $p_n(t)$ from Poisson to compound Poisson, leading to the negative binomial when $k\ln(1 + t/k)$ was taken out as independent time parameter. For present purposes, the interest is when $B(x)$ is lognormal. Convolutions are usually handled by Laplace transforms, but these are not available for the Pareto or lognormal distributions. Seal (1978) suggested that the inverse Gaussian was close to the lognormal (for confirmation see Chhikara and Folks, 1978) and compared the skewness of the two distributions fitted to claim amounts data which had appeared in the literature. The two skewnesses were mainly similar, but not in the case of fire insurance claims. Seal carried through the inversion of the Laplace transform of the inverse Gaussian, but of more direct benefit was a *Fortran* program given in his book for evaluating the convolutions of a discrete approximation to a lognormal distribution, i.e.,

$$B^{n*}(x) = \int_0^x B^{(n-1)*}(x-y)\,dB(y) \qquad n = 2, 3, 4 \ldots$$

with $B(x)$ lognormal taken at discrete values of x. Numerical results were shown for the metallurgical fire insurance claims given by Ferrara (1971) for values of n from 5 to 40 at intervals of 5. As Seal pointed out, the same expression occurs in queuing theory for the distribution of waiting times.

Beard, Pentakäinen and Pesonen (1984) also dealt with the convolution problem, providing a recursion formula for calculating progressive values when claim sizes are limited to a set of discrete values which are integer multiples of a basic claim size (a lattice distribution). The subject seems only to have been addressed in the risk literature with the lognormal case in mind.

3. THE LOGNORMAL DISTRIBUTION OF DEMAND

If firms in an industry are lognormally distributed by size, a company selling to that industry might expect sales to follow the same pattern, given that its customer accounts were a random sample of all firms in the buying industry. Even with several industries as customers, a lognormal distribution of accounts by size could occur. On the consumer goods side, Engel curve analysis shows the relationship between buying and income. The main evidence comes from work on logistics, inventory control and physical distribution systems. Marketing has surprisingly little to offer, due to the predominance of Pearson curves in model-building and analysis.

3.1 Logistics and Physical Distribution

The first full-length book on inventory problems appeared in 1931, and the scientific treatment of stock control developed in the 1950s. The essential problem is the trade-off between excess stocks resulting in costs and unproductive investment, and inadequate stocks leading to delays or lost sales. Some factors affecting the trade-off are readily quantified, such as carrying costs and reorder lot size. Two are not under control and can be treated stochastically: the level of demand, governing the outflow from inventory, and input rates dependent on the time required to replace missing items—the procurement lead time.

Normative models were developed for the control of stocks of a single item. Hadley and Whitin (1963) summarised the main lines of analysis which had developed by that time and still provide the basis for inventory control theory. They handled stochastic aspects via the Poisson, gamma and normal distributions. The lognormal was referred to in Morrell (1967), a book on stock control for industry which included a table for the adjustment to buffer stocks needed with lognormal lead times for different values of the variance.

As well as stock control for individual items, the ensemble of items which an inventory has to carry has been studied. In a *Harvard Business Review* article, Magee (1960) pointed out the importance of stock and distribution costs to companies and wrote

> Exhibit II shows the typical relationship between the number of items sold and the proportion of sales they account for. The figures are based on the records of a large number of firms in the consumer and industrial products fields. The exhibit reveals that while 10–20% of total items sold characteristically yield 80% of the sales, half of the items in the line account for less than 4% of sales. It is the bottom half of the product line that imposes a great deal of the difficulty, expense, and investment on the distribution system.

Exhibit II illustrated two approximate Lorenz curves plotting percent of sales against percent of items in a product line, passing through the 80/20 and 80/10 points respectively. Magee suggested, on the strength of his experience, that most product ranges fell between the two curves. This may have been one origin of the "80/20 rule." Equivalent lognormal σ values are 1.7 and 2.1.

Many writers on logistics acknowledged the principle in general terms but limited themselves to the idea that customers could usefully be rank-ordered by sales volume. Bowersox, Smykay and La Londe (1968), for example, suggested ABC classification. A items were defined as the top 20% of products accounting for 56% of sales activity; B items were 40% accounting for 34%; the remainder were C items. Thomas (1968) attributed ABC analysis to Ford Dickie (an article in *Factory*, July 1951) and gave an example of one particular store with 6,500 stock items. The suggested breakdown was 4% of A items covering 75% of usage value, 16% of B items covering 20% and 80% of C items covering 5% (these points would be roughly consistent with $\sigma = 2.5$ for a lognormal distribution). Ballou (1973) gave one example of company sales by size with a split over A, B and C items of 20, 30, and 50 percent respectively. Heskett (1977) used the same proportions but also gave the actual percentages of annual turnover which these items represented in firms with products such as cereal-based food, small appliances, paper, grinding wheels and chemicals. He suggested that the "80/20 relationship" could be used more systematically in developing service and distribution strategies.

The lognormal as the demand distribution over inventory items was made specific by Brown (1959, 1967) in textbooks admired and followed by Lockyer (1972). Brown was able to quote company data, showing for example the number of stock items in log-intervals of annual dollar demand for a 42,000 item inventory. Half the items represented 1% of sales. As well as showing the general fit of the lognormal, Brown was in a position to comment that it tended to underestimate the proportion of very low value items.

Morrell (1967, p. 25) said of demand rates and lead times "The distribution most commonly found in this connection is the Lognormal, as might be expected, since a negative demand is impossible." Baily (1971) mentioned the lognormal but settled for ABC analysis. Inventory control models based on lognormal assumptions were developed by Howard and Schary (1972) and Herron (1974). Crouch and Oglesby (1978) wrote " In most stockholding situations the log-normal distribution adequately represents the actual distribution of demand rates." The authors' experience was from stock holdings of sewing threads at a knitwear factory, and they derived frequency of reordering rules. Peterson and Silver (1979, p. 724) said "...empirically it has been found that annual dollar usage across a population of items tends to exhibit a pattern of remarkable regularity that can be described by a lognormal distribution," illustrated an example from a Canadian company, and gave references to work on parameter estimation in the inventory field.

Das (1983) assumed lognormal demand in the reorder point model to express the expected number of back orders per lead time, given that r is the reorder point, via

$$B(r) = \int_{r}^{\infty} (x - r)\, d\Lambda(x \mid \mu, \sigma^2)$$

and developed a quadratic to express $B(r)$ as a function of p, the probability of x exceeding r for given μ and σ^2. Aggarwal (1984) referred to the common use of lognormal assumptions for inventory items.

ABC analysis and Lorenz curves also appear in the literature of quality control. The lognormal was used by Morrison (1958) to model the deviations from prescribed standards in testing thermionic valves. Ferrell

(1958) developed quality control charts for lognormal cases. Juran and Gryna (1970) showed hypothetical curves for losses due to scrap, rework, handling complaints etc., apologizing for attributing them to Pareto rather than Lorenz (fn. p. 63). Individual figures were mentioned, such as 6% of items accounting for 78% of quality losses.

3.2 Marketing

In a textbook on statistics, Croxton and Cowden (1955) fitted the lognormal to kilowatt hours of electricity used in an Eastern city. But the distribution has been largely ignored in marketing, despite a vast literature of model-building to represent demand. Even individual points on potential Lorenz curves have seldom been identified. Using a large body of information taken from National Consumer Panel records, Fourt and Woodlock (1960) reported "Experience shows that, when buyers are grouped by purchase rates into equal thirds, the heavy buying third accounts for 65 percent of the total volume, the middle third for 25 percent, and the light third for only 10 percent" (rounded figures consistent with $\sigma = 0.8$). Twedt (1964) also used panel data to divide product users into two halves. The 'heavy half' accounted for 87–91% of consumption in such categories as beer, colas, dog food and ready-to-eat cereals. The data were still quoted to illustrate volume segmentation in a textbook twenty years later (Kotler 1984).

The first application of the lognormal beyond "80/20" points was by Brown, Hulswit and Kettelle (1956). These operational researchers studied the entire potential market for a company disguised as a large commercial printer. Seven revenue groups and the revenue from those groups were plotted on lognormal probability paper. The fit was found to be good. The 50% of smallest buyers accounted for 3% of sales. Immediate action could be taken: the small accounts received no more salesmens' calls. In conjunction with sales response curves derived from experimental results with new and existing accounts, the optimal size of the sales force was derived. This study was included in a collection of articles on mathematical models in marketing (Bass et al. 1961) with editorial comment recommending parameter estimation by two quantiles and commenting on the general applicability of the approach. However, no follow-up has occurred.

The lognormal appeared briefly in a media model developed for an advertising agency. Charnes et al. (1968) proposed that the proportions of a gross audience reading 1, 2, 3, ... copies of a periodical were lognormally distributed. But models based on the beta-binomial and other distributions were later introduced and preferred.

Easton (1975, 1980) investigated U.K. sales figures of 19 brands in separate product fields, taken from fast-moving industrial goods such as chemicals and packaging materials. Buying companies were analysed by purchase size. The LSD, Yule, gamma and lognormal distributions were fitted to the data. The lognormal gave clearly the best fit in each case. Easton also analysed purchase sizes over two consecutive time periods, establishing the transition matrix from size in one quarter to size in the next. Using a procedure suggested by Adelman (1958) to allow separate estimation of transitions from non-buyer status, he found that a stationary transition matrix and a bivariate lognormal model accounted for the purchase size records. He suggested that the complex, interacting flows of demand leading to industrial product purchase would produce a multiplicative or proportioning effect on the derived demand measured in a industrial selling organization. Easton (1974) was the first marketing writer to suggest that the 80/20 rule-of-thumb could be generalized and made complete by the use of lognormally-based Lorenz curves.

In consumer goods, Lawrence (1980) used consumer panel data on toothpaste purchasing trips by nearly 4,000 households over a three-year period. A left-truncated lognormal provided an impeccable fit, not only to all households but to breakdowns of the total by seven classifications: housewife age, income, household head education and occupation, city size, main brand used, and percentage of deal pack purchased. Log means and variance were calculated for each subgroup within a classification and were found to combine back correctly to the all households mean and variance. Left truncation was used because several hundred households were recorded as purchasing only once or twice in the three years. Lawrence argued that these households were not homogeneous with the majority of American homes practicing regular dental care, and suggested that the "ever bought" definition of a purchasing population was therefore unsatisfactory. He proposed a population of stockists, i.e., people who keep a product continuously on hand for use as required, replacing each used-up pack with a new one. The assumption defined a renewal process, with the lifetime or use-up period of the product as a random variable. It proved that when the numbers of light buyers were extrapolated from parameters calculated for households purchasing ten times or more, the lognormal fitted down to 7 purchases. For 6 or fewer purchases the lognormal under-predicted actual numbers.

Left-truncation was also applied to data selected by Ehrenberg (1959, 1972) to demonstrate the close fit of the negative binomial distribution (NBD) to consumers grouped by purchase frequency. With the truncation point at one purchase, the lognormal fitted the data almost as well, and to a high standard. Extrapolated back to the origin, the lognormal predicted the number of zero buyers as 290, compared with 1,612 who actually did not buy. Lawrence argued that the extrapolation differentiated light buyers who happened not to have bought within the six-month data period, but who as stockists were homogeneous with the buyer group, from genuine non-buyers who did not buy goods of the class in question.

In summary, the business areas most involved with sales analysis have made only modest use of the lognormal distribution. Some awareness of differences in the size of the customers—perhaps their most important attribute for a business organization—is covered mainly by references to the "80/20 rule."

REFERENCES

Adam, D. (1958). *Les Réactions du consommateur devant le prix.* SEDES, Paris. Part translated in *Pricing Strategy* (Taylor, B. and Wills, G., eds.), Staples Press, London, 75–88.

Adelman, I.G. (1958). A stochastic analysis of the size distribution of firms, *J. Amer. Statist. Assoc.*, *53*, 893–904.

Aggarwal, V. (1984). Modelling of distributions by value for multi-item inventory, *IEE Trans.*, *16*, 90–98.

Airth, A.D. (1985). The progression of wage distributions, *Eurostat News* Special number, Recent developments in the analysis of large-scale data sets, Office for Official Publications of the European Community, Luxembourg, 139–161.

Aitchison, J. and Brown, J.A.C. (1954). On criteria for descriptions of income distribution, *Metroeconomica*, *6*, 88–107.

Aitchison, J. and Brown, J.A.C. (1957). *The Lognormal Distribution*, University Press, Cambridge.

Alexander, S.S. (1964). Price movements in speculative markets: trends or random walks, No.2, *Indust. Management Rev.*, *5*, 25–46.

Amoroso, E. (1942). Nuove ricerce intorno alla distribuzione delle malattie per durata, *Atti Ist. Naz. Assic.*, *14*, 185–202.

Amoroso, L. (1934). La rappresentazione analitica delle curve di frequenza nei sinistri di infortuni e di responsabilita civile, *Atti X Cong. Intern. Attu.*, *3*, 458–472.

Bachelier, L. (1900). *Theory of Speculation.* Thesis presented to the Faculty of Sciences of the Academy of Paris. Translated in *The Random Character of Stock Market Prices* (Cootner, P.H., ed., 1964), The M.I.T. Press, Cambridge, Mass., 17–78.

Bailey, A.L. (1943). Sampling theory in casualty insurance, *Proc. Casualty Actu. Soc.*, *30*, 31–65.

Baily, P. (1971). *Successful Stock Control by Manual Systems*, Gower Press Ltd., London.

Ballou, R.H. (1973). *Business Logistics Management*, Prentice-Hall, Englewood Cliffs, N.J.

Bass, F.M., Buzzell, R.D., Greene, M.R., Lazer, W., Pessemier, E.A., Shawver, D.L., Shuchman, A., Theodore, C.A., and Wilson, G.W., eds, (1961). *Math-*

ematical Models and Methods in Marketing, Richard D. Irwin, Homewood, Ill.

Bawa, V.S. and Chakrin, L.M. (1979). Optimal portfolio choice and equilibrium in a lognormal securities market, *Portfolio Theory, 25 Years After* (Elton, E.J. and Gruber, M.J., eds.), North-Holland Publishing Co., Amsterdam–New York–Oxford, 47–62.

Bawa, V.S. and Lindenberg, E.B. (1977). Capital market equilibrium in a mean-lower partial moment framework, *J. Financial Econ., 5*, 189–200.

Beard, R.E. (1957). Analytical expressions of the risks involved in general insurance, *Trans. XV Internat. Congr. Actu.*, New York, *1*, 100–127.

Beard, R.E., Pentikäinen, T. and Pesonen, E. (1984). *Risk Theory: The Stochastic Basis of Insurance*, 3rd ed., Chapman and Hall Monographs on Statistics and Applied Probability, London-New York.

Beckers, S. (1980). The constant elasticity of variance model and its implications for option pricing, *J. Finance, 35*, 661–673.

Benckert, L.-G. (1962). The lognormal model for the distribution of one claim, *ASTIN Bull., 2*, 9–23.

Benckert, L.-G. and Jung, J. (1974). Statistical models of claim distributions in fire insurance, *ASTIN Bull., 8*, 1–25.

Bickerstaff, D.R. (1972). Automobile collision deductibles and repair cost groups: the lognormal model, *Proc. Casualty Actu. Soc., 59*, 68–102.

Black, F., Jensen, M.C. and Scholes, M. (1972). The Capital Asset Pricing Model: some empirical tests, *Studies in the Theory of Capital Markets* (M.C. Jensen, ed.), Praeger Publishers, New York, 79–121.

Black, F. and Scholes, M. (1973). The pricing of options and corporate liabilities, *J. Political Econ., 81*, 637–654.

Blattberg, R.C. and Gonedes, N.J. (1974). A comparison of the stable and Student distributions as statistical models of stock prices, *J. Business, 47*, 244–280.

Boland, L.A. (1977). Giffen goods, market prices and testability, *Austr. Econ. Papers, 16*, 72–85.

Boness, J.A., Chen, A.H. and Jatusipitak, S. (1974). Investigations in nonstationarity in prices, *J. Business, 47*, 518–537.

Bourguignon, F. (1979). Decomposable income inequality measures, *Econometrica, 47*, 901–920.

Bowersox, D.J., Smykay, E.W. and La Londe, B.J. (1968). *Physical Distribution Management*, Macmillan, New York.

Brennan, M.J. and Schwartz, E.S. (1985). On the geometric mean index: a note, *J. Financial and Quant. Anal., 20*, 119–122.

Brown, A.A., Hulswit, F.T. and Kettelle, J.D. (1956). A study of sales operations, *Operations Res., 4*, 296–308.

Brown, D. and Huang, C.F. (1981). Option pricing in a lognormal securities market with discrete trading: a comment, *J. Financial Econ., 12*, 285–286.

Brown, R.G. (1959). *Statistical Forecasting for Inventory Control*, McGraw-Hill Book Co., New York.

Brown, R.G. (1967). *Decision Rules for Inventory Management*, Holt, Rinehart and Winston, New York.

Bühlmann, H. and Hartmann, W. (1956). Anderungen in der Grundgesamtheit der Betriebsunfallkosten, *Mitt. Verein. Schweiz. Versich. Mathr.*, *56*, 303–320.

Cardozo, R.N. and Smith, D.K. Jr. (1983). Applying financial portfolio theory to product portfolio decisions: an empirical study, *J. Marketing*, *47*, 110–119.

Champernowne, D.G. (1953). A model of income distribution. *Econ. J.*, *63*, 318–351.

Champernowne, D.G. (1973). *The Distribution of Income Between Persons*, University Press, Cambridge.

Charnes, A., Cooper, W.W., Devoe, J.K., Learner, D.B. and Reinecke, W. (1968). A goal programming model for media planning, *Management Sci.*, *14*, B423–B430.

Cheng, D.C. and Karson, M.J. (1985). On the use of the geometric mean in long-term investment, *Decision Sci.*, *16*, 1–13.

Chesher, A. (1979). Testing the law of proportionate effect, *J. Indust. Econ.*, *27*, 403–411.

Chhikara, R.S. and Folks, J.L. (1978). The inverse Gaussian distribution and its statistical application–a review, *J. Roy. Statist. Soc. B*, *40*, 263–289.

Clark, P.K. (1973). A subordinated stochastic process model with finite variance for speculative prices, *Econometrica*, *41*, 135–155.

Clarke, R. (1979). On the lognormality of firm and plant size distributions: some U.K. evidence, *Appl. Econ.*, *11*, 415–433.

Cootner, P.H. (1962). Stock prices: random vs. systematic change, *Indust. Management Rev.*, *3*, 24–45.

Cowell, F.A. (1975). On the estimation of a lifetime income—a correction, *J. Amer. Statist. Assoc.*, *70*, 588–589.

Cowell, F.A. (1977). *Measuring Inequality*, Philip Allen Publishers, Oxford.

Cox, J.C. and Ross, S.A. (1976). The evaluation of options for alternative stochastic processes, *J. Financial Econ.*, *3*, 145–166.

Creedy, J. (1985). *Dynamics of Income Distribution*, Basil Blackwell Ltd., Oxford.

Crouch, R.B. and Oglesby, S. (1978). Optimization of a few lot sizes to cover a range of requirements, *J. Operational Res. Soc.*, *29*, 897–904.

Croxton, F.E. and Cowden, D.J. (1955). *Applied General Statistics*, Pitman & Sons Ltd., London.

Dale, C. (1981). Brownian motion in the Treasury Bill futures market, *Bus. Econ.*, *16*, 47–54.

Das, C. (1983). Inventory control for lognormal demand, *Computers and Operations Res.*, *10*, 267–276.

Devinney, T.M., Stewart, D.W. and Shocker, A.D. (1985). A note on the application of portfolio theory: a comment on Cardozo and Smith, *J. Marketing*, *49*, 107–112.

de Wit, G.W. and Kastelijn, W.M. (1977). An analysis of claim experience in private health insurance to establish a relation between deductibles and premium rebates, *ASTIN Bull.*, *9*, 257–266.

Dropkin, L.B. (1966). Loss distribution of a single claim, *The Mathematical Theory of Risk*, Casualty Actu. Soc., New York.

Easton, G. (1974). The 20-80 rule revisited: distribution of industrial purchase by size, *Proc. Marketing Education Group* (R.J. Lawrence, ed.), University of Lancaster, UK, 120–130.

Easton, G. (1975). Patterns of Industrial Buying, Unpublished Ph. D. thesis, University of London.

Easton, G. (1980). Stochastic models of industrial buying behaviour, *Omega*, *8*, 63–69.

Ehrenberg, A.S.C. (1959). The pattern of consumer purchases, *Appl. Statist.*, *8*, 26–41.

Ehrenberg, A.S.C. (1972). *Repeat-Buying: Theory and Applications*, North-Holland Publishing Co., Amsterdam-London-New York.

Elton, E.J. and Gruber, M.J. (1974). Portfolio theory when investment relatives are lognormally distributed, *J. Finance*, *29*, 1265–1273.

Elton, E.J., Gruber, M.J. and Gultekin, M. (1981). Expectations and share prices, *Management Sci.*, *27*, 975–987.

Epps, T.W. and Epps, M.L. (1976). The stochastic dependence of security price changes and transaction volumes: implications for the mixture-of-distributions hypothesis, *Econometrica*, *44*, 305–321.

Fama, E.F. (1965). The behavior of stock market prices, *J. Business*, *38*, 34–105.

Fama, E.F. and Blume, M.E. (1966). Filter rules and stock market trading, *J. Business*, *39*, 226–241.

Fase, M.M.G. (1971). On the estimation of lifetime income, *J. Amer. Statist. Assoc.*, *66*, 686–692.

Ferrara, G. (1971). Distributions des sinistres incendie selon leur coût, *ASTIN Bull.*, *6*, 31–41.

Ferrell, E.B. (1958). Control charts for log-normal universes, *Indust. Quality Control*, *15*, 4–6.

Finger, R.J. (1976). Estimating pure premiums by layer—an approach, *Proc. Casualty Actu. Soc.*, *63*, 34–52.

Fisk, P. (1961). The graduation of income distributions, *Econometrica*, *29*, 171–184.

Fogler, H.R. (1978). *Analyzing the Stock Market: Statistical Evidence and Methodology*, 2 ed. Grid, Inc., Columbus, Ohio.

Fouilhé, P. (1960). Evaluation subjective des prix. *Revue Française de Sociologie*, *1*: 163–172. Translated in *Pricing Strategy* (Taylor, B. and Wills, G., eds.), Staples Press, London, 89–97.

Fourt, L.A. and Woodlock, J.W. (1960). Early prediction of market success for new grocery products, *J. Marketing*, *25*, 31–38.

Fowlkes, E.B. (1979). Some methods for studying the mixture of two normal (lognormal) distributions, *J. Amer. Statist. Assoc.*, *74*, 561–575.

Frankfurter, G., Stevenson, R. and Young, A. (1979). Option spreading: theory and an illustration, *J. Portfolio Mgmt.*, *5*, 59–63.

Gabor, A. and Granger, C.W.J. (1961). On the price consciousness of consumers, *Appl. Statist.*, *10*, 170–188.

Gabor, A. and Granger, C.W.J. (1966). Price as an indicator of quality: report on an enquiry, *Economica*, *33*, 43–70.

Gibrat, R. (1931). *Les Inégalités Economiques*, Librairie du Recueil Sirey, Paris.

Granger, C.W.J. and Billson, A. (1972). Consumer attitudes toward package size and price, *J. Marketing Res.*, *9*, 239–248.

Granger, C.W.J. and Morgenstern, O. (1963). Spectral analysis of New York stock market prices, *Kuklos*, *16*, 1–27.

Grubb, D. (1985). Ability and power over production in the distribution of earnings, *Rev. Econ. and Statist.*, *67*, 188–194.

Hadley, G. and Whitin, T.M. (1963). *Analysis of Inventory Systems*, Prentice-Hall, Englewood Cliffs, N.J.

Hakansson, N.H. (1971). Capital growth and the mean-variance approach to portfolio selection, *J. Financial and Quant. Anal.*, *6*, 517–558.

Hansen, L.P. and Singleton, K.J. (1983). Stochastic consumption, risk aversion, and the temporal behavior of asset returns, *J. Political Econ.*, *91*, 249–265.

Harding, V. (1968). The calculation of premiums for excess of loss reinsurances of motor business, *Trans. XVIII Internat. Congr. Actu.*, *1*, 17–34.

Harrison, A. (1979). The upper tail of the earnings distribution: Pareto or lognormal? *Econ. Letters*, *2*, 191–195.

Hart, P.E. (1975). Moment distributions in economics: an exposition, *J. Roy. Statist. Soc. A*, *138*, 423–434.

Hart, P.E. (1982). Entropy, moments, and aggregate business concentration in the UK, *Oxford Bull. of Econ. and Statist.*, *44*, 113–126.

Hart, P.E. (1983). The size mobility of earnings, *Oxford Bull. of Econ. and Statist.*, *45*, 181–193.

Hart, P.E. and Prais, S.J. (1956). The analysis of business concentration: a statistical approach, *J. Roy. Statist. Soc. A*, *119*, 150–191.

Henry, M. (1937). Etude sur le coût moyen des sinistres en responsabilité civile automobile. *Bull. Trimestrial Inst. Actu. Franç.*, *43*, 113–178.

Hermann, R. and Montroll, E.W. (1977). Some statistical observations from a 75 year run of Sears Roebuck catalogues, *Statistical Mechanics and Statistical Methods in Theory and Application* (U. Landman, ed.), Plenum Press, New York, 785–803.

Herron, D.P. (1974). Profit oriented techniques for managing independent demand inventories, *Prod. & Inventory Mgmt.*, *15*, 57–74.

Heskett, J.L. (1977). Logistics—essential to strategy, *Harvard Bus. Rev.*, *55*, 85–96.

Hilliard, J.E. (1983). The geometric mean in investment applications: transient and steady state parameters, *Financial Rev.*, *18*, 326–335.

Hilliard, J.E. and Clayton, R.J. (1982). Obtaining and parameterizing multiperiod portfolios with desirable characteristics under lognormal returns, *Decison Sci.*, *13*, 240–250.

Howard, K. and Schary, P. (1972). Product line and inventory strategy, *Decision Sci.*, *3*, 41–58.

Hsu, D.A. (1984). The behavior of stock returns: is it stationary or evolutionary? *J. Financial and Quant. Anal.*, *19*, 11–28.

Hsu, D.A., Miller, R.B. and Wichern, D.W. (1974). On the stable Paretian behavior of stock-market prices, *J. Amer. Statist. Assoc.*, *69*, 108–113.

Ijiri, Y. and Simon, H.A. (1955). Business firms growth and size, *Amer. Econ. Rev.*, *54*, 77–89.

Iyenger, N.S. (1960). On a method of computing Engel elasticities from concentration curves, *Econometrica*, *28*, 882–891.

Jain, L.R. (1977). On fitting the three-parameter log-normal distribution to consumer expenditure data, *Sankhya C*, *39*, 61–73.

Jarrow, R. and Rudd, A. (1982). Approximate option valuation for arbitrary stochastic processes, *J. Financial Econ.*, *10*, 347–369.

Jensen, M.C. (1968). The performance of mutual funds in the period 1945–1964, *J. Finance*, *23*, 389–416.

Jones, C.P. and Litzenberger, R.H. (1970). Quarterly earnings reports and intermediate stock price trends, *J. Finance*, *25*, 143–148.

Juran, J.M. and Gryna, F.M., Jr. (1970). *Quality Planning and Analysis*, McGraw-Hill Book Co., New York.

Kakwani, N.C. (1980). *Income Inequality and Poverty*, University Press, Oxford.

Kalecki, M. (1954). On the Gibrat distribution, *Econometrica*, *13*, 161–170.

Kendall, M.G. (1953). The analysis of economic time-series. Part 1: Prices, *J. Roy. Statist. Soc. A*, *96*, 11–25.

Kotler, P. (1984). *Marketing Management: Analysis, Planning and Control*, 5th ed. Prentice-Hall, Englewood Cliffs, N.J.

Krausz, J. and Harpaz, G. (1984). An empirical investigation of the Lee, Rao and Auchmuty option pricing model, *Mid-Atlantic J. Bus.*, *22*, 1–14.

Latané H.A. (1959). Criteria for choice among risky ventures, *J. Political Econ.*, *67*, 144–155.

Latané, H.A. and Tuttle, D.L. (1967). Criteria for portfolio building, *J. Finance*, *22*, 359–373.

Latscha, R. (1956). Zur Andwendung der kollectiven Risikotheorie in der schweizerischen obligatorischen Unfallversicherung, *Mitt. Verein. Schweiz. Versich. Mathr.*, *56*, 275–302.

Lawrence, R.J. (1980). The lognormal distribution of buying frequency rates, *J. Marketing Res.*, *17*, 212–220.

Lee, W.Y., Rao, R.K.S. and Auchmuty, J.F.G. (1981). Option pricing in a lognormal securities market with discrete trading, *J. Financial Econ.*, *9*, 75–101.

Lintner, J. (1972). *Equilibrium in a random walk and lognormal securities market*, Harvard Institute of Economic Research Discussion Paper No. 235, Harvard University, Cambridge, Mass.

Little, I.M.D. and Rayner, A.C. (1966). *Higgledy Piggledy Growth Again*, Basil Blackwell, Oxford.

Lockyer, K. (1972). *Stock Control: A Practical Approach*, Cassell & Co. Ltd., London.

Lorie, J.H. and Hamilton, M.T. (1973). *The Stock Market: Theories and Evidence*, Richard D. Irwin, Homewood, Ill.

Lundberg, O. (1940). *On Random Processes and Their Application to Sickness and Accident Statistics*, Almquist and Wicksells, Uppsala.

Lydall, H.F. (1968). *The Structure of Earnings*, Clarendon Press, Oxford.

Magee, J.F. (1960). The logistics of distribution, *Harvard Bus. Rev.*, *38*, 89–101.

Mandelbrot, B. (1960). The Pareto-Levy Law and the distribution of income, *Internat. Econ. Rev.*, *1*, 79–106.

Mandelbrot, B. (1963). The variation of certain speculative prices, *J. Business*, *36*, 394–419.

Marcus, A. and Shaked, I. (1984). The relationship between accounting measures and prospective probabilities of insolvency: an application to the banking industry, *Financial Rev.*, *19*, 67–83.

Markowitz, H.M. (1952). Portfolio selection, *J. Finance*, *7*, 77–91.

Markowitz, H.M. (1959). *Portfolio Selection*, Wiley, New York.

Marlin, P. (1984). Fitting the log-normal distribution to loss data subject to multiple deductibles, *J. Risk and Insurance*, *51*, 687–701.

Marsh, T.A. and Rosenfeld, E.R. (1983). Stochastic processes for interest rates and equilibrium bond prices, *J. Finance*, *38*, 635–646.

McDonald, J.B. and Ransom, M.R. (1979). Functional forms, estimation techniques and the distribution of income, *Econometrica*, *47*, 1513–1525.

Merton, R.C. (1971). Optimum consumption and portfolio rules in a continuous-time model, *J. Econ. Theory*, *3*, 373–413.

Merton, R.C. (1976). Option pricing when underlying stock returns are discontinuous, *J. Financial Econ.*, *3*, 125–144.

Merton, R.C. (1980). On estimating the expected return on the market: an exploratory investigation, *J. Financial Econ.*, *8*, 323–361.

Mittra, S. and Gassen, C. (1981). *Investment Analysis and Portfolio Management*, Harcourt Brace Jovanovich, New York.

Monroe, K.B. (1973). Buyers' subjective perceptions of price, *J. Marketing Res.*, *10*, 70–80.

Moore, A.B. (1964). Some characteristics of changes in common stock prices, *The Random Character of Stock Market Prices* (Cootner, P.H. ed.), The M.I.T. Press, Cambridge, Mass., 139–161.

Morrell, A.J.H. (1967). *Mathematical and Statistical Techniques for Industry Monograph No. 4: Problems of Stocks and Storage*, Oliver and Boyd for Imperial Chemical Industries Ltd., Edinburgh.

Morrison, J. (1958). The lognormal distribution in quality control, *Appl. Statist.*, *7*, 160–172.

Ohlsen, J.A. and Ziemba, W.T. (1976). Portfolio selections in a lognormal market when the investor has a power utility function, *J. Financial and Quant. Anal.*, *7*, 57–71.

O'Neill, B. and Wells, W.T. (1972). Some recent results in lognormal parameter estimation using grouped and ungrouped data, *J. Amer. Statist. Assoc.*, *67*, 76–79.

Osborne, M.F.M. (1959). Brownian motion in the stock market, *Operations Res.*, *7*, 145–173.

Osborne, M.F.M. (1962). Periodic structure in the Brownian motion of stock prices, *Operations Res.*, *10*, 345–379.

Peterson, R. and Silver, E.A. (1979). *Decision Systems for Inventory Management and Production Planning*, Wiley, New York.

Praetz, P.D. (1972). The distribution of share price changes, *J. Business*, *47*, 49–55.

Prais, S.J. (1976). *The Evolution of Giant Firms in Britain*, University Press, Cambridge.

Price, K., Price, B. and Nantell, T.J. (1982). Variance and lower partial moment measures of systematic risk: some analytical and empirical results, *J. Finance*, *37*, 845–855.

Quandt, R.E. (1964). Statistical discrimination among alternative hypotheses and some economic regularities, *J. Regional Sci.*, *5*, 1–23.

Quandt, R.E. (1966). On the size distribution of firms, *Amer. Econ. Rev.*, *56*, 416–432.

Quensel, C.E. (1944). *Inkomstfördelning och skattetryck*, Sveriges Industriforbund, Stockholm.

Ramel, M. (1960). Tarification d'un traité d'excédent de pourcentage de sinistres, *C.R. XVI Internat. Congr. Actu.*, Brussels, *1*, 540–561.

Ransom, M.R. and Cramer, J.S. (1983). Income distribution functions with disturbances, *Eur. Econ. Rev.*, *22*, 363–372.

Resek, R.W. (1975). Symmetric distributions with fat tails: interrelated compound distributions estimated by Box-Jenkins methods, *Statistical Distributions in Scientific Work*, Vol. 2 (Patil, G.P., Kotz, S. and Ord, J.K., eds.) Reidel, Dordrech and Boston, 158–173.

Rhodes, E.C. (1944). The Pareto distribution of incomes, *Economica*, *11*, 1–11.

Roberts, H.V. (1959). Stock-Market "patterns" and financial analysis: methodological suggestions, *J. Finance*, *14*, 1–10.

Rogalski, R.J. (1978). Variances and option prices in theory and practice, *J. Portfolio Mgmt.*, *4*, 43–51.

Rosenberg, B. (1973). The behavior of random variables with non-stationary variance and the distribution of security prices. Unpublished paper, University of California, Berkeley.

Salem, A.B.Z. and Mount, T.D. (1974). A convenient descriptive model of income distribution: the gamma density, *Econometrica*, *42*, 1115–1127.

Samson, D. and Thomas, H. (1983). Reinsurance decision making and expected utility, *J. Risk and Insurance*, *50*, 249–264.

Seal, H.L. (1969). *Stochastic Theory of a Risk Business*, Wiley, New York.

Seal, H.L. (1978). *Survival Probabilities: The Goal of Risk Theory*, Wiley, New York.

Shaked, I. (1985). Measuring prospective probabilities of insolvency: an application to the life insurance industry, *J. Risk and Insurance*, *52*, 59–80.

Sharpe, W.F. (1963). A simplified model for portfolio analysis, *Management Sci.*, *9*, 277–293.

Sharpe, W.F. (1964). Capital asset prices: a theory of market equilibrium under conditions of risk, *J. Finance*, *19*, 425–442.

Sharpe, W.F. (1965). Risk aversion in the stock market: some empirical evidence, *J. Finance*, *20*, 416–422.

Sharpe, W.F. (1966). Mutual fund performance, *J. Business*, *39*, 119–138.

Simkowitz, M.A. and Beedles, W.L. (1980). Asymmetric stable distributed security returns, *J. Amer. Statist. Assoc.*, *75*, 306–312.

Simon, H.A. (1955). On a class of skew distribution functions, *Biometrika*, *52*, 425–440.

Simon, H.A. (1957). The compensation of executives, *Sociometry*, *20*, 32–35.

Simon, H.A. and Bonini, C.P. (1958). The size distribution of business firms, *Amer. Econ. Rev.*, *48*, 607–617.

Singh, A. and Whittington, G. (1975). The size and growth of firms, *Rev. Econ. Studies*, *42*, 15–26.

Singh, S.K. and Maddala, G.S. (1976). A function for size distribution of incomes, *Econometrica*, *44*, 963–970.

Sprenkle, C.M. (1961). Warrant prices as indicators of expectations and preferences, *Yale Econ. Essays*, *1*, 178–231.

Stapleton, R.C. and Subrahmanyam, M.G. (1984). The valuation of multivariate contingent claims in discrete time models, *J. Finance*, *39*, 207–228.

Steindl, J. (1965). *Random Processes and the Growth of Firms*, Griffin, London.

Stoetzel, J., Sauerwein, J. and de Vulpian, A. (1954). Sondages Français: études sur la consommation, *La Psychologie Economique* (Reynaud P.L., ed.) Librairie Marcel Rivière et Cie, Paris, 161–209.

Tehranian, H. and Helms, B.P. (1982). An empirical comparison of stochastic dominance among lognormal prospects, *J. Financial and Quant. Anal.*, *17*, 217–226.

Thatcher, A.R. (1968). The distribution of earnings of employees in Great Britain, *J. Roy. Statist. Soc., A*, *131*, 133–170.

Thatcher, A.R. (1976). The new earnings survey and the distribution of earnings, *The Personal Distribution of Income* (A.B. Atkinson, ed.), George Allen and Unwin, London, 227–268.

Thomas, A.B. (1968). *Stock Control in Manufacturing Industries*, Gower Press Ltd., London.

Twedt, D.W. (1964). How important to marketing strategy is the 'heavy user'? *J. Marketing*, *28*, 71–72.

Upton, D.E. and Shannon, D.S. (1979). The stable Paretian distribution, subordinated stochastic processes, and asymptotic lognormality: an empirical investigation, *J. Finance*, *34*, 1031–1039.

Van Dijk, H-K. and Kloek, T. (1980). Inferential procedures in stable distributions for class frequency data on incomes, *Econometrica*, *48*, 1139–1148.

Van Herwaarden, F.G. and de Kam, C.A. (1983). An operational concept of the ability to pay principle (with an application for the Netherlands, 1973), *Economist* (Netherlands), *13*, 55–64.

Van Herwaarden, F.G. and Kapteyn, A. (1981). Empirical comparison of the shape of welfare functions, *Eur. Econ. Rev.*, *15*, 261–286.

Van Praag, B.M.S. (1986). *Individual Welfare Functions and Consumer Behavior*, North-Holland, Amsterdam.

Van Praag, B.M.S. (1978). The perception of income inequality, *Personal Income Distribution* (Krelle, W. and Shorrocks, A.F., eds.), North-Holland, Amsterdam, 113–136.

Westerfield, R. (1977). The distribution of stock price changes: an application of transactions time and subordinated stochastic models, *J. Financial and Quant. Anal.*, *12*, 743–763.

Working, H. (1934). A random difference series for use in the analysis of time series, *J. Amer. Statist. Assoc.*, *29*, 11–24.

Working, H. (1960). Note on the correlation of first differences of averages in a random chain, *Econometrica*, *28*, 916–918.

Ying, C.C. (1966). Stock market prices and volume of sales, *Econometrica*, *34*, 676–685.

Young, W.E. and Trent, R.H. (1969). Geometric mean approximations of individual security and portfolio performance, *J. Financial and Quant. Anal.*, *2*, 179–199.

Ziai, Y. (1979). Statistical models of claim amount distributions in general insurance. Unpublished Ph.D. thesis for the Actuarial Science Division, Department of Mathematics, The City University, London.

10

Applications in Industry

S. A. SHABAN Faculty of Commerce, Economics and Political Science, Department of Insurance and Statistics, Kuwait University, Kuwait

1. INTRODUCTION

The purpose of this chapter is to review the application of the lognormal distribution as a model in some industrial problems. We include in the chapter the problem of estimating the reliability function when the underlying distribution is lognormal. The application of the lognormal distribution in certain industrial areas, such as quality control charts, optimal replacement of devices and labor turnover, will also be reviewed.

While space permits only a few instances to be discussed, a list of additional references is appended to suggest the breadth of application and possible additional reading.

2. RELIABILITY ESTIMATION

The lognormal distribution, like the Weibull and the exponential distributions, has been used widely as a life time distribution model. The distribution is specified by saying that the life time of X is lognormal if the logarithm $Y = \log X$ of the life time is normal with mean μ and variance σ^2. The reliability function at time t—in some applications it is termed the survivor function—of the lognormal is easily seen to be

$$R(t) = R(t; \mu, \sigma) = 1 - \Phi(\delta) \tag{2.1}$$

where

$$\delta = \frac{\log t - \mu}{\sigma} \tag{2.2}$$

and $\Phi(.)$ is the standard normal distribution function.

The reliability function has the appearance shown in Figures 1a and 1b.

Estimates of $\Phi(\delta)$ are discussed in the literature for both normal and lognormal life data. In this section we consider the classical and the structural approaches towards estimating (2.1). In order to avoid repetition we have omitted discussion on Bayesian estimation of (2.1) since it is already discussed in Chapter 6.

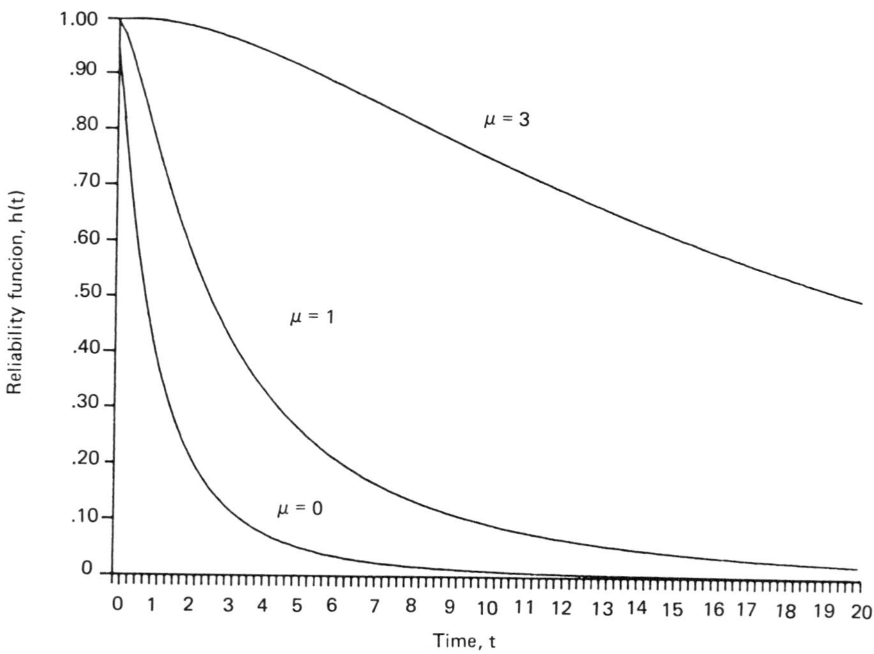

Figure 1a Lognormal reliability function with $\sigma = 1$ and $\mu = 0, 1, 3$.

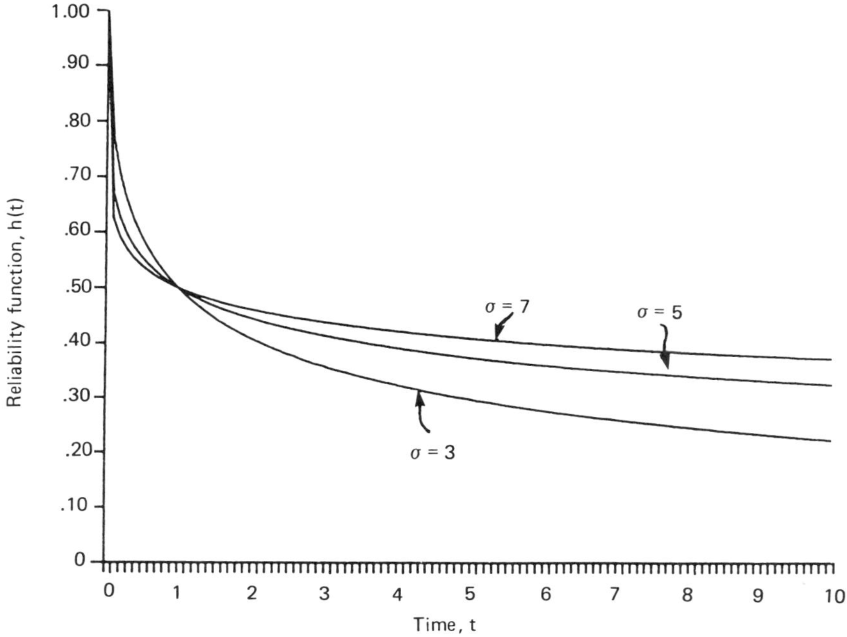

Figure 1b Lognormal reliability function with $\mu = 0$ and $\sigma = 3, 5, 7$.

2.1 Classical Estimates

2.1.1 *The MVUE and the MLE When σ is Known.* Let $X_1, X_2, \ldots, X_n$ be iid random variables having lognormal distribution $\Lambda(\mu, \sigma^2)$. If μ is unknown and σ^2 known, a complete sufficient statistic for μ is the sample mean of the transformed data, i.e.

$$\hat{\mu} = n^{-1} \sum \log X_i$$

Using a result given by Ellison (1964), the minimum variance unbiased estimator (MVUE) of (2.1) is

$$\hat{R}_{\text{MVU}}(t) = 1 - \Phi\left(\sqrt{\frac{n}{n-1}} \frac{\log t - \hat{\mu}}{\sigma}\right) \tag{2.3}$$

Since $\hat{\mu}$ is also the maximum likelihood estimator (MLE) of μ, and from the invariance property of the MLE, the MLE of (2.1) is

$$\hat{R}_{\mathrm{ML}}(t) = 1 - \Phi\left(\frac{\log t - \hat{\mu}}{\sigma}\right) \tag{2.4}$$

The expected value of $\hat{R}_{\mathrm{ML}}(t)$ is

$$E\hat{R}_{\mathrm{ML}}(t) = 1 - \Phi\left(\sqrt{\frac{n}{n+1}}\delta\right)$$

The bias of the MLE is thus

$$B(n,\delta) = \Phi(\delta) - \Phi\left(\sqrt{\frac{n}{n+1}}\delta\right)$$

The bias is positive for all $\delta > 0$, and decreases as δ increases. The mean square error (MSE) of both (2.3) and (2.4) involves the function $\Phi(u_1, u_2, \rho)$, which is the standard bivariate normal integral at the limits u_1 and u_2 and a correlation coefficient ρ. Specifically,

$$\mathrm{MSE}(\hat{R}_{\mathrm{MVU}}(t)) = \Phi\left(\delta, \delta; \frac{1}{n}\right) - \Phi^2(\delta), \tag{2.5}$$

$$\mathrm{MSE}(\hat{R}_{\mathrm{ML}}(t)) = \Phi\left(\sqrt{\frac{n}{n+1}}\delta, \sqrt{\frac{n}{n+1}}\delta, \sqrt{\frac{1}{n+1}}\right) - 2\Phi(\delta)\left[\Phi\left(\sqrt{\frac{n}{n+1}}\delta\right) - \frac{1}{2}\Phi(\delta)\right] \tag{2.6}$$

Zacks and Even (1966) studied the efficiency of both estimators in the normal case and concluded that the MVUE is more efficient than the MLE in the range of δ corresponding to the tails of the distribution. Calculating (2.5) and (2.6) requires special routines or tables for calculating $\Phi(u_1, u_2; \rho)$. A set of tables of $\Phi(u_1, u_2; \rho)$ was published by the National Bureau of Standards in 1959, for u_1, $u_2 = 0.0(0.1)4$ to six decimal places for $\rho = 0(0.05)0.95(0.01)1.0$ and to seven decimal places for $-\rho = 0(0.05)0.95(0.01)1.0$. For large enough n, asymptotic expansions of (2.5) and (2.6) to $O(n^{-3})$ are

$$E\hat{R}_{\mathrm{ML}}(t) = \Phi(-\delta) - \tfrac{1}{2}\delta\phi(\delta)\left[\frac{1}{n} + \frac{1}{4n^2}(\delta^2 - 3)\right] + O(n^{-3})$$

and

$$\mathrm{Var}(\hat{R}_{\mathrm{ML}}(t)) = \phi^2(\delta)\left[\frac{1}{n} + \frac{1}{n^2}\left(\tfrac{3}{2}\delta^2 - 1\right)\right] + O(n^{-3})$$

It is easy to show that

$$(E\hat{R}_{\mathrm{ML}}(t)) - R(t))^2 = \frac{1}{4n^2}\delta^2\phi^2(\delta) + O(n^{-3})$$

so that the MSE of $\hat{R}_{\mathrm{ML}}(t)$ is

$$E(\hat{R}_{\mathrm{ML}}(t)) - R(t))^2 = \phi^2(\delta)\left[\frac{1}{n} + \frac{1}{n^2}\left(\tfrac{7}{4}\delta^2 - 1\right)\right] + O(n^{-3}) \qquad (2.7)$$

Similarly

$$\mathrm{Var}(\hat{R}_{\mathrm{MVU}}(t)) = \phi^2(\delta)\left[\frac{1}{n-1} + \frac{1}{(n-1)^2}\left(\tfrac{3}{2}\delta^2 - 1\right)\right] + O(n^{-3}) \qquad (2.8)$$

where $\phi(.)$ is the standard normal pdf.

2.1.2 *The MVUE and the MLE When Both μ and σ Are Unknown.* The MLE of μ and σ^2 are

$$\hat{\mu} = \frac{1}{n}\sum \log X_i$$

$$\hat{\sigma}^2 = \frac{1}{n}\sum(\log X_i - \hat{\mu})^2$$

From the invariance property of the MLEs, the MLE of $R(t;\mu,\sigma)$ is

$$\hat{R}_{\mathrm{ML}}(t) = 1 - \Phi\left(\frac{\log t - \hat{\mu}}{\hat{\sigma}}\right) \qquad (2.9)$$

The MLE is biased. The bias is negative for all $\delta > 0$, and decreases as δ increases. The bias also decreases with increasing sample size.

Since $(\hat{\mu}, \hat{\sigma})$ is a minimal sufficient statistic, the MVUE of $R(t;\mu,\sigma)$ is

$$\hat{R}_{\mathrm{MVU}}(t) = \begin{cases} 1 & \text{if } W \leq 0 \\ 1 - I_W\left(\frac{n}{2} - 1, \frac{n}{2} - 1\right) & \text{if } 0 < W < 1 \\ 0 & \text{otherwise} \end{cases} \qquad (2.10)$$

where $I_Z(p,q)$ is the incomplete beta function ratio, and

$$W = \frac{1}{2}\left[1 + \frac{(\log t - \hat{\mu})}{\sqrt{n-1}\hat{\sigma}}\right]$$

The MSE of $\hat{R}_{\text{ML}}(t)$ and $\hat{R}_{\text{MVU}}(t)$ are derived by Zacks and Milton (1971). They have calculated that the variance of the MVU estimate is

$$V(\hat{R}_{\text{MVU}}(t)) = E_\delta\left[\Psi^2(\hat{\mu},\hat{\sigma})\right] - \Phi^2(\delta) \tag{2.11}$$

where $\Psi(\hat{\mu},\hat{\sigma}) = 1 - \hat{R}_{\text{MVU}}(t)$. The expectation is given as a function of the distribution function of the non-central t, $H(t \mid \nu, \varsigma)$, and the incomplete moments of the non-central t, where ν is the degree of freedom and ς is the non-centrality parameter. An expression for the expectation is

$$\begin{aligned} E_\delta\left[\Psi^2(\hat{\mu},\hat{\sigma})\right] &= H(-2m+1 \mid 2m-1, -\sqrt{2n\delta}) \\ &+ \frac{1}{B(m-1,m-1)} \sum_{i=m-1}^{2m-3} \frac{\binom{2m-3}{i} B(m-1+i, 3m-4-i)}{2^{4m-7}} \\ &\times \sum_{j=m-1+i}^{4m-6} \binom{4m-6}{j} M_j^*(2m-1, \sqrt{2m\delta}) \end{aligned}$$

where

$$M_j^*(\nu,\varsigma) = \int_{-\nu}^{\nu} \left(1+\frac{t}{\nu}\right)^j \left(1-\frac{t}{\nu}\right)^{2\nu-4-j} h(t \mid \nu,\varsigma)\, dt, j = 0,\ldots,2\nu-4$$

are the incomplete moments of the noncentral t-distribution. On the other hand the MSE of the MLE may be expressed as

$$\begin{aligned} \text{MSE}(\hat{R}_{\text{ML}}(t)) &= E\left[\Phi\left(\frac{\sqrt{n}\delta - U}{\sqrt{Q}}\right) - \Phi(\delta)\right]^2 \\ &= E\left[\Phi^2\left(\frac{\sqrt{n}\delta - U}{\sqrt{Q}}\right)\right] - 2\Phi(\delta)E\left[\Phi\left(\frac{\sqrt{n}\delta - U}{\sqrt{Q}}\right)\right] + \Phi^2(\delta) \end{aligned} \tag{2.12}$$

where $Q \sim \chi^2_{n-1}$, $U \sim N(0,1)$, and U and Q are independent. However this expression cannot be expressed in a closed form and needs to be evaluated numerically. As stated by Zacks and Milton (1971), the MVU estimator is more efficient than the MLE if $0.3 < R(t;\mu,\sigma) \leq 0.7$.

Both $\hat{R}_{\mathrm{MVU}}(t)$ and $\hat{R}_{\mathrm{ML}}(t)$ converge in law to the normal distribution. i.e.,

$$\sqrt{n}(\hat{R}_i(t) - \Phi(-\delta)) \longrightarrow N\left[0, \phi^2(\delta)\left(1 + \frac{\delta^2}{2}\right)\right], i = 1, 2$$

where

$$\hat{R}_1(t) = \hat{R}_{\mathrm{MVU}}(t) \qquad \text{and} \qquad \hat{R}_2(t) = \hat{R}_{\mathrm{ML}}(t)$$

Using results attributed to Hurt (1980), asymptotic expansions of $E\hat{R}_{\mathrm{ML}}(t)$ and the MSE's of $\hat{R}_{\mathrm{ML}}(t)$ and $\hat{R}_{\mathrm{MVU}}(t)$ are

$$E\hat{R}_{\mathrm{MVU}}(t) = \Phi(-\delta) \tag{2.13}$$

$$E\hat{R}_{\mathrm{ML}}(t) = \Phi(-\delta) + \frac{1}{4n}\phi(\delta)\delta(\delta^2 - 3) + O(n^{-2}) \tag{2.14}$$

$$\mathrm{Var}(\hat{R}_{\mathrm{MVU}}(t) = \phi^2(\delta)\left[\frac{1}{n}\left(1 + \tfrac{1}{2}\delta^2\right) + \frac{1}{8n^2}(4 + \delta^2 - 2\delta^4 + \delta^6)\right] + O(n^{-5/2}) \tag{2.15}$$

$$\mathrm{MSE}(\hat{R}_{\mathrm{ML}}(t)) = \phi^2(\delta)\left[\frac{1}{n}(1 + \tfrac{1}{2}\delta^2) + \frac{1}{16n^2}(32 - 25\delta^2 - 26\delta^4 + 7\delta^6)\right] + O(n^{-5/2}) \tag{2.16}$$

The performance of these approximations is quite satisfactory. Table 1 compares the exact variances and mean square of MVU and MLE as given in Zacks and Milton (1971) with their approximate values.

2.2 Structural Inference

The theory of structural inference differs basically from both classical and Bayesian models. In some cases there is a "numerical equivalence" between the structural solution and the classical solution, and also a Bayesian interpretation for structural density may be given.

In the structural approach, the parameters of location-scale families of distributions are treated as random variables whose joint distribution is determined by information from the sample. Dyer (1977) derived the structural distribution of the reliability function (2.1), using the relation

Table 1 Comparison Between the Exact and Approximate Values of Variance and Mean Square Errors of MVUE and MLE

$R(t) = \Phi(-\delta)$	n	$\text{Var}(\hat{R}_{\text{MVU}}(t))$ Exact equation (2.11)	Approximation equation (2.15)	$\text{MSE}(\hat{R}_{\text{ML}}(t))$ Exact equation (2.12)	Approximation equation (2.16)
0.5	6	.02938	.02874	.03549	.03537
	12	.01389	.01382	.01560	.01547
	18	.00911	.00909	.00987	.00982
0.3	6	.02497	.02466	.02680	.02782
	12	.01193	.01190	.01254	.01269
	18	.00784	.00784	.00814	.00819
0.1	6	.01013	.00985	.00768	.00677
	12	.00483	.00480	.00414	.00403
	18	.00318	.00317	.00286	.00283
0.05	6	.00481	.00461	.00334	.00256
	12	.00222	.00219	.00181	.00168
	18	.00145	.00144	.00125	.00121
0.025	6	.00215	.00208	.00152	.00136
	12	.00095	.00093	.00079	.00076
	18	.00060	.00060	.00053	.00052

between the normal and lognormal distribution and the fact that the normal distribution is a location-scale density. Consequently the structural joint density of the location and scale parameters μ and σ is

$$f(\mu, \sigma \mid a(\mathbf{x})) = C\sigma^{-n-1}(2\pi)^{-n/2} \exp \frac{-\sum(\log x_i - \mu)^2}{2\sigma^2} \tag{2.17}$$

where C is the normalizing constant and is given by

$$C^{-1} = \Gamma\left(\frac{n-1}{n}\right)\left[2\sqrt{n}(\pi n S^2)^{(n-1)/2}\right]^{-1}$$

$$nS^2 = \sum(\log X_i - \hat{\mu})^2$$

The density (2.17) is precisely the Bayes posterior distribution of μ and σ

when adopting Jeffreys' invariant prior, which is given by $\prod(\mu, \sigma) \propto \sigma^{-1}$. The structural density function of the reliability function is

$$f_R(r \mid t, a(\mathbf{x})) = K\phi\left(\sqrt{\left(1-\frac{1}{n}\right)\left(1+\frac{1}{a^2}\right)}W\right)\phi\left(\frac{W}{\sqrt{-2}}\right)$$
$$\times D_{-n+1}(W) \qquad 0 < r < 1 \qquad n \geq 2 \quad (2.18)$$

where

$$W = \sqrt{n}\Phi^{-1}(1-r)\frac{1}{\sqrt{1+a^2}}$$
$$a = \frac{\hat{\mu} - \log t}{S}$$
$$K = \frac{\sqrt{n}\pi\Gamma(n-1)}{2^{(n-5)/2}\Gamma\left(\frac{n-1}{2}\right)}(1+a^2)^{-(n-1)/2}$$

and $D_{-n+1}(w)$ is a parabolic cylinder function.

As a point estimate of the reliability function based on its structural distribution, we may use the distribution mean which is

$$\hat{R}_s(t) = E\left[R(t;\mu,\sigma)\right] = \frac{1}{2}\left\{1 + \frac{a}{(1+a^2)^{n/2}}\right.$$
$$\left.\cdot\sum_{j=0}^{\infty}\frac{\Gamma\left(j+\frac{n}{2}\right)}{\Gamma\left(j+\frac{3}{2}\right)\Gamma\left(\frac{n-1}{2}\right)}\left(\frac{a^2}{1+a^2}\right)^j F_{1,2(j+1)}(2(j+1)/n)\right\} \quad (2.19)$$

where $F_{\gamma_1,\gamma_2}(Y)$ is the F-distribution function with γ_1, γ_2 degrees of freedom.

For $|a| \leq \frac{1}{2}$ and arbitrary sample size, an approximation to $E[R(t;\mu,\sigma)]$ is

$$\tilde{E}[R(t;\mu.\sigma] = \Phi\left(\sqrt{\frac{2(2n-3)(1+a^2)a}{(2+a^2)[2(1+a^2)+n(2+a^2)]}}\right) \quad (2.20)$$

For large sample size, an approximation to $E[R(t;\mu,\sigma)]$ is

$$\tilde{\tilde{E}}[R(t;\mu.\sigma)] = \Phi\left(\frac{\lambda_1}{\sqrt{1+\lambda_2^2}}\right) - \theta\frac{\lambda_1(2+\lambda_2^2)}{\lambda_2(1+\lambda_2^2)}\phi\left(\frac{\lambda_1}{\sqrt{1+\lambda_2^2}}\right) \quad (2.21)$$

where

$$\lambda_1 = \frac{\sqrt{2(2n-3)}a\sqrt{1+a^2}}{\sqrt{n}(2+a^2)}$$

$$\lambda_2 = \sqrt{\frac{2(1+a^2)}{n(2+a^2)}}$$

and

$$\theta = \frac{\lambda_2^2}{\frac{8(2n-3)(1+a^2)}{na^2} - (\lambda_1^2 + \lambda_2^2)}$$

A numerical example given in Dyer (1977) showed that the numerical values of the structural estimate (2.19)—despite its complicated expression—is hardly distinguishable from the ML and MVUE estimates.

3. ESTIMATION OF LOGNORMAL HAZARD FUNCTION

The hazard function—in some applications it is termed the force of mortality—of the lognormal process is defined by

$$\lambda(t) = \lambda(t; \mu, \sigma) = \frac{\phi(\delta)}{t\sigma\Phi(-\delta)} \tag{3.1}$$

where δ is defined by (2.2).

Figures 2a, 2b, and 2c show lognormal hazard functions for selected values of μ and σ. For all lognormal distributions $\lambda(t)$ is zero at $t = 0$, increases to a maximum and then decreases, approaching zero at $t \to \infty$. This behavior is one of the difficulties in using the lognormal distribution in industrial applications. For many products, the failure rate does not go to zero with increasing time. However, if large values of t are not of interest, the lognormal model is found to be adequate in representing life times.

Nádas (1969) constructs a confidence region for the lognormal hazard function as follows:

Let $\underline{\mu} < \bar{\mu}$ and $0 \leq \underline{\sigma} < \bar{\sigma}$ be statistics such that the rectangle

$$W = \left\{(\mu, \sigma); \underline{\mu} \leq \mu \leq \bar{\mu}, \underline{\sigma} \leq \sigma \leq \bar{\sigma}\right\}$$

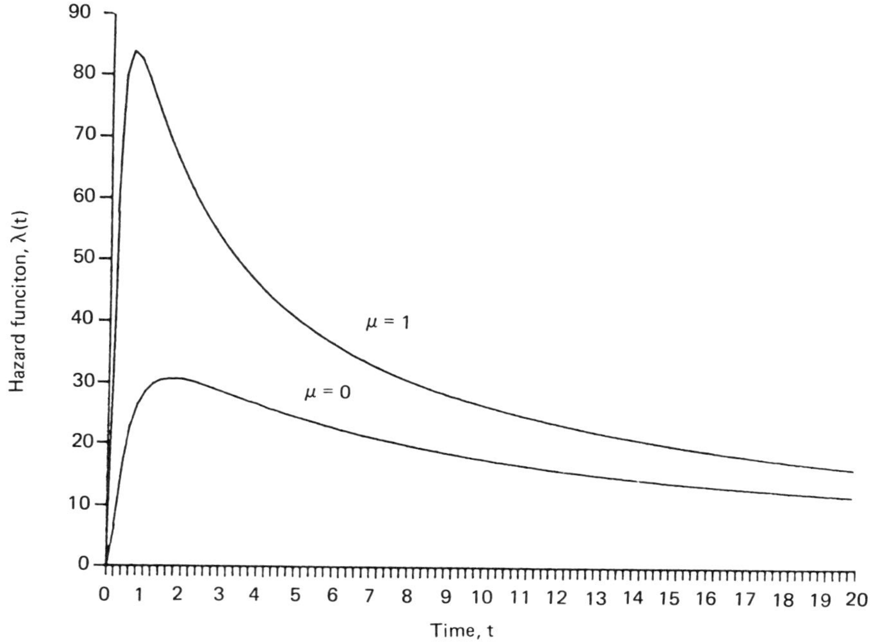

Figure 2a Lognormal hazard function with $\sigma = 1$ and $\mu = 0, 1$.

is a $100(1-\alpha)\%$ confidence set for the true parameter value (μ_0, σ_0). Then the $100(1-\alpha)\%$ confidence region for $\lambda(t; \mu_0, \sigma_0)$ defined by

$$p\{\lambda_l \leq \lambda(t; \mu_0, \sigma_0) \leq \lambda_u, \text{ all } t \geq 0\} \geq 1 - \alpha \tag{3.2}$$

is given by

$$\lambda_l = \min\{\lambda(t; \bar{\mu}, \underline{\sigma}), \lambda(t; \bar{\mu}, \bar{\sigma})\} \qquad t \geq 0 \tag{3.3}$$

and

$$\lambda_u = \begin{cases} \lambda(t; \underline{\mu}, \bar{\sigma}) & \log t < \underline{\mu} - c\bar{\sigma} \\ \lambda(t; \underline{\mu}, (\underline{\mu} - \log t/c)) & \underline{\mu} - c\bar{\sigma} \leq \log t \leq \underline{\mu} - c\underline{\sigma} \\ \lambda(t; \underline{\mu}, \underline{\sigma}) & \underline{\mu} - c\underline{\sigma} < \log t \end{cases} \tag{3.4}$$

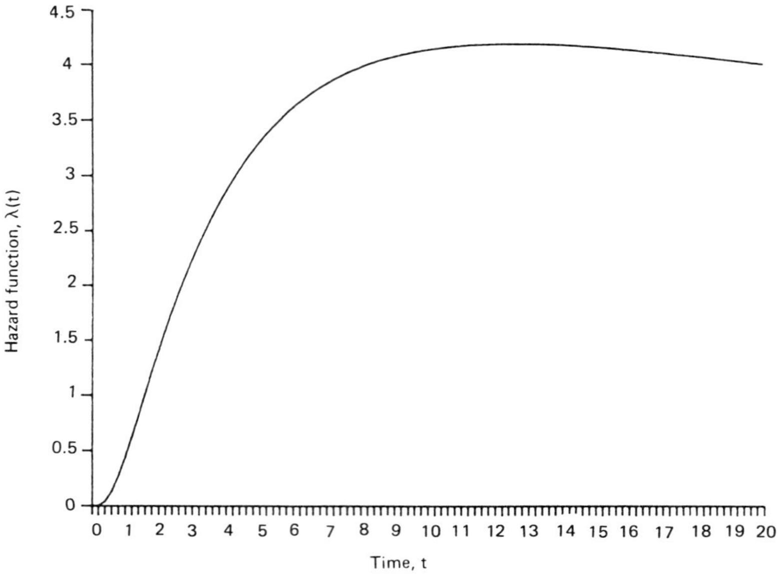

Figure 2b Lognormal hazard function with $\sigma = 1$ and $\mu = 3$.

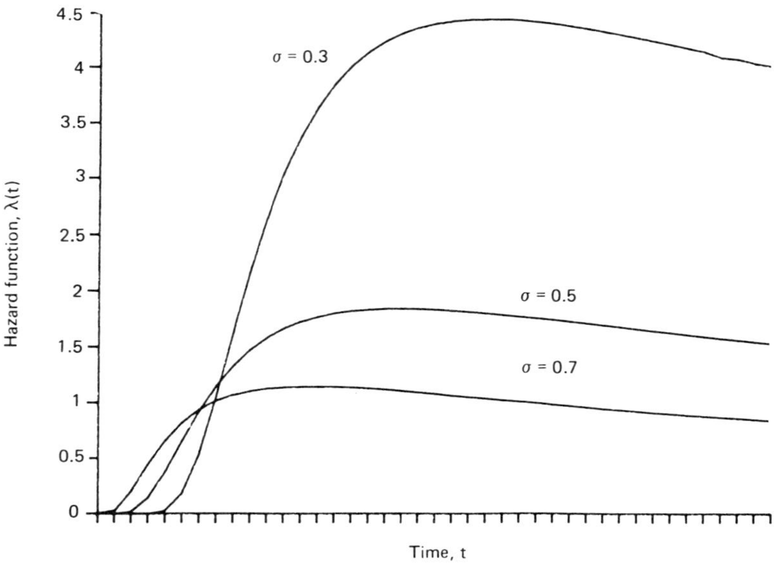

Figure 2c Lognormal hazard function with $\mu = 0$ and $\sigma = 0.3, 0.5, 0.7$.

where c (approximately 0.84) is the unique root of $Z(S) - S = S^{-1}$, and $Z(.)$ is the standard normal hazard function.

Jones (1971) suggested an approximate confidence interval for $\lambda(t)$, for large sample size, utilizing the asymptotic property of maximum likelihood estimates. His confidence interval is

$$\hat{\lambda}(t) \pm Z_{\alpha/2}\sqrt{V(\hat{\lambda})} \tag{3.5}$$

where $Z_{\alpha/2}$ is the $100\alpha/2$ fractile of the standard normal distribution. The asymptotic variance of $\hat{\lambda}(t)$ obtained from the inverse of Fisher's information matrix for $\hat{\mu}$, $\hat{\sigma}$, is

$$V(\hat{\lambda}) = \lambda^2(t)\left[D^2(\delta_0) + \frac{(\delta_0 D(\delta_0) + 1)^2}{2}\right]$$

where $\delta_0 = (\log t_0 - \mu)/\sigma$ and $D(S) = Z(S) - S$. By estimating δ_0 and $\lambda(t)$ using $\hat{\mu}, \hat{\sigma}$, we obtain an estimator of $V(\hat{\lambda})$:

$$\hat{V}(\hat{\lambda}) = \hat{\lambda}^2(t)\left[D^2(\hat{\delta}_0) + \frac{(\hat{\delta}_0 D(\hat{\delta}_0) + 1)^2}{2}\right]$$

Jones (1971) provided tables for

$$B^2(\delta) = D^2(\delta) + \frac{(\delta D(\delta) + 1)^2}{2} \qquad \text{for } -5 \leq \delta \leq 5$$

He expected that his interval (3.5) should be shorter than the interval (3.2), because of the efficiency property of the maximum likelihood method.

4. QUALITY CONTROL CHARTS

In standard industrial quality control charts, the normality of the industrial process is assumed. When a skew distribution appears, it is taken to indicate a lack of control. Morrison (1958) explains the falsity of this assumption in some applications in a thermionic valve system, where the appearance of a skew distribution is typical of the process and thus is not an indication of the lack of control. Another example was given by Joffe and Sichel (1968), who advocated the use of lognormal theory in controlling the dust levels in South African gold mines.

One way to deal with such processes is to apply the logarithmic transformation to the original data to achieve normality, and use the well-known normal theory. The disadvantage of this procedure is the additional numerical work and the difficulty in interpreting the charts. To overcome this difficulty, both Morrison (1958) and Joffe and Sichel (1968) proposed control charts dependent on using the original lognormal data.

Morrison's scheme depends on the relationship between the normal and lognormal to derive control limits for the lognormal. Two control charts are usually used, one to control the average and the second to control the variability. To control the average, we may use the geometric mean, which corresponds to the arithmetic mean in normal charts, but this complicates the procedure. Alternatively, we can use the arithmetic mean within modified normal limits, and then calculate the geometric mean only for the samples whose arithmetic means fall beyond the control limits. To control the variability, corresponding to the range used in normal charts, a sensible measure of variability for the lognormal data is to use the ratio: $r = X_{(n)}/X_{(1)}$, where $X_{(1)}$, $X_{(n)}$ are the smallest and the largest ordered values of the observations. The control limit factors for the lognormal scheme can be obtained from the existing factors of a normal scheme. The limits should be modified to preserve the correspondence between the normal and the lognormal. For example, the sample mean target in a normal case is calculated by $Y_t = \frac{1}{2}(Y_{\min} + Y_{\max})$, whereas in the lognormal chart it will be $X_t = \sqrt{X_{\min} \cdot X_{\max}}$.

Joffe and Sichel (1968) suggested charts for sequentially testing observed arithmetic means from a lognormal population when σ^2 is known. Let $X_1, \ldots, X_n$ be independent observations from a lognormal distribution such that $Y_i = \log X_i \sim N(\mu, \sigma^2)$ where σ^2 is known. Their development is based on approximating the sampling distribution of the arithmetic mean, $\overline{X}$, of the lognormal—which is unknown—by defining a new variable t'', say, such that

$$t'' = \exp\left\{\overline{Y} + \left(\frac{n-1}{2n}\right)\sigma^2\right\}$$

where $\overline{Y} = \sum Y_i/n$. The variable t'' has the same mean as $\overline{X}$ and for a small σ^2, approximately the same variance. Hence the control limits for $\overline{X}$ will depend on the fact that (approximately)

$$\log \overline{X} \sim N\left(\mu + \frac{n-1}{2n}\sigma^2, \frac{\sigma^2}{n}\right) \tag{4.1}$$

It is required to test

$$H_0 : \mu = \mu_0 + 2\sigma$$

against

$$H_1 : \mu = \mu_0 - 2\sigma$$

where μ_0 is a predetermined safety level. The application of SPRT (sequential probability ratio test) requires the continuation of sampling as long as

$$T_l < \overline{Y} = \sum \frac{\log X_i}{n} < T_u \tag{4.2}$$

where

$$T_u = \mu_0 + \frac{\sigma}{4n} \log\left(\frac{1-\alpha}{\alpha}\right)$$

$$T_l = \mu_0 - \frac{\sigma}{4n} \log\left(\frac{1-\alpha}{\alpha}\right)$$

and α is the probability of rejecting H_0 when H_1 is true and also the probability of rejecting H_1 when H_0 is true. Using (4.1) and the limits of (4.2) will yield the procedure in terms of $\overline{X}$: continue sampling whenever

$$S_l < \overline{X} < S_u \tag{4.3}$$

where

$$S_u = \theta_0 \exp\left\{-\frac{\sigma^2}{2n} + \frac{\sigma}{4n} \log\left(\frac{1-\alpha}{\alpha}\right)\right\}$$

$$S_l = \theta_0 \exp\left\{-\frac{\sigma^2}{2n} - \frac{\sigma}{4n} \log\left(\frac{1-\alpha}{\alpha}\right)\right\}$$

and $\theta_0 = \exp(\mu_0 + \sigma^2/2)$. The limits S_u and S_l are symmetric about $\theta_0 \exp(-\sigma^2/2n)$.

5. REPLACEMENT STRATEGY OF DEVICES WITH LOGNORMAL FAILURE DISTRIBUTIONS

As stated in Cheng (1977), reliability studies in many semi-conductor devices, such as LED, IMPATT diodes, lasers and APD (avalanche photodiode detector), have life-time distributions well represented by the lognormal. There are differences between the lognormal and the conventional exponential distribution in the replacement rate strategy. In the exponential distribution, the devices have no memory, which implies that the failure rate is time independent, and if the devices are repaired or replaced immediately, the replacement rate remains constant. On the other hand the replacement rate in the lognormal model is time dependent, and is a function of both t_m, the mean time to fail (MTTF), and σ. It is desired to estimate this replacement rate function. If n, N denote the number of replacements and the number of devices in the system respectively, and t_1, t_2, ..., t_n, $t_i < t_{i+1}$, are "quasi-average" replacement times, it can be shown that

$$\frac{1}{N} = \Phi\left(\frac{\log t_1 - \mu}{\sigma}\right)$$

$$\frac{n}{N} = \Phi\left(\frac{\log t_n - \mu}{\sigma}\right) + \frac{1}{N}\sum_{i=1}^{n-1} \Phi\left(\frac{\log(t_n - t_i) - \mu}{\sigma}\right) \qquad n = 2, 3, \ldots$$

These equations need to be solved successively for t_i, $i = 1, \ldots, n$, which can be performed numerically. The replacement rate at $t = t_i$ is defined as

$$r(t_i) = (\Delta t_i)^{-1} = (t_i - t_{i-1})^{-1}$$

and the normalized replacement rate, $r_N(t_i/t_m)$, is defined as

$$r_N\left(\frac{t_i}{t_m}\right) = \frac{t_m}{N(t_i - t_{i-1})}$$

A study of the replacement rate curves with different σ indicates that when the lognormal devices are introduced randomly the replacement rate approaches asymptotically, for large times, 1 cycle/t_m, which is the value predicted by the exponential model. For extreme values of σ ($\sigma \ll 1$, and $\sigma \gg 1$), the peak replacement rates become much larger than 1 cycle/t_m; for intermediate values of σ, typically $0.25 \le \sigma \le 1.5$, the lognormal scheme requires that the new device should have an MTTF larger than that of an exponential scheme with the same service objectives.

6. THE LOGNORMAL DISTRIBUTION AS A MODEL OF LABOR TURNOVER

The lognormal distribution has been suggested as a model of labor turnover by different authors, for example Lane and Andrew (1955) and McClean (1976). The lognormal model relates an employee's probability of leaving to his length of service. In the model the completed length of service has the lognormal distribution; therefore the probability of leaving in t years is

$$q_t = \frac{1}{\sqrt{2\pi}} \int_{-\infty}^{(\log t-\mu)/\sigma} e^{-\frac{1}{2}x^2}\, dx = \Phi\left(\frac{\log t - \mu}{\sigma}\right)$$

Hence, $\Phi^{-1}(q_t) = A + B \log t$, where $A = -\mu/\sigma$ and $B = 1/\sigma$. A and B can be estimated using linear regression techniques, and $\hat{p}_t = 1 - \hat{q}_t$, can be found for all t. Another model suggested in the labor turnover studies is the transition model, which considers the different grades in the company and the state of having left to be a Markovian process. The comparison between the two models given by McClean (1976) indicated that the lognormal model is a better estimator of leaving patterns, while for the prediction of leavers, no general conclusion can be made and each company should be examined to see which model is more appropriate.

REFERENCES

Cheng, S.S. (1977). Optimal replacement rate of devices with lognormal failure distribution, *IEEE Trans. Reliability*, *R-26*, 174–178.

Dyer, D. (1977). Structural inference on reliability in a lognormal model, *The Theory and Application of Reliability*, *Vol. 1* (Tsokos, C.P. and Shimi, I.N., eds.), Academic Press, Inc., N.Y., 277-306.

Ellison, B.E. (1964). Two theorems for inference about the normal distribution with applications in acceptance sampling, *J. Amer. Stat. Assoc.*, *59*, 89–95.

Hurt, J. (1980). Estimates of reliability for the normal distribution, *Appl. Math.*, *25*, 432–444.

Jones, C.F. (1971). A confidence interval for the lognormal hazard, *Technometrics*, *13*, 885–888.

Joffe, A.D. and Sichel, H.S. (1968). A chart for sequentially testing observed arithmetic means from lognormal populations against a given standard. *Technometrics*, *10*, 605–612.

Lane, K.F. and Andrew, J.E. (1955). A method of labour turnover analysis, *J. Royal Statist. Soc.*, *A.*, *118*, 196–323.

McClean, S. (1976). A comparison of the lognormal and transition models of wastage, *Statistican*, *25*, 281–294.

Morrison, J. (1958). The lognormal distribution in quality control, *Applied Statistics*, *7*, 160–172.

Nádas, A. (1969). A confidence region for the log-normal hazard function, *Technometrics*, *11*, 387–388.

National Bureau of Standards (1959). *Tables of the Bivariate Normal Distribution Function and Related Functions*, Applied Mathematics Series 50, U.S. Government Printing Office, Washington, D.C.

Zacks, S. and Even, M. (1966). The efficiencies in small samples of the maximum likelihood and best unbiased estimators of reliability functions, *J. Amer. Statist. Assoc.*, *61*, 1033–1051.

Zacks, S. and Milton, R.C. (1971). Mean square errors of the best unbiased and maximum likelihood estimators of tail probabilities in normal distributions, *J. Amer. Statist. Assoc.*, *66*, 590–593.

ADDITIONAL REFERENCES

Chhikara, R.S. and Folks, J.L. (1977). The inverse Gaussian distribution as a lifetime model, *Technometrics*, *19*, 461–468.

Davidson, F. and Romanski, J. (1976). Experimental performance of point process estimation procedures for lognormal fading. *IEEE Trans. Inform. Theory.*, *I T 22*, 366–372.

Fertig, K.W. and Mann, N.R. (1975). A new approach to the determination of exact and approximate one-sided prediction intervals for normal and lognormal distributions, with tables, *Reliability and Fault Tree Analysis*, SIAM, Philadelphia., 533–556.

Feinleib, M. (1960). A method of analyzing lognormally distributed survival data with incomplete follow-up, *J. Amer. Statist. Assoc.*, *55*, 534–545.

Goldthwaite, L.T. (1961). Failure rate study for the lognormal life time model, *Proceedings of the 7th National Symposium on Reliability and Quality Control*, 208–213.

Gupta, S.S. (1962). Life test sampling plans for normal and lognormal distributions, *Technometrics*, *4*, 151–175.

Hurt, J. (1976). Asymptotic expansions of functions of statistics. *Appl. Math.*, *21*, 444–456.

Jackson, O.A.Y. (1969). Fitting a gamma or lognormal distribution to fiber-diameter measurements on wool tops, *Applied Statistics*, *18*, 70–75.

King, T.L., Antle, C.E. and Kappenman, R.F. (1979). Sample size for selecting most reliable of K normal (lognormal) populations, *IEEE Trans. Reliability*, *R-28*, 44–46.

Mann, N.R. (1977). An F approximation for two parameter Weibull and lognormal tolerance bounds based on possibly censored data, *Naval Res. Logist. Quart.*, *24*, 187–195.

Mann, N.R., Schafer, R.E., and Singpurwalla, N.D. (1974). *Methods for Statistical Analysis of Reliability and Life Data*, John Wiley & Sons, New York.

McCulloch, A.J. and Walsh, J.E. (1967). Life-testing results based on a few heterogeneous lognormal observations, *J. Amer. Statist. Assoc.*, *62*, 45–47.

Nelson, W. and Kielpinski, T.J. (1976). Theory for optimum censored accelerated life tests for normal and lognormal life distributions, *Technometrics*, *18*, 105–114.

Nelson, W. and Schmee, J. (1979). Inferences for (log) normal life distributions from small singly censored samples and BLUEs, *Technometrics*, *21*, 43–54.

Nowick, A.S. and Berr, B.S. (1961). Lognormal distribution function for describing anelastic and other relaxation process, *I.B.M. Journal of Research and Development*, *5*, 297–320.

Owen, W.J. and DeRouen, T.A. (1980). Estimation of the mean for lognormal data containing zeroes and left-censored values, with applications to the measurement of worker exposure to air contaminants, *Biometrics*, *36*, 707–719.

Padgett, W.J. and Wei, L.J. (1977). Bayes estimation of reliability for the two-parameter lognormal distribution. *Commun. Statist. Theor. Meth.*, *A6(5)*, 443–457.

Padgett, W.J. and Tsokos, C.P. (1977). Bayes estimation of reliability for the lognormal failure model, *The Theory and Application of Reliability*, *Vol. II* (Tsokos and Shimi, ed.). Academic Press, Inc. N.Y., 133–161.

Padgett, W.J. and Wei, L.J. (1978). Bayesian lower bounds on reliability for the lognormal model, *IEEE Trans. Reliability*, *R-27*, 161–165.

Padgett, W.J. and Robinson, J.A. (1978). Empirical Bayes estimators of reliability for lognormal failure model, *IEEE Trans. Reliability*, *R-27*, 332–336.

Padgett, W.J. (1978). Comparison of Bayes and maximum likelihood estimators in the lognormal model for censored samples, *Metron*, *36*, 79–98.

Schmee, J. and Nelson, W. (1979). Predicting from early failures the last failure time of a (log) normal sample, *IEEE Trans. Reliability*, *R-28*, 23–26.

Stoica, P. (1980). Prediction of autoregressive lognormal processes, *IEE Trans. on Automatic Control*, *AC-25*, 292–293.

Upton, D.E. and Shannon, D.S. (1979). The stable paretian distribution, subordinated stochastic process, and asymptotic lognormality: an empirical investigation, *The Journal of Finance*, *34*, 1031–1039.

11

Applications in Biology: Simple Growth Models

JAMES E. MOSIMANN AND GREGORY CAMPBELL Laboratory of Statistical and Mathematical Methodology, Division of Computer Research and Technology, National Institutes of Health, Bethesda, Maryland

1. INTRODUCTION

Applications of parametric statistical distributions in biology may be divided into two broad classes. First, there are those applications in which a specific underlying distributional model is not clearly specified. There is only a vague understanding of the random processes generating the observed data, often coupled with coarse accuracy for the experimental measurements. In such situations it is more with hope than confidence that the investigator may posit the existence of many independent additive sources of variation and invoke a central limit theorem to justify the use of statistical inference based on normal theory. Many uses of methods such as regression and ANOVA seem implicitly to rely on such a justification. However, the investigator may perceive that such methods give more stable or interpretable results when the data are transformed by logarithms. Such a perception is often based on the "straightening" of an apparent curvilinear regression or on the "attaining of more homogeneous variances" across groups after the log transformation. In any event, if the investigator posits the existence of many independent sources of variation, but now multiplicative rather than additive, then the use of a central limit theorem on the log-data provides a justification for the use of statistical methods based on the lognormal distribution. In biology there are so many applications of this nature, where data are converted to logarithms and subsequently analyzed using normal distributional methods, that it would be hopeless to attempt even a partial list.

But there is a second class of applications, albeit in the minority, in which specific assumptions about the real-world processes generating the data may be tenable, and lead to specific distributional models for describing the random variation observed. It is this class to which the present discussion is devoted. By concentrating on positive random variables, and the characteristic properties of various distributions for these variables, it is hoped to demonstrate the generality and limitations of the multivariate lognormal distribution for describing random variation.

2. TISSUE GROWTH MODELS

2.1 General Tissue Growth Models

Consider the growth of some living tissue, where the word "tissue" is meant in a very broad sense. Suppose there are observations on the size (or amount) of this tissue, W_k, for equally-spaced times $k = 0, 1, \ldots, n$. A finite logarithmic growth rate from time k to $k+1$ may simply be defined as $[\log(W_{k+1}) - \log(W_k)] = \log(W_{k+1}/W_k)$. Call the antilog $V_{k+1} = W_{k+1}/W_k$ the growth ratio. It is assumed that $W_k > 0$ for $k = 0, 1, \ldots, n$; hence, the growth ratio is always defined. Now let $W_0 = u_0$ be the initial fixed amount of tissue at time 0. Since $W_k = V_k W_{k-1}$ holds for successive times to n,

$$W_k(\mathbf{V}) = u_0 \prod_{i=1}^{k} V_i \qquad \text{for } k = 1, \ldots, n$$

Now consider a model in which the vector of growth ratios $\mathbf{V} = (V_1, \ldots, V_n)'$ is a positive random vector. Its coordinates may or may not be statistically independent, and may or may not have identical marginal distributions, but they are all physically dimensionless, the logarithms of which are rates of change (in the log scale) per the unit of time.

Note that, conditional upon the finite collection of equally-spaced time points $\{0, 1, \ldots, n\}$, it is always possible to define an equivalent "constant-ratio" model from time period 0 to k which would yield the same intermediate size W_k, conditional on the initial size u_0. Thus

$$W_k(\mathbf{V}) = u_0 \left[\prod_{i=1}^{k} V_i^{1/k} \right]^k = u_0 \, [M_k]^k$$

Here the equivalent constant growth ratio for W_k is the geometric mean, $M_k = \prod_{i=1}^{k} V_i^{1/k}$, of the initial k growth ratios. The value of this equivalent constant ratio reflects growth from time 0 if > 1, decay if < 1 and stability if $= 1$. Of course the same constant ratio model will not apply simultaneously for all k unless the vector of growth ratios has the same distribution as the random vector $(V_1, \ldots, V_1)' = V_1(1, \ldots, 1)'$, degenerate on the equiangular ray, where V_1 is the growth ratio from time 0 to 1. It should be emphasized that in all these models the growth ratios are random variables.

At this point the models presented are very general, conditional upon the time points $\{0, 1, \ldots, n\}$. The only requirement is that the size be positive at all points.

2.2 The Active Tissue Model

Consider biological growth in which the tissue resulting from growth at a previous stage itself actively contributes to the production of new tissue at the next stage. Such "active" growth models commonly involve multiplicative "errors", or in more apt terms, the products of random variables. Start with u_0, a fixed amount of actively reproducing tissue at time 0. Consider, as above, points in time $k = 0, 1, \ldots, n$, and let X_k be the specific growth ratio of the tissue at time k, so for example, $W_1 = X_1 u_0$. Since the tissue produced at time 1 as well as the tissue existing at time 0 actively participates in the production of tissue at time 2 and so forth, then

$$W_k = X_k W_{k-1} = u_0 \prod_{i=1}^{k} X_i,$$

$k = 1, \ldots, n$. In this model it is clear that the vector $\mathbf{X} = (X_1, \ldots, X_n)'$ of specific tissue growth ratios is simply the vector $\mathbf{V}$ of growth ratios. Since this is so, the equivalent constant ratios are simply the geometric means of the specific growth ratios of the tissue; namely,

$$M_k = M_k(\mathbf{X}) = \prod_{i=1}^{k} X_i^{(1/k)} \qquad k = 1, \ldots, n$$

Contiguous ratios of these constant ratios are the multiplicative size ratios, M_{k+1}/M_k, for $k = 1, \ldots, n-1$, of the vector $\mathbf{X}$.

2.3 The Inert Tissue Model (Constant Source Tissue)

In contrast to the above model for active tissue, consider a simple biological growth model in which the product of growth from a previous stage is "inert" tissue (say a fingernail) which itself does not contribute to the production of new tissue at the next stage. If it is also assumed that the amount of productive tissue, say u_0, is constant over the relevant time period (as might be approximately true with tissue at the base of the fingernail in an adult), then a simple "additive" error model applies. Let $\mathbf{X} = (X_1, \ldots, X_n)'$ be the positive random vector of tissue growth rates (not ratios) for times $1, \ldots, n$. Then $W_1 = X_1 u_0$ and

$$W_k = X_k u_0 + W_{k-1} = u_0 \sum_{i=1}^{k} X_i = u_0 S_k \qquad k = 2, \ldots, n$$

where S_k denotes the partial sum

$$S_k = S_k(\mathbf{X}) = \sum_{i=1}^{k} X_i \qquad k = 1, \ldots, n$$

In this inert tissue model the growth ratios, V_{k+1}, are the additive size ratios S_{k+1}/S_k, for $k = 1, \ldots, n-1$. For each k, the equivalent constant rate is the arithmetic mean S_k/k of the tissue growth rates from time 1 to k. Contiguous ratios of these constant rates are scalar multiples of the additive size ratios $(k/(k+1))(S_{k+1}/S_k)$, for $k = 1, \ldots, n-1$. Hence, in this inert tissue model with a constant amount of productive tissue, the statistical independence of the growth ratios is equivalent to the independence of the contiguous ratios of constant rates. This was not the case in the active tissue model.

Note that unlike the active tissue model, this model allows only for growth and not for decay. It is true that, by reversing time (or, equivalently, by allowing all negative X's), monotone decay instead of monotone growth can be modeled. However, unlike the active tissue model, the inert tissue model does not allow growth and decay at different times in history of the tissue.

Suppose the amount of source tissue is not constant but is changing by random growth ratios; this would result if the source tissue produced active (productive) tissue as well as inert tissue. In this situation multiplicative and additive effects are intermingled. This model will not be examined carefully here, but it should be noted that even with inert tissue models, multiplicative effects may be important.

2.4 Finite Growth Models and Infinite Divisibility

The above discussion relies on fixed, equally-spaced time points. The ability to change the spacing, to further subdivide the time intervals and yet preserve distributions, suggests the use of infinitely divisible distributions to model such growth. Two important such distributions are the normal and the gamma, which for the tissue models lead to the lognormal and the gamma. These latter distributions play an especially important role in the study of relative growth.

3. LOGNORMAL AND GAMMA DISTRIBUTIONS IN FINITE GROWTH MODELS

3.1 Size and Shape

In the preceding section, simple models for tissue growth lead naturally to the consideration of random vectors $\mathbf{X} = (X_1, \ldots, X_n)'$ for tissue growth, as well as to certain functions of $\mathbf{X}$ (called size variables) such as the partial geometric means, M_k, and partial sums, S_k, for $k = 1, \ldots, n$. Recall the definitions of size variables and shape vectors (Mosimann, 1970, 1975a, 1975b). Let R_+ denote the positive real numbers and R_+^n its n-dimensional analog. A *size variable* $G : \mathsf{R}_+^n \to \mathsf{R}_+$ is a positive-valued function with the homogeneity property $G(\alpha\mathbf{x}) = \alpha G(\mathbf{x})$ for all $\mathbf{x} \in \mathsf{R}_+^n$ and all $\alpha \in \mathsf{R}_+$. Examples of size variables are $S_n(\mathbf{X})$, X_i, $(\sum X_i^2)^{1/2}$, $M_n(\mathbf{X})$, and $\max_{1 \le i \le n} X_i$. A *shape vector* is a vector-valued function $\mathbf{Z}_G$ given by $\mathbf{Z}_G(\mathbf{x}) = \mathbf{x}/G(\mathbf{x})$, all $\mathbf{x} \in \mathsf{R}_+^n$, where G is a size variable. Associated shape vectors for the size variable examples are proportions $\mathbf{X}/S_n(\mathbf{X})$, ratios $\mathbf{X}/X_i$, direction cosines $\mathbf{X}/(\sum X_i^2)^{1/2}$, $\mathbf{X}/M_n(\mathbf{X})$, and $\mathbf{X}/\max_{1 \le i \le n} X_i$. For any two shape vectors, because there exists a one-to-one map from one onto the other, either shape vector may represent shape. Further, if a random variable T is independent of one such random shape vector of $\mathbf{X}$, it is independent of all others as well. Thus, one may speak unambiguously of the "n-dimensional shape" of $\mathbf{X}$ and of its independence from T (Mosimann, 1975a).

3.2 Characterizations of Size and Shape

The independence of the random shape of $\mathbf{X}$ and the size variable $G(\mathbf{X})$ is called *isometry with respect to G*. If $\mathbf{X}$ is isometric with respect to the size variable $G(\mathbf{X})$, then no other size variable $H(\mathbf{X})$ can be independent of shape unless the ratio $G(\mathbf{X})/H(\mathbf{X})$ is degenerate at a point (Mosimann, 1970). In that isometry is a very strong condition, it is not surprising that it has lead to characterizations of statistical distributions. The following characterization of the lognormal distribution was given by Mosimann (1970):

Theorem 1 Let $X_1, \ldots, X_n$ be positive, statistically independent, nondegenerate random variables (not necessarily identically distributed) and let $Y_i = \log(X_i)$. If n-dimensional shape is statistically independent of size M_n, then each X_i has a lognormal distribution with the same $\mathrm{Var}(Y_i)$.

There is a parallel result in which partial sums replace partial geometric means (Mosimann, 1970):

Theorem 2 Let $X_1, \ldots, X_n$ be positive, statistically independent, nondegenerate random variables (not necessarily identically distributed). If n-dimensional shape is statistically independent of size S_n, then each X_i has a gamma distribution with the same scale parameter.

Both $(M_1, \ldots, M_n)$ and $(S_1, \ldots, S_n)$ are *regular sequences of size variables* as defined by Mosimann (1975a). As seen above, $(M_2/M_1, \ldots, M_n/M_{n-1})$ and $(S_2/S_1, \ldots, S_n/S_{n-1})$ occur naturally as growth ratios of equivalent constant ratios: the ratios of geometric means in the active tissue model, and the ratios of partial sums in the inert tissue (constant source) model. For the consecutive ratios of any regular sequence of size variables, there is a one-to-one onto function from the size ratios to any shape variable. The independence of the $(n-1)$-dimensional shape of $(X_1, \ldots, X_{n-1})'$ and the contiguous ratio $G_n(\mathbf{X})/G_{n-1}(\mathbf{X})$ where the G's are from a regular sequence is called *neutrality with respect to* G_{n-1}, G_n. If the $(n-1)$-dimensional shape of $(X_1, \ldots, X_{n-1})'$ is neutral with respect to G_{n-1}, G_n, then it is not neutral with respect to H_{n-1}, H_n for any other regular sequence $H_1, \ldots, H_n$, unless the ratio G_n/H_n is degenerate at a point (Mosimann, 1975a, 1975b).

Two characterization results concerning neutrality are stated before considering the implications of these theorems for models of tissue growth. The following was given by Mosimann (1975b):

Theorem 3 Let $\mathbf{X}$ be a positive random vector constrained so that $M_n(\mathbf{X}) = 1$ with probability 1. Let $\mathbf{X}^* = (X_n, \ldots, X_1)'$; that is $\mathbf{X}$ with coordinates written in reverse order. Suppose
(1) $M_2(\mathbf{X})/M_1(\mathbf{X}), \ldots, M_n(\mathbf{X})/M_{n-1}(\mathbf{X})$ are statistically independent

(2) The shape of $(X_n, \ldots, X_2)'$ is statistically independent of $M_n(\mathbf{X}^*)/M_{n-1}(\mathbf{X}^*)$,

then $\mathbf{X}$ has a multivariate lognormal distribution, with the parameter matrix $\Sigma = (\sigma_{ij})$ (the covariance matrix of $\mathbf{Y} = (\log(X_1), \ldots, \log(X_n))')$ necessarily of the form: for some $\beta \geq 0$, $\sigma_{ii} = (n-1)\beta$, $i = 1, \ldots, n$; $\sigma_{ij} = -\beta$, $i \neq j$.

Both conditions of the above theorem concern shape. Condition (1) is equivalent to *complete multiplicative neutrality* of the vector $\mathbf{X}$. These conditions imply complete multiplicative neutrality with respect to any permutation of $\mathbf{X}$.

A companion theorem, but for the partial sums rather than the partial geometric means, was proved in James and Mosimann (1980):

Theorem 4 Let $\mathbf{X}$ be a positive random vector constrained so that $S_n(\mathbf{X}) = 1$ with probability 1. Let $\mathbf{X}^* = (X_n, \ldots, X_1)'$; that is, $\mathbf{X}$ with coordinates written in reverse order. Suppose:

(1) $S_2(\mathbf{X})/S_1(\mathbf{X}), \ldots, S_n(\mathbf{X})/S_{n-1}(\mathbf{X})$ are statistically independent,

(2) the shape of $(X_n, \ldots, X_2)'$ is statistically independent of $S_n(\mathbf{X}^*)/S_{n-1}(\mathbf{X}^*)$,

then the distribution of $\mathbf{X}$ is Dirichlet.

Condition (1) is equivalent to *complete additive neutrality* for the vector $\mathbf{X}$ and, with (2), implies complete additive neutrality for any permutation of $\mathbf{X}$. If the constraint that $S_n(\mathbf{X}) = 1$ is removed from the hypotheses of this theorem, then the conclusion becomes that the distribution of $\mathbf{X}/S_n(\mathbf{X})$ is Dirichlet. (If the restriction on $\mathbf{X}$ is changed from positive to nonnegative, then the conclusion becomes that the distribution is Dirichlet or a limit of Dirichlets, as in James and Mosimann (1980).)

3.3 Relationships of the Characterizations with Growth Models

Consider the above characterizations for the tissue growth models. In the active tissue model it may be that the tissue growth ratios, X_k, are mutually independent. In addition, if the vector $(M_2/M_1, \ldots, M_n/M_{n-1})'$ of contiguous ratios is independent of the equivalent constant ratio M_n (this is isometry), then the conditions of Theorem 1 are met. Each X_i must then be lognormal with the same $\mathrm{Var}(Y_i)$. In this case the distribution of $\mathbf{X}/M_n$ is multivariate lognormal with parameter matrix as in Theorem 3. Conditional on the initial size $U_0 = u_0$, $\mathbf{W} = (W_1, \ldots, W_n)'$ has a multivariate lognormal distribution as well. The unconditional distribution of $\mathbf{W}/U_0$ is multivariate lognormal.

In the active tissue model, suppose that the X_i's are not necessarily independent and consider the size ratios M_{k+1}/M_k (ratios of equivalent growth ratios). If these M_{k+1}/M_k are positively correlated, then the paths or trajectories of growth are more unstable or variable than if the ratios were independent. Further, consider the time-reversed vector $\mathbf{X}^* = (X_n, \ldots, X_1)'$. If the $(n-1)$-dimensional shape of $(X_n, \ldots, X_2)'$ is independent of $M_n(\mathbf{X}^*)/M_{n-1}(\mathbf{X}^*)$, then $\mathbf{X}/M_n$ must have the multivariate lognormal distribution of Theorem 3.

In the inert tissue model with constant source, if the tissue growth rates, X_k, $k = 1, \ldots, n$, are mutually independent and shape is independent of $S_n(\mathbf{X})$, then each tissue rate must have a gamma distribution with common scale parameter; further, the shape $\mathbf{X}/S_n(\mathbf{X})$ has a Dirichlet distribution. In terms of the vector $\mathbf{W}$, the vector $(W_1/W_n, (W_2 - W_1)/W_n, \ldots, (W_n - W_{n-1})/W_n)' = \mathbf{X}/S_n$ has a Dirichlet distribution, or alternatively, $\mathbf{W}/W_n$ has an ordered Dirichlet distribution.

In the inert tissue model with constant source, if the growth ratios, the V_k's rather than the X_k's, are mutually independent (i.e., the contiguous ratios S_{k+1}/S_k of constant rates are mutually independent), this is just the complete additive neutrality of the tissue rate vector $\mathbf{X}$. Also, if for the time-reversed vector $\mathbf{X}^*$, the $(n-1)$-dimensional shape of $(X_n, \ldots, X_2)'$ is independent of $S_n(\mathbf{X}^*)/S_{n-1}(\mathbf{X}^*)$, then Theorem 4 shows that the distribution of $\mathbf{X}/S_n$ must be Dirichlet. Here, if S_n (or W_n) has a gamma distribution with the corresponding parameter, then the tissue growth rates are independent gammas as above. If not, then the X's are not independent, but size S_n can still be independent of shape.

In summary, under fairly simple assumptions of independence, lognormal and gamma distributions arise naturally in the two growth models considered. Of the two models, the active tissue model associated with the lognormal would seem to have far wider application in growth studies.

4. MORE GENERAL SIZE VARIABLES AND THEIR RATIOS

4.1 Linear and Loglinear Size Variables

In this section generalizations of M_n and S_n are considered. Ratios of these general size variables are used to define families of distributions that naturally contain size variables and ratios of these size variables. This allows the study of the covariances of these ratios in order to investigate partial isometry and neutrality.

Consider the following generalizations of M_k and S_k. Let $\mathbb{R}_0$ denote the nonnegative real numbers, with A a subset of $\mathbb{R}_0$ and A^n the set of n-dimensional vectors with components from A. For $\mathbf{a} \in A^n - \mathbf{0}$ (the notation

$-\mathbf{0}$ excludes $\mathbf{0}$), define

$$M(\mathbf{X};\mathbf{a}) = \prod_{i=1}^{n} X_i^{a_i}$$

and

$$L(\mathbf{X};\mathbf{a}) = \sum_{i=1}^{n} a_i X_i$$

Note that, for $\mathbf{X}$ a positive random vector, both newly defined variables are positive. In fact, the linear L is always a size variable and, provided $\sum a_i = 1$, the loglinear $M(\mathbf{X};\mathbf{a})$ is as well. That M_k and S_k are special cases of the above is easy to see (the appropriate $\mathbf{a}$-vectors are $(1/k,\ldots,1/k,0,\ldots,0)'$ and $(1,\ldots,1,0,\ldots,0)'$, respectively).

Consider the situation in which the X_i's in $\mathbf{X}$ are independent but not necessarily identically distributed; this may arise in the tissue models if the time points are not equally spaced or if the parameters are not homogeneous. If the X_i are independent lognormals with $\mathrm{Var}(Y_i) = \sigma_i^2$, then $M(\mathbf{X};\mathbf{a})$ with $a_i = \tau/\sigma_i$ and $\tau = 1/\sum_{i=1}^{n}(1/\sigma_i)$ is a size variable that is independent of shape. Further, if size $G(\mathbf{X})$ is also independent of shape, then $M(\mathbf{X};\mathbf{a})/G(\mathbf{X})$ must be degenerate; i.e., $M(\mathbf{X};\mathbf{a})$ is unique up to a positive multiplier c. If the X_i's are independent gammas with scale parameter β_i (such that X_i/β_i has scale parameter 1), then $L(\mathbf{X};\mathbf{a})$ with $a_i = c/\beta_i$ is the size variable, unique up to the positive constant c, that is independent of shape (Mosimann, 1970, 1975b).

4.2 Associated Families of Distributions

Based on the loglinear variable M just defined, the following is a more general definition of the multivariate lognormal distribution which is equivalent in the full rank case to the density representation of Chapter 1.

Definition For $n \geq 2$, a random vector $\mathbf{X} = (X_1,\ldots,X_n)'$ has a *multivariate lognormal (MLN) distribution* if there exists m (≥ 1) independent (univariate) lognormal random variables $U_1,\ldots,U_m$, with common $\mathrm{Var}(\log(U_i))$, and vectors $\mathbf{a}_i, \mathbf{b}_i \in \mathrm{R}_0^m - \mathbf{0}$ ($i = 1,\ldots,n$), such that, for $\mathbf{U} = (U_1,\ldots,U_m)'$, $\mathbf{X}$ has the same distribution as

$$(M(\mathbf{U};\mathbf{a}_1)/M(\mathbf{U};\mathbf{b}_1),\ldots,M(\mathbf{U};\mathbf{a}_n)/M(\mathbf{U};\mathbf{b}_n))$$

Note that for $\mathbf{c}_i = \mathbf{a}_i - \mathbf{b}_i \in \mathsf{R}^m$, this is merely the "log" version of Definition II of the multivariate normal in Rao (1973, p. 522), without the zero mean vector adjustment. The representation is not unique. The use of $\mathbf{a}$'s and $\mathbf{b}$'s rather than $\mathbf{c}$'s is a device to suggest the following analog based on linear gamma size ratios.

For the parallel development for linear size variables, the following univariate family is introduced.

Definition A random variable X is in the *gamma-ratio-gamma (GRG) family of distributions* for the set $A \subseteq \mathsf{R}_0$ if there exist m (≥ 1) independent gamma random variables $U_1, \ldots, U_m$ with common scale parameter such that, for $\mathbf{U} = (U_1, \ldots, U_n)'$, X has the same distribution as either $L(\mathbf{U}; \mathbf{a})$ or $L(\mathbf{U}; \mathbf{a})/L(\mathbf{U}; \mathbf{b})$ for $\mathbf{a}$, $\mathbf{b} \in A^m - \mathbf{0}$.

This representation is not unique. It is quite easy to characterize all the distributions of the members of this GRG family for $A = \{0, 1\}$. It consists of the gamma (in the case of the denominator degenerate at 1), the beta, the inverted beta, the distribution degenerate at 1 (for $\mathbf{a} = \mathbf{b}$), the distribution of the inverted beta plus 1, and the sum of an inverted beta and an independent beta (with common parameter). The more general family for $A = \mathsf{R}_0$ is much larger. If the GRG family is used without reference to the set A, assume $A = \mathsf{R}_0$.

This univariate family is now extended to a multivariate family:

Definition For $n \geq 2$, a random vector $\mathbf{X} = (X_1, \ldots, X_n)'$ is in the *multivariate gamma-ratio-gamma (MGRG) family of distributions* for the set $A^n \subseteq \mathsf{R}_0^n$ if there exist m (≥ 1) independent gamma random variables $U_1, \ldots, U_m$ with common scale parameter such that, for $\mathbf{U} = (U_1, \ldots, U_n)'$, $\mathbf{X}$ has the same distribution as

$$(L_1(\mathbf{U})/L_1^*(\mathbf{U}), \ldots, L_n(\mathbf{U})/L_n^*(\mathbf{U}))$$

where $L_i(\mathbf{U}) = L(\mathbf{U}; \mathbf{a}_i)$ and $L_i^*(\mathbf{U}) \equiv 1$ or $= L(\mathbf{U}; \mathbf{b}_i)$, where $\mathbf{a}_i$, $\mathbf{b}_i \in \mathsf{R}_0^n - \mathbf{0}$.

The representation is not unique. Note that all marginal distributions are in the GRG family. As earlier, if A is unspecified, it is understood that $A = \mathsf{R}_0$.

This MGRG family is quite large. If $A = \{0, 1\}$ and the denominator is degenerate at 1, then the independent gammas and the multivariate gamma I of Johnson and Kotz (1972, p. 216) are in the family. If $A = \{0, 1\}$ and $\mathbf{b}_i = \mathbf{b}$ for $i = 1, \ldots, n$, then the Dirichlet, the inverted Dirichlet and the ordered Dirichlet are also included. If different denominators are permitted, the generalized Dirichlet is also in the family. If A is expanded to R_0, this allows ratios of gamma random variables with different scale parameters, among others.

4.3 Characterizations and the Families

Attention is now focused on the more realistic situation in which neither the X_i's nor their size ratios are independent. For the MLN family the covariance of X_i and X_j has the same sign (+ or −) as that of $\log(X_i)$ and $\log(X_j)$; further, if one covariance is zero, so is the other, in which case X_i and X_j are independent. With the above definition of MLN, there exists an m-dimensional vector $\mathbf{Z} = (Z_1, \ldots, Z_m)'$ of independent normal random variables with common variance and $\mathbf{c}_i = \mathbf{a}_i - \mathbf{b}_i$ $(i = 1, \ldots, n)$ such that $\mathbf{Y} = \log(\mathbf{X})$ has the same distribution as $C\mathbf{Z} = (\mathbf{c}_1'\mathbf{Z}, \ldots, \mathbf{c}_n'\mathbf{Z})$ where $C = (\mathbf{c}_1', \ldots, \mathbf{c}_n')'$ is an $n \times m$ matrix. Let $D = C^-$ denote an $m \times n$ generalized inverse of C. If $D = (\mathbf{d}_1', \ldots, \mathbf{d}_m')'$ then $M(\mathbf{X}; \mathbf{d}_1), \ldots, M(\mathbf{X}; \mathbf{d}_m)$ are independent lognormal random variables with the same log variance and $(\prod_{i=1}^{m} M(\mathbf{X}; \mathbf{d}_i))^{1/m}$ is a variable independent of m-dimensional shape (but not necessarily n-dimensional shape if $n > m$). Furthermore, properly normalized, it is the unique such size variable (up to a constant multiplier) that is concentrated on the subset that is the support of the distribution. This extends the result of Sampson and Siegel (1985). Other sequences of size variables such as M_k can lead to size ratios that are positively or negatively correlated. Further, Mosimann (1975b) has shown that, for any MLN distribution not degenerate on a single ray, S_n cannot be independent of shape.

If $\mathbf{X}$ has a distribution in the MGRG family with some dependent components but $\mathbf{X}/S_n$ is not Dirichlet, one can examine the special case in which $\mathbf{X}$ has the same distribution as $A\mathbf{U}$, where $\mathbf{U} = (U_1, \ldots, U_m)'$ consists of independent gammas with common scale and $A = (\mathbf{a}_1', \ldots, \mathbf{a}_n')'$, $\mathbf{a}_i \in \mathsf{R}_0^n - \mathbf{0}$. This restriction on A, necessary to ensure $A\mathbf{U}$ is positive, implies that correlation of X_i and X_j is nonnegative. If A^- denotes a generalized inverse of A, then the variable $V = \mathbf{1}'A^-\mathbf{X}$ is a size variable (unique up to a positive constant on the subset that is the support of $\mathbf{X}$) independent of m-dimensional shape (but not necessarily n-dimensional shape for $n > m$) for the vector $\mathbf{X}$ since $\sum U_i$ is independent of $\mathbf{U}/\sum Ui$ if and only if $\sum U_i$ is independent of $A\mathbf{U}/\sum U_i$.

These two families MLN and MGRG permit the investigations of isometry and neutrality and the exploration of the relationships of size and shape and of size ratios. For the size variables $M(\mathbf{X}; \mathbf{a})$ and $L(\mathbf{X}; \mathbf{a})$, in particular, for sequences M_k and S_k, the relationships of size and shape and of consecutive size ratios are easily studied, as all relevant distributions are in the families MLN and MGRG.

4.4 Applications to Growth Models

The MGRG family is more limiting than MLN for the growth models. Models based on MLN can model positive and negative correlations in $\mathbf{X}$ and in the vector of size ratios; MGRG cannot model both growth and decay and is limited in correlations of the size ratios. Whereas amalgamation of X_1 and X_2 using $(X_1X_2)^{1/2}$ is lognormal, the variable $(X_1 + X_2)/2$ is not necessarily GRG unless the denominators in the independent gamma representation are equal. Furthermore, it is a simple matter to investigate partial isometry and neutrality with inference on the covariance matrix of $\log(X_i)$ in MLN.

An important issue in applications is the choice of the size variable. The uniqueness of the size variable independent of shape underlines the crucial nature of this decision. In particular, if the geometric mean (or some loglinear size variable M) is independent of shape, then the sum cannot also be. (In fact, Mosimann (1975b) has shown that no member of the MLN family (except the uninteresting case of those degenerate on a single ray) can have $(n-1)$-dimensional shape independent of the additive ratio S_n/S_{n-1}.) In some problems, the choice between the sum and the geometric mean may not be as crucial in that either may appear to be uncorrelated with shape in the sample statistics. In practice, it may be reasonable to choose more than one size variable for a particular experiment; for example, measurements by weight and by volume in an experiment may suggest different sizes.

5. SOME CURRENT APPLICATIONS IN ALLOMETRY

As seen above, simple but quite general random models for growth lead readily to multiplicative errors. It has been noted that the multivariate lognormal distribution is the only distribution for which both the growth ratios may themselves be independent and have their proportions (shape) independent of the geometric mean of the growth ratios. Of course one of the major advantages of the multivariate lognormal distribution is that it allows the vector of growth ratios to exhibit all sorts of dependencies, both positive and negative. Given its remarkable flexibility with respect to ratios and products of positive random variables, it is hardly surprising that the multivariate lognormal distribution has played a central role in the field of biology known as allometry, or relative growth, the study of changes in proportions of various parts of organisms with change in size (cf. Gould (1966, 1977), Huxley (1932), Reeve and Huxley (1945), Griffiths and Sandland (1982), Jolicoeur (1963), Mosimann (1958), Jungers (1985)). In this section some current applications of the multivariate lognormal distribution in such studies are indicated.

In allometric studies prior to the work of Mosimann (1970), there was no emphasis on the need to specify a particular (unique) size variable for the study of the association of size and shape. Jolicoeur (1963) had proposed that the major axis of the ellipse of concentration of the log measurements be used to represent the allometric relationship. He considered isometry to be the condition that this axis be the equiangular line. This is completely consistent with Mosimann's results for the multivariate lognormal distribution, where shape is shown to be independent of the geometric mean if and only if the equiangular vector is an eigenvector of the covariance matrix of the log data. Thus, Jolicoeur's definition implicitly selects the geometric mean of the measurements as size. See, most recently, Jolicoeur (1984). Mosimann and Malley (1979) discuss a variety of definitions of size.

Hopkins (1966) and Sprent (1968, 1972) presented models in which the observed covariance matrix is the sum of a "model" or structural matrix plus an error matrix. They cautioned that tests of certain allometric hypotheses may be considered, but only as applied to the structural portion of the model. The presence of an error matrix often has destroyed the ability to test certain isometric hypotheses, but Mosimann, Malley, Cheever and Clark (1978), using multivariate lognormal models, were able to study some hypotheses in the presence of error. Their data was for the distribution of schistosome eggs in human organs at autopsy. Using mean vector as well as covariance information, they concluded for some cases that there was no association of infection intensity (a geometric mean) with the proportional distribution of parasite eggs in various organs (shape).

In another paper dealing with disruptive errors, Jolicoeur and Heusner (1971) make interesting use of plots of antilogs of fitted equal-frequency ellipses, and show in a convincing way that the bivariate lognormal distribution, rather than the bivariate normal distribution, is more applicable to their data on oxygen consumption as related to body size in rats.

Recent interesting applications of the lognormal distribution may be found in the work of Seebeck (1983a, 1983b) who uses linear models in the analysis of size and shape. He studies growth and body composition in cattle and other livestock. In (1983a), he uses a variety of covariance analyses of log shape and demonstrates that taking size into account, faster growing animals give lower yields of carcass. His studies exploit in an interesting way the use of loglinear size variables embedded in experimental designs which utilize the normal theory underlying the general linear model.

In another interesting application based on loglinear size variables, Darroch and Mosimann (1985) present canonical discriminant (and principal component) analyses for log shape which are invariant when applied to any log shape vector formed from a loglinear size variable. They provide new insight for discrimination of the classic iris data studied by Fisher (1936); namely, that the species of iris are as well discriminated on scale-free (log shape) information alone, in comparison with the analysis using scale and scale-free (log size and log shape) information.

Veitch (1978) gives a brief but thoughtful summary of the history of allometry, discusses the transformation to logarithms, and uses a multivariate linear model to study changes in the proportions of claws and appendages of two species of fiddler crabs. The size variable chosen is the geometric mean of carapace length and breadth.

In a paper based explicitly on the multivariate lognormal distribution, Sampson and Siegel (1985) define as a "residual" size variable the unique loglinear size variable $M(\mathbf{X}; \mathrm{a})$ independent of shape. They then give an allometric interpretation of the the huge antlers in the Irish elk. They conclude that for its shape (the antler/shoulder heigth ratio) the Irish elk had a larger than expected body size, a fact consistent with the enormous amount of tissue (antlers) which had to be regenerated on a (presumably) annual basis in this extinct species.

Mosimann and James (1979) also use an explicit multivariate lognormal assumption in analyzing geographic variation in Florida red-winged blackbirds. They consider variance component analyses of various loglinear size variables, and conclude that winglength is the best of those considered in that its between/within ratio was highest.

The above sampling of the current literature of allometry is sufficient to indicate the active interest in the multivariate lognormal model in studies of allometry. A number of other articles, from other fields could also be cited, for this distribution is of fundamental importance wherever multiplicative errors play a basic role in random processes.

REFERENCES

Darroch, J. N. and Mosimann, J. E. (1985). Canonical and principal components of shape, *Biometrika*, *72*, 241–252.

Ferguson, T. S. (1962). Location and scale parameters in exponential families of distributions, *Ann. Math. Statist.*, *33*, 986–1001.

Fisher, R. A. (1936). The use of multiple measurements in taxonomic problems, *Ann. Eugen.*, *7*, 179–188.

Ford, S. M. and Corruccini, R. S. (1985). Intraspecific, interspecific, metabolic, and phylogenetic scaling in platyrrhine primates, *Size and Scaling in Primate Biology* (W. L. Jungers, ed.), Plenum Press, New York, pp. 401–435.

Gould, S. J. (1966). Allometry and size in ontogeny and phylogeny, *Biological Reviews*, *41*, 587–640.

Gould, S. J. (1977). *Ontogeny and Phylogeny*, The Belknap Press of Harvard University Press, Cambridge, Massachusetts.

Griffiths, D. A. and Sandland, R. L. (1982). Allometry and multivariate growth revisited, *Growth*, *46*, 1–11.

Hopkins, J. W. (1966). Some considerations in multivariate allometry, *Biometrics*, *22*, 747–760.

Huxley, J. S. (1932). *Problems of Relative Growth*, The Dial Press, New York.

James, I. R. (1979). Characterization of a family of distributions by the independence of size and shape variables, *Ann. Statist.*, *7*, 869–881.

James, I. R. (1981). Distributions associated with neutrality properties for random proportions, *Statistical Distributions in Scientific Work*, *4*, (C. Taillie, G. P. Patil, and B. A. Baldessari, eds), D. Reidel Publishing Company, Dordrecht-Holland, pp. 125–136.

James, I. R. and Mosimann, J. E. (1980). A new characterization of the Dirichlet distribution through neutrality, *Ann. Statist.*, *8*, 183–189.

Johnson, N. L. and Kotz, S. (1972). *Distributions in Statistics: Continuous Multivariate Distributions*, Wiley, New York.

Jolicoeur, P. (1963). The multivariate generalization of the allometry equation, *Biometrics*, *19*, 497–499.

Jolicoeur, P. (1984). Principal components, factor analysis, and multivariate allometry: a small sample direction test, *Biometrics*, *40*, 685–690.

Jolicoeur, P. and Heusner, A. A. (1971). The allometry equation in the analysis of the standard oxygen consumption and body weight of the white rat, *Biometrics*, *27*, 841–855.

Jungers, W. L. (1985). Body size and scaling of limb proportions in primates, *Size and Scaling in Primate Biology* (W. L. Jungers, ed.), Plenum Press, New York, pp. 345–381.

Lukacs, E. (1955). A characterization of the gamma distribution, *Ann. Statist.*, *26*, 319–324.

Mosimann, J. E. (1958). An analysis of allometry in the chelonian shell, *Revue Canadienne de Biologie*, *17*, 137–228.

Mosimann, J. E. (1962). On the compound multinomial distribution, the multivariate β distribution, and correlations among proportions, *Biometrika*, *49*, 66–82.

Mosimann, J. E. (1970). Size allometry: size and shape variables with characterizations of the lognormal and generalized gamma distributions, *J. Amer. Statist. Assoc.*, *65*, 930–945.

Mosimann, J. E. (1975a). Statistical problems of size and shape. I. Biological applications and basic theorems, *Statistical Distributions in Scientific Work*, *2* (G. P. Patil, S. Kotz, and K. Ord, eds.), D. Reidel Publishing Company, Dordrecht-Holland, pp. 187–217.

Mosimann, J. E. (1975b). Statistical problems in size and shape, II, Characterizations of the lognormal, gamma and Dirichlet distributions, *Statistical Distributions in Scientific Work*, *2* (G. P. Patil, S. Kotz, and K. Ord, eds.), D. Reidel Publishing Company, Dordrecht-Holland, pp. 219–239.

Mosimann, J. E. and James, F. C. (1979). New statistical methods for allometry with application to Florida red-winged blackbirds, *Evolution*, *33*, 444–459.

Mosimann, J. E. and Malley, J. D. (1979). Size and shape variables, *Statistical Ecology*, *7* (L. Orloci, C. R. Rao, and W. M. Stiteler, eds.), International Co-operative Publishing House, Fairland, Maryland, pp. 175–189.

Mosimann, J. E. and Malley, J. D. (1981). The independence of size and shape before and after scale change, *Statistical Distributions in Scientific Work*, *4* (C. Taillie, G. P. Patil, and B. A. Baldessari, eds). D. Reidel Publishing Company, Dordrecht-Holland, pp. 137–145.

Mosimann, J. E., Malley, J. D., Cheever, A. W., and Clark, C. B. (1978). Size and shape analysis of schistosome egg-counts in Egyptian autopsy data, *Biometrics*, *34*, 341–356.

Rao, C. R. (1973). *Linear Statistical Inference and Its Applications*, (2nd ed.) Wiley, New York.

Reeve, E. C. R. and Huxley, J. S. (1945). Some problems in the study of allometric growth, *Essays on Growth and Form* (W. E. Le Gros Clark and P. B. Medawar, eds.), Oxford Univ. Press, Oxford, pp. 121–156.

Reyment, R. A., Blackith, R. E., and Campbell, N. A. (1984). *Multivariate Morphometrics*, (2nd ed.) Academic Press, New York.

Sampson, P. D. and Siegel, A. F. (1985). The measure of "size" independent of "shape" for multivariate lognormal populations, *J. Amer. Statist. Assoc.*, *80*, 910–914.

Seebeck, R. M. (1983a). Factors affecting patterns of development and their assessment, *Animal Production*, *37*, 53–66.

Seebeck, R. M. (1983b). The dependence of lean carcass composition on carcass fat, as assessed by multivariate shape/size methods, *Animal Production*, *37*, 321–328.

Sprent, P. (1968). Linear relationships in growth and size studies, *Biometrics*, *24*, 639–656.

Sprent, P. (1972). The mathematics of size and shape, *Biometrics*, *28*, 23–27.

Veitch, L. G. (1978). Size, shape and allometry in *Uca*; a multivariate approach, *Math. Scientist*, *3*, 35–45.

12
Applications in Ecology

BRIAN DENNIS College of Forestry, Wildlife, and Range Sciences, University of Idaho, Moscow, Idaho

G. P. PATIL* Center for Statistical Ecology and Environmental Statistics, Department of Statistics, The Pennsylvania State University, University Park, Pennsylvania

1. INTRODUCTION

Perhaps the lognormal distribution finds the widest variety of applications in ecology. Ever since Malthus and Darwin, biologists have been acutely aware that populations of animals and plants grow multiplicatively. Studying the consequences arising from the enormous potential for increase possessed by most species on earth forms a major component of modern ecological research. Whenever quantities grow multiplicatively, the lognormal becomes a leading candidate for a statistical model of such quantities.

In this chapter, we discuss some of the theoretical and descriptive modeling studies in ecology that have featured the lognormal. We focus primarily on the lognormal as a model of the abundances of species and not as a model of the size growth of individual organisms. We review and critique several of the more important ecological modeling approaches related to the lognormal; in some cases, we display new results or offer thoughts

*Prepared in part as a Visiting Professor of Biostatistics, Department of Biostatistics, Harvard School of Public Health and Dana-Farber Cancer Institute, Harvard University, Boston, Massachusetts.

on future statistical and ecological research problems. The material in this paper is divided into three sections, for which we here provide the following summaries.

In Section 2, we reexamine the lognormal as a theoretical model of population abundance. The traditional multiplicative growth model is recast as a stochastic differential equation. Population size then becomes a diffusion process, that is, a Markov process with sample paths that are continuous functions of time. The transition distribution of the process is lognormal; various other statistical properties such as time-dependent measures of central tendency are obtained. The model is of limited practical usefulness in ecology, as it is essentially just a stochastic version of exponential growth. The model could only describe growth of a species for a short time interval, since all species eventually encounter environmental limits to growth.

However, we also describe a different stochastic growth model leading to the lognormal. The model is a stochastic differential equation based on the Gompertz growth equation. The model contains an underlying deterministic stable equilibrium for population size, representing the outcome of growth regulated by limiting environmental resources. The transition distribution for population size, as well as the equilibrium distribution, is lognormal in form. Thus, ecologists can regard the lognormal not only as a model of unbounded exponential growth, but also as a model of population regulation in the presence of an environmental carrying capacity.

An interpretive problem arises when using stochastic differential equations. A given stochastic differential equation represents two different diffusion processes, depending on whether Ito or Stratonovich stochastic integrals are used. A main reason for using stochastic differential equations is to approximate more complicated stochastic models. The details of the approximation process determine which type of stochastic integral to use. We show that the statistical properties of both lognormal models under the Stratonovich interpretation are easily recovered from those properties under the Ito interpretation, and vice versa, using the concept of weighted distributions.

We review in Section 3 the role of the lognormal as a model of species frequencies. The lognormal is confined to representing a single species in Section 2; by contrast, in Section 3, the lognormal represents patterns displayed by ecological communities with dozens, even hundreds, of species. Ecological and statistical research on quantitative species abundance patterns began in earnest with the introduction of the logseries model in the early 1940s. Shortly afterward, the lognormal model was proposed in reaction to the logseries, since many data sets did not appear J-shaped when plotted on a logarithmic scale. Numerous ecological studies have incorporated the lognormal model. Unfortunately, ecologists have not paid enough attention to sampling considerations and proper inference methods in these studies. As a result, whether the logseries, the lognormal, or some other distribution will be more widely applicable in species abundance studies is an open question. We try to clarify the problems of making statistical inferences for species frequency models, and we describe one promising inference approach that has been proposed but seldom used.

One intriguing aspect of the lognormal species frequency distribution is the so-called Canonical Hypothesis of species abundance. This hypothesis arose from an empirical pattern that had been noticed on logarithmic species frequency plots. The plots indicated that a randomly selected individual organism in the community would most likely come from a species whose log-abundance was in the same class as that of the largest species. The hypothesis received considerable attention in the ecological literature, and a "canonical" lognormal distribution was proposed having parameters constrained in such a fashion as to fix this abundance relationship. We review this hypothesis in Section 3, and we raise a cautionary note to the effect that the Canonical Hypothesis has seldom been formally tested in any way known to be statistically valid. The hypothesis has been studied more recently in the statistics literature, though. The results, which we summarize in Section 3, suggest that a new level of statistical awareness should be injected into the empirical studies of species frequency patterns.

Possibly the main role of the lognormal in ecology is simply to serve as the handiest adjustable wrench in the toolbox of statistical distributions. Ecological abundance data are intrinsically positive, with a few enormously high data points typically arising in every study. The lognormal distribution is an ideal descriptor of such data, with a positive range, right skewness, heavy right tail, and easily computed parameter estimates.

Ecological data sets, however, sometimes contain complicating factors which rule out the use of a simple two-parameter lognormal. In Section 4 we review three typical modifications of the lognormal. First, ecological data sets often consist of count data. The Poisson-lognormal distribution represents a discrete version of the lognormal potentially applicable to such cases. Second, ecological abundance surveys often contain an overly large number of samples with abundances of zero. The delta-lognormal, formed as a finite mixture of an ordinary lognormal distribution and a degenerate (spike) distribution at zero, offers advantages when estimating mean abundance is the objective of the surveys. Third, ecological abundances observed in samples sometimes grew from random numbers of initial propagules in each sample. We review a compound distribution structure recently proposed for such data; the structure also provides a degenerate component for added zeros.

We adopt the following notation throughout the paper. If $X = \log N$ has a normal distribution with probability density function (pdf) given by

$$f_X(x) = \frac{1}{(\sigma^2 2\pi)^{1/2}} \exp\left[-\frac{(x-\mu)^2}{2\sigma^2}\right] \tag{1.1}$$

where $-\infty < x < +\infty$, that is, if

$$X \sim \text{normal}(\mu, \sigma^2) \tag{1.2}$$

then $N = e^X$ has a lognormal distribution with pdf

$$f_N(n) = \frac{1}{n(\sigma^2 2\pi)^{1/2}} \exp\left[-\frac{(\log n - \mu)^2}{(2\sigma^2)}\right] \tag{1.3}$$

and we write

$$N \sim \text{lognormal}(\mu, \sigma^2) \tag{1.4}$$

For additional information on discrete and continuous statistical distributions, including those appearing in this paper, we refer the reader to Patil et al. (1984a, 1984b, 1984c).

2. POPULATION GROWTH MODELS

2.1 Multiplicative Population Growth

MacArthur (1960) quantified for ecologists the notion that the abundance of a single species should have, under certain circumstances, a lognormal distribution. His reasoning was more intuitive than mathematical, but was nonetheless adopted by ecologists as a principal explanation of observed lognormal abundance patterns (May, 1975). MacArthur assumed that the growth rate of a species could be represented by an *ordinary differential equation* (ODE) of the form

$$\frac{dn(t)}{dt} = r(t)n(t) \tag{2.1}$$

where $n(t)$ is population abundance (typically measured in numbers of individuals or biomass per unit area or volume) at time t, and $r(t)$ is the per individual (or per unit biomass) growth rate. This ODE integrates to

$$\log n(t) = \log n_0 + \int_0^t r(\tau)\, d\tau \tag{2.2}$$

where $n_0 = n(0)$. MacArthur noted that the function $r(t)$ might vary randomly in time for some species due to fluctuations of environmental factors. The integral in (2.2) could then be regarded as the accumulated sum of random variables. MacArthur invoked the Central Limit Theorem to predict that $\log n(t)$ would have a normal distribution.

Two features of MacArthur's intuitive derivation are noteworthy: (a) the idea of random fluctuations in the per individual growth rate, and (b) the time-dependence of the normal distribution for $\log n(t)$. The fluctuations, for the derivation to hold, must be of such a nature that the sum of random variables given by $X_1 + X_2 + \cdots + X_k$, where

$$X_i = \int_{a_{i-1}}^{a_i} r(\tau)\, d\tau \tag{2.3}$$

and $0 = a_0 < a_1 < a_2 < \ldots < a_k = t$, conforms to one of the various Central Limit Theorem schemes. The resulting normal distribution for $\log n(t)$ would have a mean that essentially grows linearly with t and a variance that grows proportional to t. Thus, two ecological conditions underlying this derivation become apparent: (a) Any autocovariance of the fluctuations must decay rapidly for the Central Limit Theorem to hold. (b) The time t must be relatively early in the population's growth trajectory, before state-dependent changes in r, due to crowding or food limitation, become important. MacArthur pointed out that the model would only apply to opportunistic species, or species colonizing unutilized resources.

2.2 Stochastic Differential Equations

It is useful to derive various statistical properties for stochastic models such as MacArthur's, in order to test them with ecological data. The analysis is greatly simplified by using *stochastic differential equations* (SDEs). SDEs, known also as diffusion processes, can serve as approximations to many stochastic processes, including stochastic difference equations, branching processes, and birth-death processes (see Karlin and Taylor, 1981, p. 168). The approach to SDEs and lognormal growth models taken here follows that of Dennis and Patil (1984). See also Patil (1984).

An SDE model for the growth of a single species may be written as

$$dN(t) = N(t)g(N(t))\,dt + \sigma N(t)\,dW(t) \tag{2.4}$$

Here $N(t)$ is population abundance (now in upper case to denote a stochastic process) at time t, and $g(N(t))$ is the per unit abundance growth rate, which in general may depend on the population abundance. Also, $W(t)$ is a standard Wiener process ($W(t) \sim \text{normal}(0,t)$; $dW(t) \sim \text{normal}(0,dt)$) and σ is a positive scale constant. The form of (2.4) arises from an assumption that the per unit abundance growth rate, $g(N(t))$, is perturbed by unpredictable environmental factors. Mathematically, the differential $dN(t)$ is defined in terms of an Ito or a Stratonovich stochastic integral (e.g., Karlin and Taylor, 1981, p. 346). The quantity $N(t)$ becomes a diffusion process, a type of Markov process having continuous sample paths with probability one. Two functions, the infinitesimal mean and the infinitesimal variance, characterize most of the statistical properties of a diffusion process. They are defined respectively by

$$m_N(n) = \lim_{h\to 0}\frac{1}{h}E[\Delta N \mid N(t) = n] \tag{2.5}$$

$$v_N(n) = \lim_{h\to 0}\frac{1}{h}E[(\Delta N)^2 \mid N(t) = n] \tag{2.6}$$

where $\Delta N = N(t+h) - N(t)$. A standard result is that the infinitesimal mean and variance of the process $N(t)$ defined by the SDE (2.4) are

$$m_N(n) = ng(n) + \omega n \tag{2.7}$$

$$v_N(n) = \sigma^2 n^2 \tag{2.8}$$

Here ω is an indicator variable which depends on the type of stochastic integral being used to define the SDE (2.4): $\omega = 0$ if the Ito integral is used, and $\omega = \sigma^2/2$ if the Stratonovich integral is used.

Whether the Ito or Stratonovich integral is appropriate depends on the interpretation of (2.4) as an approximation to some underlying stochastic process. If (2.4) is seen as an approximation to a stochastic difference equation with uncorrelated noise, then the Ito interpretation should be used. If, however, $N(t)$ is viewed as an approximation to some process produced by integrating along a sample path of a smooth Gaussian process, then the Stratonovich interpretation of (2.4) should apply. These points are developed further by Ricciardi (1977), Karlin and Taylor (1981), and Horsthemke and Lefever (1984).

The fact that the Ito and Stratonovich interpretations of (2.4) produce different quantitative predictions has caused some consternation in the ecological literature (Feldman and Roughgarden, 1975; Turelli, 1977). The controversy has diminished in more recent years, as ecological modelers now do not take SDEs of form (2.4) too literally, but merely regard the SDEs as mathematically convenient approximations to more detailed, underlying processes. We show below that two SDE models producing lognormal distributions have qualitatively similar predictions under the Ito and Stratonovich interpretations, and that any differences are easily sorted out using weighted distributions.

One of the most useful features of diffusion processes is the transformation property. If $N(t)$ is a diffusion process, then $X(t) = h(N(t))$ is also a diffusion process, provided h is a continuous, strictly increasing (decreasing) function. The infinitesimal mean and variance of $X(t)$ are given by

$$m_X(x) = \frac{v_N(n)h''(n)}{2} + m_N(n)h'(n), \tag{2.9}$$

$$v_X(x) = v_N(n)[h'(n)]^2 \tag{2.10}$$

(if h' and h'' are uniformly continuous functions) with $n = h^{-1}(x)$ (Karlin and Taylor, 1981, p. 173). This property often permits the transformation of a novel diffusion process into a known process with well-studied statistical properties.

2.3 Stochastic Exponential Growth Model

MacArthur's intuitive model can be recast in statistical terms as an SDE of the form (2.4) with a constant per unit abundance growth rate:

$$dN(t) = rN(t)\,dt + \sigma N(t)\,dW(t) \tag{2.11}$$

The Stratonovich version of this model was extensively analyzed by Capocelli and Ricciardi (1974) (see also Tuckwell, 1974). A discrete time stochastic version of the exponential growth model was studied by Lewontin and Cohen (1969), and is essentially recaptured in the Ito version of (2.11). The differences in the Ito and Stratonovich versions were studied by Gray and Caughey (1965), Feldman and Roughgarden (1975), Ricciardi (1977), and Braumann (1983). The main properties of the model are found using the transformation $X(t) = \log N(t)$ and the formulas (2.9) and (2.10), producing

$$m_X(x) = r + \omega - \frac{\sigma^2}{2} \tag{2.12}$$

$$v_X(x) = \sigma^2 \tag{2.13}$$

These are the infinitesimal moments of a Wiener process with drift. A well-known result (e.g. Ricciardi, 1977, p. 58) gives a normal transition distribution for $X(t)$:

$$X(t) \sim \text{normal}\left(x_0 + \left(r + \omega - \frac{\sigma^2}{2}\right)t, \sigma^2 t\right) \tag{2.14}$$

Equivalently, the distribution of $N(t)$ becomes lognormal:

$$N(t) \sim \text{lognormal}\left(\log n_0 + \left(r + \omega - \frac{\sigma^2}{2}\right)t, \sigma^2 t\right) \tag{2.15}$$

It is interesting to compare various measures of central tendency with the deterministic solution of the ODE $dn(t)/dt = rn(t)$ given by

$$n(t) = n_0 e^{rt} \tag{2.16}$$

The mean, geometric mean, and harmonic mean of $N(t)$ are, respectively,

$$E[N(t)] = n_0 e^{(r+\omega)t} \tag{2.17}$$

$$\exp\{E[\log N(t)]\} = n_0 e^{[r+\omega-(\sigma^2/2)]t} \tag{2.18}$$

$$1/E[1/N(t)] = n_0 e^{(r+\omega-\sigma^2)t} \tag{2.19}$$

The expectations are conditioned on $N(0) = n_0$. Also, the median and mode of the distribution of $N(t)$ are found to be

$$\text{median}(N(t)) = n_0 e^{[r+\omega-(\sigma^2/2)]t} \tag{2.20}$$

$$\text{mode}(N(t)) = n_0 e^{[r+\omega-(3\sigma^2/2)]t} \tag{2.21}$$

For smaller values of r, some of the central tendency measures increase exponentially while others decay to zero. In fact, if $r < (\sigma^2/2) - \omega$, the probability that $N(t)$ is arbitrarily close to zero approaches 1 as t becomes large:

$$\begin{aligned} P[0 < N(t) \leq \varepsilon] &= P[-\infty < X(t) \leq \log \varepsilon] \\ &= P\left\{-\infty < \frac{X(t) - E[X(t)]}{[\text{var}(X(t))]^{1/2}} \leq \frac{\log \varepsilon - E[X(t)]}{[\text{var}(X(t))]^{1/2}}\right\} \longrightarrow 1 \end{aligned} \tag{2.22}$$

as $t \to \infty$, since the last expression is the probability that a standard normal random variable is less than or equal to $[\log \varepsilon - x_0]/(\sigma\sqrt{t}) - [r + \omega -$

$(\sigma^2/2)]\sqrt{t}/\sigma$, which increases without limit as t increases. A similar result was described by Lewontin and Cohen (1969) for a discrete time process, and by Capocelli and Ricciardi (1974) for the Stratonovich version of this model.

Note that in the above central tendency measures, the geometric mean equals the median, which is a general property of the lognormal distribution. Also, the deterministic trajectory (2.16) equals the arithmetic mean for the Ito version, while for the Stratonovich version the deterministic trajectory equals the geometric mean. This point was stressed by Braumann (1983) in asserting that a main practical difference in the Ito and Stratonovich versions is the semantic interpretation of (2.16) as a mean. In fact, a cascade of interrelationships between the central tendency measures of the two versions exists (Dennis and Patil, 1984). Let $N_I(t)$ and $N_S(t)$ denote, respectively, the Ito and Stratonovich versions of $N(t)$. Then,

$$E[N_I(t)] = \exp\{E[\log N_S(t)]\} = n(t) \tag{2.23}$$

$$\exp\{E[\log N_I(t)]\} = 1/E[1/N_S(t)] \tag{2.24}$$

$$1/E[1/N_I(t)] = \text{mode}(N_S(t)) \tag{2.25}$$

More generally, the generalized means of the Ito and Stratonovich versions are related. The θth moment about the origin of $N(t)$ is

$$E[(N(t))^\theta] = n_0^\theta e^{[r+\omega-\sigma^2/2]\theta t+(\sigma^2/2)\theta^2 t} \tag{2.26}$$

The generalized mean of $N(t)$ is then:

$$\left\{E[(N(t))^\theta]\right\}^{1/\theta} = n_0 e^{[r+\omega+(\theta-1)(\sigma^2/2)]t} \tag{2.27}$$

The Ito-Stratonovich interrelationship becomes

$$\left\{E[(N_I(t))^\theta]\right\}^{1/\theta} = \left\{E[(N_S(t))^{\theta-1}]\right\}^{1/(\theta-1)} \tag{2.28}$$

These interrelationships are derived from a general property of the exponential growth SDE: the Stratonovich transition lognormal distribution is a weighted Ito transition lognormal distribution. The statistical concept of weighted distributions was defined by Rao (1965) and has been investigated by Patil and Ord (1976), Patil and Rao (1977, 1978) and Mahfoud and Patil (1982). As pointed out by Dennis and Patil (1984), the Ito and

Stratonovich lognormal transition pdfs are related by

$$f_S(n,t \mid n_0) = \frac{n^{1/2} f_I(n,t \mid n_0)}{E[(N_I(t))^{1/2}]} \tag{2.29}$$

We will further point out here that the Stratonovich pdf can be obtained in this model as a scale-transformed Ito pdf. It is a property of the lognormal distribution that

$$\frac{n^{\beta} f(n)}{E[N^{\beta}]} = e^{-\lambda\beta} f(e^{-\lambda\beta} n) \tag{2.30}$$

where $f(n)$ is the pdf of a lognormal (μ, λ) random variable, N. Thus, letting $\beta = 1/2$, the Stratonovich pdf can be obtained from the Ito pdf through the scale transformation $N_S(t) = \exp(\sigma^2 t/2) N_I(t)$.

2.4 Stochastic Gompertz Growth Model

The stochastic growth model given by the SDE (2.11) could not be used indefinitely to represent a population's abundance. An increasing population would eventually encounter limits on nutrient supply, space, or other resources necessary for growth. This situation is frequently modeled in ecology with an ODE of the form

$$\frac{dn(t)}{dt} = n(t) g(n(t)) \tag{2.31}$$

(e.g. Freedman, 1980), where $g(n(t))$ is assumed to be a decreasing function of $n(t)$ with the following properties:

$$g(\bar{n}) = 0 \tag{2.32}$$

for some $\bar{n}$ such that $0 < \bar{n} < \infty$, and

$$g'(\bar{n}) < 0 \tag{2.33}$$

Population abundance for such models increases (or decreases) from n_0 to the stable equilibrium value given by $\bar{n}$.

One particular form of (2.31) is the Gompertz growth model:

$$\frac{dn(t)}{dt} = an(t) \log \left[\frac{\bar{n}}{n(t)} \right] \tag{2.34}$$

This ODE integrates readily to

$$n(t) = \exp\left\{\log \bar{n} + \left[\log \frac{n_0}{\bar{n}}\right] e^{-at}\right\} \tag{2.35}$$

This growth trajectory is a sigmoid curve with an inflection point at $\bar{n}/e$.

Such models can often be approximated quite well by the logistic growth model. The procedure approximates $g(n)$ with a linear function using a Taylor series expansion around $\bar{n}$ (Dennis and Patil, 1984):

$$\frac{dn(t)}{dt} = n(t)\left[r - \frac{r}{\bar{n}}n(t)\right] \tag{2.36}$$

Here $r = -\bar{n}g'(\bar{n})$. For the Gompertz model, the logistic approximation has $r = a$. This approximation has a trajectory that starts at n_0, and levels off at $\bar{n}$ (like the Gompertz), but the inflection point occurs at $\bar{n}/2$.

Stochastic versions of such models can be built as SDEs of the form (2.4). In the resulting stochastic models, population abundance does not level off at a stable equilibrium. Rather, the distribution for $N(t)$ may approach a limiting stationary distribution that is independent of the initial conditions as well as t. The stationary distribution, when it exists, has the following pdf (see Dennis and Patil, 1984):

$$f(n) = \psi \exp\left\{\frac{2}{\sigma^2}\int \frac{g(n)}{n}\,dn - 2\left(1 - \frac{\omega}{\sigma^2}\right)\log n\right\} \tag{2.37}$$

The constant ψ is evaluated by setting the area under $f(n)$ equal to one.

This stationary pdf is a member of the log-exponential family of pdfs defined by Patil and Ord (1976). A feature of this family is the "form-invariance" property: the size-biased version always retains the same form as the original pdf. Patil and Rao (1978) provide a general discussion of the properties of size-biased distributions.

One implication of this property is that the Ito and Stratonovich versions of the SDE (2.4) have stationary distributions of the same form (Dennis and Patil, 1984). From (2.37), the Stratonovich stationary pdf is found to be a size-biased Ito stationary pdf:

$$f_S(n) = \frac{n f_I(n)}{E[N_I]} \tag{2.38}$$

Thus, if the Ito version of (2.4) predicted a certain type of stationary distribution, such as a lognormal or a gamma, then the Stratonovich version would predict the same type (provided both pdfs exist). An immediate consequence of this relationship (2.38) is that the harmonic mean of the stationary Stratonovich pdf is the mean of the stationary Ito pdf:

$$1/E[1/N_S] = E[N_I] \tag{2.39}$$

The stochastic version of the Gompertz model is in the form (2.4), with $g(N(t)) = a \log[\bar{n}/N(t)]$. The stationary pdf is found from (2.37) to be that of a lognormal random variable:

$$N(\infty) \sim \text{lognormal}\left(\log \bar{n} - \frac{[(\sigma^2/2) - \omega]}{a}, \frac{\sigma^2}{2a}\right) \tag{2.40}$$

This result considerably extends the conceptual use of the lognormal in ecology as a population growth model. The lognormal under the MacArthur-type scenarios was strictly a time-dependent, transient model for a population in the early phase of its growth. By contrast, the lognormal (2.40) is a model for a population fluctuating around a stable equilibrium value.

A stochastic version of the logistic model (2.36) takes the form (2.4), with $g(N(t)) = r - (r/\bar{n})N(t)$. The stationary distribution is a gamma distribution with the following pdf:

$$f(n) = \frac{\alpha^\beta}{\Gamma(\beta)} n^{\beta-1} e^{-\alpha n} \tag{2.41}$$

where $\alpha = 2r/(\sigma^2 \bar{n})$ and $\beta = 2r/\sigma^2 - 1 + 2\omega/\sigma^2$. If $\beta > 1$, the gamma has roughly the same shape as the lognormal: unimodal and right-skewed. In many practical instances, ecologists would not be able to distinguish between the lognormal and the gamma on the basis of fit to a given data set. Just as the logistic (2.36) is an approximation to growth models with a stable equilibrium, the gamma can serve as an approximation to the stationary distributions of stochastic growth models of the form (2.4). Dennis and Patil (1984) provide further discussion of the ecological role of the gamma as an abundance model.

The stochastic Gompertz model has the convenient feature that the complete transition pdf can be obtained (Ricciardi, 1977). The transformation $X(t) = \log N(t)$ yields, with the help of formulas (2.9) and (2.10), a diffusion process with infinitesimal moments given by

$$m_X(x) = a \log \bar{n} - \frac{\sigma^2}{2} + \omega - ax \qquad v_X(x) = \sigma^2 \tag{2.42}$$

These are the infinitesimal moments of the well-known Ornstein-Uhlenbeck process (e.g. Karlin and Taylor, 1981, p. 170). The transition distribution for $X(t)$ is normal:

$$X(t) \sim \text{normal}(\mu(t), \lambda(t)) \tag{2.43}$$

where

$$\begin{aligned} \mu(t) &= \log \bar{n} - \frac{(\sigma^2/2) - \omega}{a} \\ &\quad + \left\{ \log n_0 - \log \bar{n} + \frac{(\sigma^2/2) - \omega}{a} \right\} e^{-at} \end{aligned} \tag{2.44}$$

$$\lambda(t) = \frac{\sigma^2}{2a}(1 - e^{-2at}) \tag{2.45}$$

Thus, the transition distribution of $N(t)$ is lognormal:

$$N(t) \sim \text{lognormal}(\mu(t), \lambda(t)) \tag{2.46}$$

We have here a time-dependent lognormal distribution that could represent a population's abundance for large values of t as well as small values. The limiting stationary distribution (2.40) is recovered from (2.46) as $t \to \infty$.

Various time-dependent measures of central tendency can be written down:

$$\left\{ E[(N(t))^\theta] \right\}^{1/\theta} = e^{\mu(t) + \theta\lambda(t)/2} \tag{2.47}$$

$$\text{mode}(N(t)) = e^{\mu(t) - \lambda(t)} \tag{2.48}$$

The arithmetic mean, geometric mean (= median), and harmonic mean are found from (2.47) by setting $\theta = 1$, 0, and -1, respectively. In particular, the geometric mean of $N(t)$ for the Stratonovich version of the Gompertz SDE equals the deterministic growth trajectory given by (2.35).

The relationship between the Ito and Stratonovich versions of the stochastic Gompertz model is manifested in terms of a weighted distribution, as was the case for the stochastic exponential growth model. A property of the lognormal distribution relates the lognormal random variable N with its β-weighted version N_β. Specifically, if

$$N \sim \text{lognormal}(\mu, \lambda) \tag{2.49}$$

with pdf $f(n)$, and if N_β has a pdf given by

$$f_\beta(n) = \frac{n^\beta f(n)}{E[N^\beta]} \tag{2.50}$$

then

$$N_\beta \sim \text{lognormal}(\mu + \beta\lambda, \lambda) \tag{2.51}$$

This β-weighted lognormal is also the distribution of a scale transformation of the original random variable given by $\exp(\beta\lambda)N$ (see (2.30)). The property (2.51) is related to the variance-invariance characterization theorem of the lognormal due to Mahfoud and Patil (1982): N is lognormal iff $\text{Var}(\log N) = \text{Var}(\log N_\beta)$ for all $\beta > 0$. From (2.44), we find that the means of the log-transformed Ito and Stratonovich variables are related by

$$\mu_S(t) = \mu_I(t) + \frac{\sigma^2}{2a}(1 - e^{at}) \tag{2.52}$$

where $\mu_I(t) = E[\log N_I(t)]$, etc. Thus, the Stratonovich and Ito lognormal distributions share precisely the weighted relationship (2.50), where β is a function of time. Specifically,

$$f_S(n, t \mid n_0) = \frac{n^{\beta(t)} f_I(n, t \mid n_0)}{E[(N_I(t))^{\beta(t)}]} \tag{2.53}$$

where

$$\beta(t) = \frac{\sigma^2}{2a}(1 - e^{-at})/\lambda(t) \tag{2.54}$$

As t becomes large, the relationship (2.53) of the transition pdfs approaches the stationary size-biased relationship (2.38), since $\beta(t) \to 1$.

3. SPECIES FREQUENCY MODELS

3.1 Fisher's Logseries Models

A pivotal, three-part paper by Fisher, Corbet, and Williams (1943) launched four decades of ecological research on quantitative patterns of species abundance. Though it was not used by Fisher et al., the lognormal distribution has been one of the main tools used by ecologists in this research.

C. B. Williams had been studying samples of moths from Great Britain, and A. S. Corbet had been studying butterfly samples from Malaya. Many species were represented in the samples. Corbet and Williams noted that any particular sample usually had a large number of species represented by only a single individual apiece, while a less number of species were represented by two individuals, even less by three individuals, and so on. Corbet had observed that these species frequencies appeared to follow a harmonic series pattern: if N was the number of species with one individual in the sample, then $N/2$ was approximately the number of species with two individuals, $N/3$ was approximately the number of species with three individuals, and so on. Unfortunately, this mathematical model had the inconvenient property that it diverged: the sum given by $N(1+\frac{1}{2}+\frac{1}{3}+\ldots)$, if continued indefinitely, would predict an infinite number of species in the sample.

R. A. Fisher proposed a modification of the model. Fisher supposed that the expected numbers of species in the sample might be proportional to the terms of a negative binomial distribution. Upon analyzing the data of Corbet and Williams, Fisher found that the estimated values of the parameter k in the negative binomial were invariably small, usually quite close to zero. Fisher reduced the number of parameters in the model by taking the limit $k \to 0$, $s \to \infty$ in such a way that

$$sk \longrightarrow \alpha \tag{3.1}$$

in the negative binomial model. The quantity s is the proportionality constant in the negative binomial terms, representing the number of species in the ecological community being sampled (we must note that the limit (3.1) never explicitly appears in Fisher's discussion (Fisher et al., 1943), but rather is more or less implied). The result was that the expected number of species with r individuals in the sample, m_r, became proportional to the terms of a logseries distribution:

$$m_r = \frac{\alpha q^r}{r} \qquad r = 1, 2, 3, \ldots \tag{3.2}$$

where $\alpha > 0$ and $0 < q < 1$.

The logseries distribution has been used extensively since Fisher et al. (1943) to describe species frequencies, most notably by C. B. Williams

(1964) and R. A. Kempton (Kempton and Taylor, 1974, 1979; Kempton, 1975; Taylor et al., 1976).

3.2 Preston's Lognormal Model

F. W. Preston published an influential objection to the logseries model a few years after the Fisher et al. paper appeared (Preston, 1948). Preston worked with data sets on bird communities as well as moth communities, including some of Williams' data. Preston grouped the data into logarithmic abundance intervals which he called "octaves": the number of species with 1-2 individuals in the samples, with 2-4 individuals, with 4-8 individuals, etc., were displayed as a frequency histogram. Preston observed that the histograms, when drawn on such a logarithmic scale, tended to have modes, and in fact, tended to look quite Gaussian. Preston fitted a (left-truncated) normal curve to the histograms, which seemed to describe the data sets very well. Since Preston's paper, the normal curve has been widely used to "graduate," in Preston's words, species frequency data grouped into logarithmic abundance intervals (see reviews by Whittaker, 1972, and May, 1975).

3.3 Sampling Considerations

These applications of the lognormal as a species frequency model have unfortunately been marred by a lack of statistical rigor. Preston and subsequent investigators in many cases fit the Gaussian curves to the histograms by eye. Later, ecologists employed nonlinear regression routines to find the least-squares fits of the Gaussian curves to the histograms (Gauch and Chase, 1974). Such procedures ignore any probabilistic content of the Gaussian curve, ignore the intrinsically discrete nature of the data, and ignore sampling mechanisms. Lacking an explicit likelihood function, the ecologists are unable to provide valid confidence intervals for the parameters, test for goodness of fit, or tests for differences between samples.

Statisticians, in fact, have been unable to agree on the appropriate sampling model to use in conjunction with either the logseries or the lognormal models (see, for instance, Rao, 1971; Watterson, 1974; Kempton, 1975; Engen, 1979; Lo and Wani, 1983). It is unclear whether Fisher originally had an explicit sampling model in mind for the logseries. It is the authors' opinion that Kempton's (1975) sampling model is likely to find the widest use in species frequency studies, though more statistical and ecological research on this question certainly remains to be done. We will briefly describe Kempton's sampling model here, with attention to the role of the lognormal distribution in this approach.

Let N_r be the number of species with r representatives in the sample, $r = 1, 2, 3, \ldots$. The numbers $N_1, N_2, \ldots$, are assumed to be independent, but not identically distributed, Poisson random variables. The total number of species in the sample is assumed to be a Poisson random variable with mean s. Also, the number of individuals in the sample of a particular species is assumed to be a Poisson random variable with mean λ. The values of λ differ among species; it is assumed that the λ values arise from a continuous distribution on the positive real line with pdf $f(\lambda)$. The result of these assumptions is that

$$E[N_r] \equiv m_r = s \int_0^\infty \frac{e^{-\lambda}\lambda^r}{r!} f(\lambda)\, d\lambda \tag{3.3}$$

The pdf $f(\lambda)$ would typically be that of either a lognormal or a gamma distribution. Thus, the observed species frequencies, $n_1, n_2, \ldots$, are realized values of independent Poisson variables, $N_1, N_2, \ldots$, whose means, $m_1, m_2, \ldots$, contain a common set of unknown parameters. The unknown parameters are found in (3.3) and consist of s plus the parameters in the pdf $f(\lambda)$.

If the λ values arise from a lognormal (μ, σ^2) pdf, then

$$m_r(s, \mu, \sigma^2) = \frac{s}{r!(\sigma^2 2\pi)^{1/2}} \int_0^\infty \lambda^{r-1} \exp[-\lambda - (\log\lambda - \mu)^2/(2\sigma^2)]\, d\lambda \tag{3.4}$$

In other words, the expected values m_r are proportional to the terms of a discrete Poisson-lognormal distribution (Holgate, 1969; Bulmer, 1974; Kempton and Taylor, 1974; Shaban, this volume). For the gamma model, with $f(\lambda) = [\beta^k/\Gamma(k)]\lambda^{k-1}e^{-\beta\lambda}$, the m_r values are proportional to the terms of a negative binomial distribution:

$$m_r(s, k, \beta) = s\binom{k+r-1}{r} q^r p^k \tag{3.5}$$

where $q = 1 - p = 1/(1+\beta)$. Taking Fisher's limit $s \to \infty$, $k \to 0$, and $sk \to \alpha$ here produces the logseries:

$$m_r(\alpha, q) = \frac{\alpha q^r}{r} \tag{3.6}$$

Let the unknown parameters be denoted by the vector θ. The likelihood function becomes the product of Poisson probabilities:

$$l(\theta) = \prod_{r=1}^{\infty} \frac{\exp[-m_r(\theta)][m_r(\theta)]^{n_r}}{n_r!} \tag{3.7}$$

With product-Poisson sampling, iteratively reweighted least squares could be used for calculating maximum likelihood estimates (Jennrich and Moore, 1975). Using the lognormal model, though, requires an additional routine for numerical integration in order to evaluate the Poisson-lognormal terms (3.4).

3.4 Preston's Canonical Hypothesis

Preston (1962) noticed a curious pattern in his lognormal curves of species frequencies. The pattern formed the basis of "Preston's Canonical Hypothesis" of species abundance. The Canonical Hypothesis (CH) essentially states that the species frequency curves observed in nature will be predominantly lognormal, and that the parameter values observed will be found only in a small, constrained region of the parameter space. The CH has attracted considerable attention in the ecological literature (see May, 1975).

Specifically, the CH consists of a lognormal distribution with the following structure. We must first define the so-called individuals curve. If $f(\lambda)$ is the species abundance pdf in (3.3), then the expected number of species with abundances greater than λ would be

$$s \int_{\lambda}^{\infty} f(u)\, du \tag{3.8}$$

It would follow that the expected total abundance of all those species with abundance greater than λ would be

$$s \int_{\lambda}^{\infty} u f(u)\, du \tag{3.9}$$

Because of (3.8) and (3.9), $sf(\lambda)$ is called the species curve, and $s\lambda f(\lambda)$ is called the individuals curve. In economics, if $f(\lambda)$ represents a distribution of wealth among individuals, then (3.8) is the number of individuals with wealth greater than λ, and (3.9) is the total amount of wealth these individuals have cornered. On a logarithmic scale, with $\tau = \log \lambda$, these curves become $se^{\tau} f(e^{\tau})$ and $se^{2\tau} f(e^{\tau})$, respectively. When $f(\lambda)$ is a lognormal pdf, these logarithmic species and individuals curves are of course Gaussian.

Preston found, through examining many of his logarithmic histogram diagrams, that the mode of the logarithmic individuals curve tended to fall in the octave of the largest species. In other words, a randomly picked individual (dollar) would most likely come from the logarithmically largest species (wealthiest individual), rather than from, say, a group of species with intermediate logarithmic abundance (middle class). This pattern occured repeatedly in Preston's eye-fitted curves, leading Preston to propose a "canonical" lognormal distribution in which the parameters are constrained so as to fix this mode = max relationship.

Ecologists, judging from their literature, have practically come to regard the CH as an established empirical law of nature. Sugihara (1980), for instance, states: "Few propositions in ecology have as much empirical support as Preston's (1962) canonical hypothesis of species abundance." Sugihara goes on to propose a refinement of the lognormal sequential breakage model (see Aitchison and Brown, 1957; Pielou, 1975) which produces a canonical lognormal distribution: the pieces being broken are niches in a multidimensional niche space, and a breakage corresponds to the evolutionary splitting of a species or a successful invasion of a niche occupied by another species. Preston himself regards departures from the canonical lognormal distribution as indicative of defective, nonrandom sampling, of sampling heterogeneous ecological communities, or of sampling overpacked communities with more species than niches (Preston, 1980).

The enthusiasm ecologists have for this hypothesis must be judged from a statistical standpoint as premature. The studies supporting the CH are based on data sets analyzed with dubious parameter estimation methods having no known statistical validity. By contrast, Kempton's extensive analyses of British moth communities incorporated the explicit sampling model and likelihood function described earlier. These studies reported no evidence that the canonical lognormal is the best fitting distribution; in fact, the logseries model tended to outperform the full lognormal model for many of the moth collections (Kempton and Taylor, 1974; Taylor et al., 1976).

3.5 Statistics of Preston's Canonical Hypothesis

Patil and Taillie (1979a) have defined the CH in statistically precise terms. Their work provides formal statistical hypotheses concerning the CH that potentially could be tested for any data set on species frequencies.

Patil and Taillie define the predicted abundance of the largest species as

$$\lambda_{\max} = \bar{F}^{-1}\left(\frac{1}{s+1}\right) \tag{3.10}$$

where $\bar{F}(\lambda) = 1 - F(\lambda)$, and F is the cumulative distribution function given by

$$F(\lambda) = \int_0^\lambda f(u)\,du \tag{3.11}$$

The idea arises from the fact that $E[\bar{F}(\Lambda_{\max})] = 1/(s+1)$, where $\Lambda_{\max}$ is the largest observation from a random sample of size s from $f(\lambda)$. Then $\lambda_{\max}$ is a convenient, tractable approximation to $E[\Lambda_{\max}]$. The mode, $\tilde{\tau}$ of the logarithmic individuals curve is found by setting $d\log[se^{2\tau}f(e^\tau)]/d\tau = 0$. The CH is then formally stated by Patil and Taillie as $\log\lambda_{\max} \approx \tilde{\tau}$, or

$$\bar{F}^{-1}\left(\frac{1}{1+s}\right) \approx e^{\tilde{\tau}} \tag{3.12}$$

Using a lognormal (μ, σ^2) pdf for $f(\lambda)$, the CH becomes

$$\left[\bar{\Phi}^{-1}\left(\frac{1}{s+1}\right)\right]^2 \approx \sigma^2 \tag{3.13}$$

where $\bar{\Phi}$ is the right tail of a standard normal distribution. This statement of the CH amounts to a constraint on the parameters s and σ^2. If s is large, as is the case for most species abundance studies, then

$$2\log s - \log\log s - \log(4\pi) \approx \sigma^2 \tag{3.14}$$

provides a very good approximation to the relationship (3.13).

Patil and Taillie further point out that other distributions besides the lognormal could be used for $f(\lambda)$ in the CH (3.12). For instance, the gamma model with $f(\lambda) = [\beta^k/\Gamma(k)]\lambda^{k-1}e^{-\beta\lambda}$ yields the following version of the CH:

$$\frac{\Gamma(k, k+1)}{\Gamma(k)} \approx \frac{1}{s+1} \tag{3.15}$$

Here $\Gamma(k, x)$ is the incomplete gamma function defined by

$$\Gamma(k, x) = \int_x^\infty t^{k-1} e^{-t}\, dt \tag{3.16}$$

Thus, the CH for the gamma implies a constraint between the parameters s and k.

The parameters k in the gamma and σ^2 in the lognormal are related to the degree of evenness of the species abundances. The coefficient of variation in the gamma is $1/\sqrt{k}$, while in the lognormal it is $[\exp(\sigma^2)-1]^{1/2}$. Large k, or small σ^2, corresponds to a small coefficient of variation in the species abundances. The species would tend to have similar λ values under such circumstances, resulting in greater evenness of the abundances in the community. Taillie (1979) and Patil and Taillie (1979b) have formalized this notion of evenness in species curves using the concept of Lorenz ordering from economics. They have shown that k and σ^2 completely determine the Lorenz ordering for the gamma and lognormal models.

The CH constraints (3.13) and (3.15), as pointed out by Patil and Taillie (1979a), imply an inverse relationship exists between species richness and evenness in an ecological community. For the gamma model, in fact, the relationship (3.15) between k and s is well-approximated by

$$sk \approx 4.56 \tag{3.17}$$

for large s and small k. This is found by dividing both sides of (3.15) by k, taking the limit $s \to \infty$, $k \to 0$, $sk \to \alpha$, and then numerically evaluating the integral. So Preston's CH applied to the gamma model turns out to be a special case ($\alpha \approx 4.56$) of Fisher's limiting logseries!

A type of limiting lognormal model can be derived using the CH, in analogy with the logseries as a limiting form of the gamma model. The CH constraint (3.13) is approximately a linear relationship between σ and s for large $s : \sigma \approx a + bs$. In the Poisson-lognormal model (3.4) for species frequencies, one can substitute $\sigma = a + bs$ and take the limit as $s \to \infty$, producing

$$\lim_{s \to \infty} m_r(s, \mu, (a + bs)^2) = \frac{\gamma}{r} \tag{3.18}$$

for $r = 1, 2, \ldots$, with $\gamma = 1/[b(2\pi)^{1/2}]$. In a sense, we have come full circle in recovering Corbet's original harmonic series model for species frequencies as a limiting lognormal model. We point out that Patil and Taillie (1979a) studied a somewhat different divergent series as a limiting lognormal model obtained using a different limiting scheme.

The topic of species frequency distributions, from an ecological standpoint, would now benefit from some large-scale, serious data analysis. There is presently no reason to draw any more sweeping conclusions based on makeshift estimation techniques and eyeball testing. Claims concerning which distributions fit best, changes in distributions or parameters following ecological disturbance, or the CH should now be rigorously examined through careful attention to appropriate statistical modeling of sampling procedures. It is exciting to contemplate what patterns in nature remain to be discovered through a healthy injection of statistical thinking into species abundance studies.

4. MODIFIED LOGNORMAL MODELS AS DESCRIPTIVE ABUNDANCE MODELS

The lognormal is commonly used in ecology in a purely descriptive role as a model of abundance of a single species present in different samples. If many samples are taken across time or space, the abundance of a species typically varies greatly from sample to sample. The lognormal is used to describe these abundances mostly for convenience. Parameter estimates for the lognormal are easy to compute; and, an added attraction for ecologists is the theoretical underpinning of the lognormal as a single species growth model. (Section 2).

However, ecological data are frequently not so cooperative. Ecological studies can contain complicated factors, and the lognormal distribution often requires some modification for use as a descriptive model of abundance. We will not dwell here on reviewing standard descriptive uses of the lognormal in ecology. Rather, we will mention here a few of the typical modifications to the lognormal that are in use.

4.1 Poisson-Lognormal

When plankton are sampled using replicated net hauls or other methods, the frequency distribution of sampled abundances tends to be a unimodal, right-skewed distribution resembling a lognormal (Barnes and Marshall, 1951; Barnes, 1952). However, plankton samples are typically count data, representing numbers of particles suspended in a unit volume of water. A given sample could be assumed to have a Poisson distribution with mean parameter λ. Additional between-sample variability could then be induced by a mixing distribution with pdf $f(\lambda)$. The plankton count distribution is Poisson-lognormal if the mixing distribution is lognormal. The probabilities would then be

$$P[X = x] = \frac{1}{x!(\sigma^2 2\pi)^{1/2}} \int_0^\infty \lambda^{x-1} \exp[-\lambda - (\log \lambda - \mu)^2/(2\sigma^2)]\, d\lambda \quad (4.1)$$

where X is the number of particles in a unit volume of water. This distribution was discussed earlier in the entirely different context of species frequency models. Cassie (1962) gives an extensive discussion of the Poisson-lognormal as a plankton abundance model, with particular attention to its differences from the negative binomial. Further statistical properties and applications as a plankton model are developed by Reid (1981). Readers are also referred to the article in this volume by S. A. Shaban on the Poisson-lognormal distribution.

4.2 Delta-Lognormal

Data from surveys on abundances of marine organisms, including plankton, often contain a large proportion of zeros. The lognormal distribution typically provides a reasonable description of abundances for samples in which organisms are present. The spatial distribution of marine organisms tends to be patchy, though; samples are drawn from a mosaic of areas where organisms are present and areas where organisms are absent. When the objective of such surveys is to estimate mean abundance, there are advantages to using a modified lognormal distribution with an added discrete probability mass at zero (Pennington, 1983). Such a distribution is called a delta-distribution by Aitchison and Brown (1957) and a delta-lognormal distribution by K. Shimizu in Chapter Two of the present volume. The delta-lognormal has "pdf" given by

$$g(x) = \alpha\delta(x) + (1 - \alpha) f(x) \quad (4.2)$$

where $f(x)$ is a lognormal pdf, and $\delta(x)$ is the Dirac delta function defined by

$$\int_a^b \delta(x)\, dx = \begin{cases} 1 & a < 0 < b \\ 0 & \text{otherwise} \end{cases} \quad (4.3)$$

and $0 \leq \alpha \leq 1$. Estimation for distributions of the general form (4.2) was studied by Aitchison (1955).

The purpose of the marine abundance surveys often is to estimate $E[X] = \kappa$, which for the delta-lognormal becomes

$$\kappa = E[X] = (1 - \alpha)e^{\mu+(\sigma^2/2)} \tag{4.4}$$

Suppose a random sample of size n drawn from the delta-lognormal has m nonzero values, and suppose $\bar{y}$ and s^2 are the sample mean and sample variance, respectively, of the log-transformed nonzero values in the sample. One unbiased estimate of κ is of course the ordinary sample mean of the observations, zeros and all. However, $\bar{y}$, s^2, and m/n are joint complete sufficient statistics for μ, σ^2, and α, and this fact can be exploited to produce a much better estimate. Aitchison (1955) obtained the minimum variance unbiased estimate of κ; Pennington (1983) obtained the MVUE for the variance of the estimate and applied the results to fish and plankton survey data. Pennington noted that the MVUE for κ is considerably more efficient than the ordinary sample mean under the high variability conditions encountered in marine abundance surveys. These results and generalizations are contained in Section 3.1 of Chapter Two of the present volume.

We might remark here that it would be useful to study the addition of extra zeros to the Poisson-lognormal, in connection with marine surveys involving count data.

4.3 Delta-Compound-Lognormal

When terrestial plant communities are sampled with quadrats, the data often consist of large proportions of quadrats with no plants, and continuous, right-skewed distributions of plant abundances among quadrats where plants are present. Plant abundance in such studies is typically measured in terms of cover. The situation is more difficult than the preceding marine surveys in which the delta-lognormal could be used, for two reasons: (a) Plant cover present in a quadrat arises from a random number of initial propagules (seeds, rhizomes, etc.). (b) Plant cover typically grows as a function of time.

Steinhorst et al. (1985) proposed distribution models to describe plant cover development in forest communities following clearcutting and burning. The models consist of a randomly stopped sum of continuous iid random variables, plus an additional probability mass at zero. Such a model would have a Laplace-Stieltjes transform given by

$$\phi_Y(s) = \alpha + (1 - \alpha)\phi_N(-\log \phi_X(s)) \tag{4.5}$$

where Y is the total cover on a quadrat, N is the number of initial propagules on the quadrat (a discrete random variable on the non-negative integers), X is the size attained by a plant at the time of sampling, and $\phi_X(s) = E[e^{-sX}]$, $\phi_N(s) = E[e^{-sN}]$, $\phi_Y(s) = E[e^{-sY}]$. The parameters in the distribution of X were assumed to be functions of time such that $E[X]$ would follow a growth law like the logistic. Steinhorst et al. consider models in which N is either Poisson or negative binomial, and X is either gamma or normal. They were able to compute maximum likelihood estimates for data sets on various species of shrubs. The estimates were computed using the EM algorithm for those species which grow from rhizomes, since N is then an unobservable variable ("individual" plants not being distinguishable).

While Steinhorst et al. (1985) did not explicitly discuss using the lognormal as a distribution model for X, they have now investigated its use and are studying statistical inference problems for models in the form (4.5) in more generality (Steinhorst, manuscript in preparation). The lognormal would seem to be a promising candidate for modeling the size attained by a plant at a given time. The stochastic Gompertz model (see Section 2.4), for instance, would provide an explicit lognormal model with a mean that evolves according to a well-known growth law. The lognormal, however, does not have a convenient Laplace-Stieltjes transform, and so writing the model in the form (4.5) may not be very useful.

REFERENCES

Aitchison, J. (1955). On the distribution of a positive random variable having a discrete probability mass at the origin, *J. Amer. Statist. Assoc.*, *50*, 901–908.

Aitchison, J. and Brown, J. A. C. (1957). *The Lognormal Distribution*, Cambridge University Press, Cambridge, Massachusetts.

Barnes, H. and Marshall, S. M. (1951). On the variability of replicate plankton samples and some applications of 'contagious' series to the statistical distribution of catches over restricted periods, *J. Marine Biol. Assoc.*, *30*, 233–263.

Barnes, H. (1952). The use of transformations in marine biological statistics, *J. Cons. Explor. Mer.*, *18*, 61–71.

Braumann, C. A. (1983). Population growth in random environments, *Bull. Math. Biol.*, *45*, 635–641.

Bulmer, M. G. (1974). On fitting the Poisson lognormal distribution to species-abundance data, *Biometrics*, *30*, 101–110.

Capocelli, R. M. and Ricciardi, L. M. (1974). A diffusion model for population growth in random environment, *Theor. Popul. Biol.*, *5*, 28–41.

Cassie, R. M. (1962). Frequency distribution models in the ecology of plankton and other organisms, *J. Anim. Ecol.*, *31*, 65–92.

Dennis, B. and Patil, G. P. (1984). The gamma distribution and weighted multimodal gamma distributions as models of population abundance, *Math. Biosci.*, *68*, 187–212.

Engen, S. (1979). Abundance models: sampling and estimation, *Statistical Distributions in Ecological Work* (J. K. Ord, G. P. Patil, and C. Taillie, eds.), International Co-operative Publishing House, Fairland, Maryland, pp. 313–332.

Feldman, M. W. and Roughgarden, J. (1975). A population's stationary distribution and chance of extinction in a stochastic environment with remarks on the theory of species packing, *Theor. Popul. Biol.*, *7*, 197–207.

Fisher, R. A., Corbet, A. S., and Williams, C. B. (1943). The relation between the number of species and the number of individuals in a random sample of an animal population, *J. Anim. Ecol.*, *12*, 42–58.

Freedman, H. I. (1980). *Deterministic Mathematical Models in Population Ecology*, Marcel Dekker, New York.

Gauch, H. G. and Chase, G. B. (1974). Fitting the Gaussian curve to ecological data, *Ecology*, *55*, 1377–1381.

Gray, A. H. and Caughey, T. K. (1965). A controversy in problems involving random parametric excitation, *J. Math. and Phys.*, *44*, 288–296.

Holgate, P. (1969). Species frequency distributions, *Biometrika*, *56*, 651–660.

Horsthemke, W. and Lefever, R. (1984). *Noise-Induced Transitions: Theory and Applications in Physics, Chemistry, and Biology*, Springer-Verlag, Berlin.

Jennrich, R. I. and Moore, R. H. (1975). Maximum likelihood estimation by means of nonlinear least squares, *Am. Statist. Assoc. Proc. Statist. Comput. Sec.*, 57–65.

Karlin, S. and Taylor, H. M. (1981). *A Second Course in Stochastic Processes*, Academic Press, New York.

Kempton, R. A. (1975). A generalized form of Fisher's logarithmic series, *Biometrika*, *62*, 29–37.

Kempton, R. A. and Taylor, L. R. (1974). Log-series and log-normal parameters as diversity discriminants for the Lepidoptera, *J. Anim. Ecol.*, *43*, 381–399.

Kempton, R. A. and Taylor, L. R. (1979). Some observations on the yearly variability of species abundance at a site and the consistency of measures of diversity, *Contemporary Quantitative Ecology and Related Ecometrics* (G. P. Patil and M. Rosenzweig, eds.), International Co-operative Publishing House, Fairland, Maryland, pp. 3–22.

Lewontin, R. C. and Cohen, D. (1969). On population growth in a randomly varying environment, *Proc. Nat. Acad. Sci.*, *62*, 1056–1060.

Lo, H. and Wani, J. K. (1983). Maximum likelihood estimation of the parameters of the invariant abundance distributions, *Biometrics*, *39*, 977–986.

MacArthur, R. H. (1960). On the relative abundance of species, *Am. Nat.*, *94*, 25–36.

Mahfoud, M. and Patil, G. P. (1982). On weighted distributions, *Statistics and Probability: Essays in Honor of C. R. Rao* (G. Kallianpur, P. R. Krishnaiah, and J. K. Ghosh, eds.), North-Holland, pp. 479–492.

May, R. M. (1975). Patterns of species abundance and diversity, *Ecology and Evolution of Communities* (M. L. Cody and J. M. Diamond, eds.), Belknap Press, Cambridge, Massachusetts, pp. 81–120.

Patil, G. P. (1984). Studies in statistical ecology involving weighted distributions, in *Proceedings of the Indian Statistical Institute Golden Jubilee International Conference in Statistics: Applications and New Directions* (J. K. Ghosh and J. Roy, eds.), Statistical Publishing Society, Calcutta, India, pp. 478–503.

Patil, G. P., Boswell, M. T., Joshi, S. W., and Ratnaparkhi, M. V. (1984a). *Dictionary and Bibliography of Statistical Distributions in Scientific Work. Vol. 1: Discrete Models*. International Co-operative Publishing House, Fairland, Maryland.

Patil, G. P., Boswell, M. T., and Ratnaparkhi, M. V. (1984b). *Dictionary and Bibliography of Statistical Distributions in Scientific Work. Vol. 2: Continuous Univariate Models*. International Co-operative Publishing House, Fairland, Maryland.

Patil, G. P., Boswell, M. T., Ratnaparkhi, M. V., and Roux, J. J. (1984c). *Dictionary and Bibliography of Statistical Distributions in Scientific Work. Vol. 3: Multivariate Models*. International Co-operative Publishing House, Fairland, Maryland.

Patil, G. P. and Ord, J. K. (1976). On size-biased sampling and related form-invariant weighted distributions, *Sankhya B*, *38*, 48–61.

Patil, G. P. and Rao, C. R. (1977). The weighted distributions: a survey of their applications, *Applications of Statistics* (P. R. Krishnaiah, ed.), North-Holland, pp. 383–405.

Patil, G. P. and Rao, C. R. (1978). Weighted distributions and size-biased sampling with applications to wildlife populations and human families, *Biometrics*, *34*, 179–189.

Patil, G. P. and Taillie, C. (1979a). Species abundance models, ecological diversity, and the canonical hypothesis, *Bull. Int. Statist. Inst.*, *44*, 1–23.

Patil, G. P. and Taillie, C. (1979b). An overview of diversity, *Ecological Diversity in Theory and Practice* (J. F. Grassle, G. P. Patil, W. Smith, and C. Taillie, eds.), International Co-operative Publishing House, Fairland, Maryland, pp. 3–27.

Pennington, M. (1983). Efficient estimators of abundance, for fish and plankton surveys, *Biometrics*, *39*, 281–286.

Pielou, E. C. (1975). *Ecological Diversity*, John Wiley, New York.

Preston, F. W. (1948). The commonness, and rarity, of species, *Ecology*, *29*, 254–283.

Preston, F. W. (1962). The canonical distribution of commonness and rarity, *Ecology*, *43*, 185–215 and 410–432.

Preston, F. W. (1980). Noncanonical distributions of commonness and rarity, *Ecology*, *61*, 88–97.

Rao, C. R. (1965). On discrete distributions arising out of methods of ascertainment, *Classical and Contagious Discrete Distributions* (G. P. Patil, ed.), Statistical Publishing Soc., Calcutta, pp. 320–332.

Rao, C. R. (1971). Some comments on the logarithmic series distribution in the analysis of insect trap data, *Spatial Patterns and Statistical Distributions* (G. P. Patil, E. C. Pielou, and W. E. Waters, eds.), The Pennsylvania State University Press, University Park, Pennsylvania pp. 131–142.

Reid, D. D. (1981). The Poisson lognormal distribution and its use as a model of plankton aggregation, *Statistical Distributions in Scientific Work, Vol. 6* (C. Taillie, G. P. Patil, and B. A. Baldessari, eds.), D. Reidel, Dordrecht, Holland, pp. 303–316.

Ricciardi, L. M. (1977). *Diffusion Processes and Related Topics in Biology*, Springer-Verlag, Berlin.

Steinhorst, R. K., Morgan, P., and Neuenschwander, L. F. (1985). A stochastic-deterministic simulation model of shrub succession, *Ecological Modelling*, *29*, 35–55.

Sugihara, G. (1980). Minimal community structure: an explanation of species abundance patterns, *Am. Nat.*, *116*, 770–787.

Taillie, C. (1979). Species equitability: a comparative approach, *Ecological Diversity in Theory and Practice* (J. F. Grassle, G. P. Patil, W. Smith, and C. Taillie, eds.), International Co-operative Publishing House, Fairland, Maryland, pp. 51–62.

Taylor, L. R., Kempton, R. A., and Woiwood, I. P. (1976). Diversity statistics and the log-series model, *J. Anim. Ecol.*, *45*, 255–272.

Tuckwell, H. C. (1974). A study of some diffusion models of population growth, *Theor. Popul. Biol.*, *5*, 345–357.

Turelli, M. (1977). Random environments and stochastic calculus, *Theor. Popul. Biol.*, *12*, 140–178.

Watterson, G. A. (1974). Models for the logarithmic species abundance distributions, *Theor. Popul. Biol.*, *6*, 217–250.

Whittaker, R. H. (1972). Evolution and measurement of species diversity. *Taxon*, *21*, 213–251.

Williams, C. B. (1964). *Patterns in the Balance of Nature*, Academic Press, New York.

13

Applications in Atmospheric Sciences

EDWIN L. CROW Institute for Telecommunication Sciences, National Telecommunications and Information Administration, U.S. Department of Commerce, Boulder, Colorado

1. INTRODUCTION

Lognormal distributions have been used extensively in atmospheric sciences to describe phenomena that take on non-negative values, such as storm, daily, and longer-period rain, snow, and hail amounts, particle (aerosol) size distributions, pollutant concentrations, cloud dimensions, air velocity fluctuations, flood frequencies, and radio wave amplitude fluctuations. In most cases these are empirically observed and tested fits, and often other models, such as the gamma distribution, are also fitted, or at least cannot be excluded as being consistent with the data. However, there are also refrences to logical derivation from the law of proportionate effect in the breakup of particles (Lopez, 1977; Patterson and Gillette, 1977).

Because of the complexity of atmospheric phenomena and the difficulty of conducting controlled experiments, there is much modeling to derive results numerically or analytically based on a variety of assumptions, which often include lognormal distributions of basic variables (e.g., Giorgi, 1986).

A considerable tool in this survey paper was the computer searching of Meteor/Astrogeophysical Abstracts by Gayl Gray of the National Center for Atmospheric Research. Citations in the period 1970-May 1985 were covered in which "lognormal," "log-normal," or "log normal" appeared in the title, descriptors, or abstract. About 200 references were obtained, 15 of which had the term in the title.

Often the papers go beyond description, to statistical inference comparing distributions arising under different conditions. An important instance of this is the comparison of precipitation distributions from clouds that have been seeded with silver iodide with clouds of the same general type that have not been seeded (controls) to determine whether there has been significant change. This has been discussed in more detail than other applications in Sec. 5, with numerical calculations, including examples of using the elegant results of Chapters 2 and 3 to obtain unbiased estimates of parameters and optimal confidence intervals for the mean.

In other sections applications of lognormal distributions are discussed for precipitation amounts (Sec. 4), particle size distributions (Sec. 2), pollutant concentrations (Sec. 3), hydrology (Sec. 6), and quite a few other phenomena (Sec. 7).

It is worth mentioning here the not uncommon use of the term "spectrum" in meteorology as a synonym for "distribution," especially in referring to particle or drop size distributions. For example, Neiburger and Weickmann (1974) have Fig. 3.22 entitled "Raindrop distributions..." and Fig 3.23 "Raindrop spectra...," both with number concentrations plotted against drop diameter. See also Kelkar and Joshi (1977) and Feingold and Levin (1986). It seems desirable to reserve the term "spectrum" for its classical use, a decomposition into a series of wavelengths or frequencies such as for electromagnetic waves or turbulence or time series analysis, where "spectral density" is distinctly different from "probability density."

2. PARTICLE SIZE DISTRIBUTIONS

2.1 Introduction and Definitions

Aitchison and Brown (1957, Sec. 10.2) summarize the early history (1929 on) of application of lognormal distributions to particle sizes, such as produced naturally in sediments or mechanically by grinding. They point out the early recognition in this field that size (i.e., diameter) distributions obtained by counting particles and size (still diameter) distributions obtained by weighing the amounts through nested sieves can be related lognormal distributions. The distributions of particles in the air have been of vital and increasing importance because of the concern with air pollution and radioactive fallout. "Particles influence weather and climate in two ways:

First, by absorbing, radiating, and scattering radiant energy; and second, by acting as nuclei for cloud elements" (Machta and Telegadas, 1974). A dispersion of solid or liquid particles in a gas (usually air) is now called an "aerosol."

A basic reference for small particle statistics is the book by Herdan (1960). It introduces a large number of moment parameters as a result of the physical methods of measuring the distributions, such as sieving, but we may take as its primary parameters for the two-parameter lognormal distribution the *geometric mean* x_g and the *geometric standard deviation* σ_g (eq. (6.15a), page 81), a pair different from the (μ, σ) of equation (2.1) in Chapter 1:

$$x_g = e^{\mu}, \qquad \sigma_g = e^{\sigma} \tag{2.1}$$

Thus x_g is the median of the lognormal distribution $\Lambda(\mu, \sigma^2)$, whereas the mean is, by equation (4.2) of Chapter 1,

$$E(X) = \mu_1'(X) = \exp\left(\mu + \tfrac{1}{2}\sigma^2\right) \tag{2.2}$$

"The emergence of the geometric mean in particle size analysis is due to the empirical finding that many particle size distributions conform satisfactorily to the log normal law, in which case the geometric mean represents the value with the greatest frequency of particles, and is, therefore, better suited to serve as a representative of particle size than the arithmetic mean (Herdan, page 31)."

Herdan should have justified the geometric mean as the median rather than as "the value with the greatest frequency of particles," since the mode of $\Lambda(\mu, \sigma^2)$ is $\exp(\mu - \sigma^2)$ (equation (4.5) in Chapter 1). The use of x_g and σ_g is well established in particle statistics, with justification, but confusion does arise over the several kinds of means and moments. Allen's large book, now in its third edition (1981), states incorrectly (page 136) that "the mean of a log-normal distribution is the geometric mean, i.e., the arithmetic mean of the logarithms." Likewise, Herdan states (page 81) that "σ^2 ... is the standard deviation of the distribution of ratios around the geometric mean and whose logarithm is the standard deviation of $\log x$," whereas, from equation (4.3) in Chapter 1, the variance of X/x_g is

$$\operatorname{Var}\left(\frac{X}{x_g}\right) = \exp(\sigma^2)\left\{\exp(\sigma^2) - 1\right\} = \sigma_g^{\sigma}(\sigma_g^{\sigma} - 1)$$

so

$$SD\left(\frac{X}{x_g}\right) \doteq \sigma\left(1+\tfrac{3}{4}\sigma^2\right) \qquad \text{for very small } \sigma$$

and the mean square deviation of X/x_g from 1 is

$$\exp(2\sigma^2) - 2\exp(\sigma^2/2) + 1$$

the square root of which is approximately $\sigma(1 + 7\sigma^2/8)$ for very small σ. Thus neither one of these possibilities is the geometric standard deviation.

If we denote by X the diameter of a random spherical particle and assume it has the lognormal distribution $\Lambda(\mu, \sigma^2)$, then the particle surface S is πX^2 and its volume V is $\pi X^3/6$. By Section 4.3 of Chapter 1, S has the lognormal distribution $\Lambda(1n\pi + 2\mu, 4\sigma^2)$ and V has the lognormal distribution $\Lambda(\ln(\pi/6) + 3\mu, 9\sigma^2)$. Hence the geometric means and the geometric standard deviations of S and V are, by the definition of Herdan's (6.15a),

$$S_g = \pi e^{2\mu} = \pi x_g^2 \qquad \sigma_g(S) = e^{2\sigma} = \sigma_g^2 \tag{2.3}$$

$$V_g = \frac{\pi}{6}e^{3\mu} = \frac{\pi}{6}x_g^3 \qquad \sigma_g(V) = e^{3\sigma} = \sigma_g^3 \tag{2.4}$$

On the other hand, the (true) mean surface and volume are, by (2.2) above,

$$E(S) = \mu_1'(S) = \pi\exp(2\mu + 2\sigma^2) \tag{2.5}$$

$$E(V) = \mu_1'(V) = \frac{\pi}{6}\exp\left(3\mu + \tfrac{9}{2}\sigma^2\right) \tag{2.6}$$

The diameters corresponding to these,

$$x_S = e^{\mu+\sigma^2} \qquad x_V = e^{\mu+(3/2)\sigma^2} \tag{2.7}$$

are called the *surface mean diameter* and *volume mean diameter* respectively (Herdan, page 33). (If a weight mean diameter is defined, it would, out of consistency, be identical with the volume mean diameter.) Hence

$$\ln x_S = \ln x_g + \sigma^2 = \ln x_g + (\ln\sigma_g)^2 \tag{2.8}$$

$$\ln x_V = \ln x_g + \tfrac{3}{2}\sigma^2 = \ln x_g + \tfrac{3}{2}(\ln\sigma_g)^2 \tag{2.9}$$

where a notation for x_g consistent with x_S and x_V would be x_D or x_L, for diameter or length, say $x_g \equiv x_D$. Likewise, an unambiguous term for particle size distribution would be particle *diameter* distribution, but the physically pertinent property may often be surface (in optical studies) or volume (in pollution control) (Patterson and Gillette, 1977).

However, the surface and weight distributions considered in the particle size literature are defined quite differently than above (Herdan, pages 34, 84; Allen, pages 122, 137). They still refer to *diameter* distributions but with frequencies (or numbers) replaced by total surface or weight of particles with each diameter. Aitchison and Brown (1957, pages 12, 100) generalized this to the *moment distributions* of $\Lambda(\mu, \sigma^2)$, which are of the form $\Lambda(\mu + j\sigma^2, \sigma^2)$ for order j. Thus the definitions of these terms in particle statistics are different from those in mathematical statistics. For example, on page 84 Herdan states: "If the number distribution is log normal, the weight distribution is also log normal and with the same log standard deviation," whereas we saw the weight (volume) distribution above to have the log standard deviation tripled. What Herdan, Allen (1981, page 137), and others call the number distribution or distribution by count is the probability distribution of diameters (or radii), which may be considered the primary size distribution. Thus the number mean diameter (radius) is the (true) mean diameter (radius). Feingold and Levin (1986) (who find the lognormal a better squared-error fit to their raindrop size distributions than the gamma family) also refer to the distributions of moments, but they interpret their expression (14) incorrectly as the distribution of the power of a lognormal variable.

The origin of the different definitions lies in the frequent measurement of the particle distribution by the *weight* (or also surface) of particles in the various sieve sizes, the *number* n_i of particles being unknown. Let the weight of particles for sieve size i be w_i, corresponding to spheres of diameter x_i. Then the weight mean diameter is calculated as

$$\bar{x}_w = \frac{\sum w_i x_i}{\sum w_i} \tag{2.10}$$

But

$$w_i = n_i \rho \frac{\pi}{6} x_i^3$$

where ρ is the particle density, assumed uniform. Hence

$$\bar{x}_w = \frac{\sum n_i x_i^4}{\sum n_i x_i^3} \tag{2.11}$$

The analogous theoretical mean is

$$\frac{\int_0^\infty x^4 f\left(x \mid \mu, \sigma^2\right)\, dx}{\int_0^\infty x^3 f\left(x \mid \mu, \sigma^2\right)\, dx} \tag{2.12}$$

where $f(x \mid \mu, \sigma^2)$ is the probability density function of the diameters. Hence, by (4.1) of Chapter 1, the theoretical mean that $\bar{x}_w$ is estimating equals

$$\frac{E(X^4)}{E(X^3)} = \exp\left(\mu + \tfrac{7}{2}\sigma^2\right) = x_g \exp\left(\tfrac{7}{2}\sigma^2\right) \tag{2.13}$$

This is what Herdan refers to as "the weight [or volume] average particle size," (size being defined as diameter) on pages 34 and 84, but we see that it differs from the volume mean diameter (2.7) derived from Herdan's page 33 definition. The implication in Herdan is that they are the same, but they are not. Equation (2.13) is consistent with Herdan's (6.20) and Allen's (4.125).

Equation (2.11) is essentially the same as the equation preceding equation (4.124) on page 159 of Allen (1981), and (2.13) is consistent with his (4.125). One must understand that what Allen calls "the mean size of a weight distribution" (Page 159) is the mean diameter weighted by physical weights, as in (2.10) multiplied by ρ.

The cumulative size distribution by weight obtained by plotting the cumulative weights $w_i / \sum w_j$ as cumulative frequencies against x_i on lognormal probability paper is often used to obtain the median and log geometric standard deviation graphically. The estimation methods of Chapters 2, 3, 4, and 5 might well be used.

2.2 Aerosol Studies

Aerosols enter the atmosphere in a great variety of ways from the Earth's surface, and about 10 percent of the total is anthropogenic (Monin, 1986, pages 61–68). The particles have a mean radius r ranging from 10^{-3} to $10^3 \mu$m (1 μm = 1 micron = 10^{-6} meter). Those important in the formation of clouds (cloud condensation nuclei) have radii from 10^{-2} to 10^2 μm. Monin goes on to state that, in the range of radii from 0.01 to 2.5 μm, the global aerosol can be described on the average by a lognormal distribution, the number in $(r, r + dr)$ being given by

$$\frac{N_0}{\sqrt{2\pi \ln \sigma_0}} \exp\left\{-\frac{[\ln(r/r_0)]^2}{2\ln \sigma_0}\right\} \frac{dr}{r}$$

with $N_0 = 10\ \mathrm{cm}^{-3}$, $r_0 = 0.0725\ \mu\mathrm{m}$, and $\sigma_0 = 1.86$, or also by an exponential distribution. Thus r_0 is the geometric mean, so $\mu = \ln r_0 = -2.62$, but σ_0 is not the geometric standard deviation;

$$\sigma = \ln \sigma_g = (\ln \sigma_0)^{1/2} = 0.788 \qquad \sigma_g = 2.20$$

Using σ_0 as above rather than σ_g seems unfortunate. By (2.2) the (true) mean radius is $0.099\ \mu\mathrm{m} \doteq 0.1\ \mu\mathrm{m}$.

The above lognormal distribution is an approximation to the physical distribution and becomes systematically different over cities, continents, and oceans. The particles lead to light scattering and lowered visibility that is typically only one-fourth that of a clean atmosphere (with no aerosol) but eight times the average visibility over an American city, according to Monin (page 63).

Patterson and Gillette (1977) provide an interesting summary and comparison of several sets of measurements of the size distributions of tropospheric aerosols having a soil-derived component. They identify three components, or modes, A, B, and C, and believe that C, the smallest particles, is probably anthropogenic rather than soil-derived. Thus they fit a mixture of two normal distributions to the logarithms of measured particle radii; the method of fitting is not stated. The methods of collecting and measuring the particles (e.g., collecting on filters and optical or electron microscopy) appear demanding and subject to error, especially for smaller particles. Thus the distributions are in fact truncated.

"Log normal distributions were chosen [state Patterson and Gillette] rather than other distributions for several reasons: we have found empirically that the fit is good for soil-derived aerosols; the choice is reasonable theoretically because soil-derived aerosols are produced as a result of comminution processes (e.g., sandblasting) in the soil, and such processes do result in log normal distributions [Epstein, 1947]; and finally, the log normal distribution is convenient mathematically because the parameters of the volume and surface moments of the distribution are related in a simple way to the parameters of the number distribution. This relation is expressed by the formulae

$$\ln r_v = \ln r_n + 3\ln^2 \sigma_g \tag{2a}$$

$$\ln r_s = \ln r_n + 2\ln^2 \sigma_g \tag{2b}$$

with r_n the number radius, r_s the surface mean radius, and r_v the volume mean radius."

Patterson and Gillette have thus cited the breakage theory derivation of the lognormal distribution (Sec. 3.2, Chapter 1) and referred to three alternative mean radii. Their equations (2a) and (2b) are consistent with Herdan's (6.21) and Allen's (4.129) and (4.135), where the mean radii are the theoretical equivalent of (2.10) above with w_i taken as taken as u_i, $n_i x_i^2$, and $n_i x_i^3$ respectively for number, surface, and volume mean radius. Although Patterson and Gillette had defined r_n as "mean radius," which is the number mean radius (Sec. 2.1), its position in their equation (1) showed that it is the geometric mean (or median) radius, for which (2a) and (2b) would not be correct. The point has been belabored, but it has seemed necessary to do so to show that the proliferation of possible distributions and means thereof has led to confusion and that writers should be explicit in their definitions or references.

A valuable series of stratospheric aerosol measurements and comparison with fitted parametric distributions is provided by Gras and Laby (1981). They found agreement between two measuring techniques—in situ single-particle counting and jet impaction—and equally good fits to the data by either lognormal distributions or "zero-order log distributions," which have probability density functions (pdfs) of the same form (aside from a constant) as the lognormal pdf multiplied by the independent variable. The maximum difference between the corresponding pdf's for the 16–22 km altitude data was stated to be 1 part in 10^5. A nonlinear least sqaures procedure (Leavenberg algorithm) was used to fit these functions, and standard errors are given, but no reference is cited.

The assumption of lognormal distribution of the size (radius) of aerosols is a key component of a recent study (Giorgi, 1986) of the "effects of airborne particulate material," in particular, the global effects of massive aerosol injections caused by a full-scale nuclear war (the "nuclear winter" effect). Based on theoretical and observational evidence (Whitby, 1978), Giorgi assumed that the aerosol distribution can be approximated by a mixture of three lognormal distributions and that the dynamical evolution of the distribution can be described by the evolution of the geometric mean radius r_g, the geometric standard deviation σ_g, and the total number, N_0, of particles per unit volume of each component lognormal distribution. (What Giorgi calls the geometric standard deviation is actually the σ of (2.1), not the geometric standard deviation.) The three component distributions have physical origins: the smallest particles are associated with urban pollution, the intermediate with disintegration of the solid earth surface, and the largest with mechanical disintegration of the ocean surface. At successive instants of time at each location the frequency of each particle size is determined numerically by a "prognostic equation" that adds up contributions from sources (such as dust stirred up by a nuclear explosion), transport, coagulation, sedimentation, dry deposition, and wet (rain) removal. Then r_g, σ_g, and N_0 are determined from the 0th, 1st, and 3rd moments using the equivalent of equation (4.1) of Chapter 1. It should be recalled that the moments are not statistically efficient in estimating the lognormal parameters (Sec. 2.6 of Chapter 2). In this case we have a modeling situation in which essentially the entire distribution is available rather than estimation from sample data, but it appears that the component distributions at each instant and location are not necessarily lognormal, so an element of estimation is involved. Applying the model to the entire globe involves immense computations, so computational efficiency is a prime consideration. Using three lognormal distributions with three parameters each was a major improvement by Giorgi over the two previous methods of (1) integrating a nonparametric distribution bin by bin and (2) using a single parametric distribution.

Some of the many other references to fitting or modeling aerosol data with a lognormal distribution or a mixture of two or three lognormal distributions are: Whitby (1978), Low (1979), Renoux et al. (1982), Tanaka et al. (1983), Fitch (1983), and Harrison and Pio (1983).

3. POLLUTANT CONCENTRATIONS

Of primary interest in the quality of air is the amount of six pollutants, carbon monoxide, nitrogen dioxide, ozone, sulfur dioxide, lead, and total suspended particulates, measured in milligrams or micrograms per cubic meter or parts per million (Council on Environmental Quality (CEQ), 1984, page 13). These are measured daily in U.S. urban areas and combined into a Pollutant Standard Index (PSI) by taking the maximum of several subindices (Ott, 1978, page 147). Ott (1978, page 175) plotted the 1974 cumulative frequencies for five U.S. cities on lognormal probability paper and obtained approximately parallel straight lines. The CEQ shows a general downward trend in the PSI by plotting the number of days per year in which it registered in the "unhealthful," "very unhealthful," and "hazardous" ranges. "The downward trends in the PSI in most of these cities have been mainly influenced by the drop in CO levels from 1976 to 1981. More recently, from 1981 to 1984, the index is mainly driven by O_3 [ozone]."

Legrand (1974) summarized the distributions of daily concentrations of sulfur dioxide and smoke measured at each of 180 semiautomatic stations scattered throughout Belgium. The cumulative distributions plot as straight lines on lognormal probability paper and are thus characterized by their medians and slopes. Maps showing contours of constant values of these two characteristics were drawn for each of five urban areas.

Much of the following material is taken from a comprehensive review by Tsukatani (1981). Larsen (1969) introduced a mathematical model, called the arrowhead chart, of pollutant concentration quantiles for various averaging times from 1 hour to 1 year. He assumed the distributions to be lognormal for all averaging times and empirically fitted the median and the logarithmic standard deviation as a function of averaging time. The derived quantiles are confined between empirically fitted straight lines on log paper that represent the maximum and minimum concentrations and converge to a point at (usually) 1 year—hence the "arrowhead chart." As the result of work by Mage and Ott, Larsen (1977) added a third parameter, an "increment," apparently the threshold parameter of Chapters 1, 4, and 5 of this book, fitted simply by trial and error. (See Chapters 4 and 5 for an effective method.) However, Mage and Ott (1978) call their model the censored three-parameter lognormal model, criticize Larsen's 1977 paper, and advise the analyst not to "automatically use one model." (Larsen calls his first two parameters the geometric mean and the "standard geometric deviation," but the latter appears to be the same as the geometric standard deviation of the particle size scientists.) Shoji and Tsukatani improved the theoretical basis of the arrowhead chart, as well as the fit for Japanese data shown in Tsukatani (1981), by using an autocorrelation function to get the logarithmic standard deviation for any averaging time.

Air pollution has spawned a vast literature, which has been barely touched on here with respect to particle size distributions and concentration distributions. The interested reader should consult recent issues of the journals *Atmospheric Environment* and *Journal of the Air Pollution Control Association.*

4. PRECIPITATION AMOUNTS, DURATIONS, AND RATES

Lognormal distributions have been fitted to the amounts of precipitation over various periods, including daily and monthly, to the extremes of such amounts, to the durations of rainfall, and to rain rates. On the whole, however, gamma distributions are fitted to rainfall amounts more often than the lognormal, the alternative not necessarily being rejected; see, for example, Brier (1974).

Volynets (1982) categorized summer rain in the Ukraine as single-cell (single-cloud) and multicell and analyzed the distributions of weights, durations, and areas of each type. All of these distributions were stated to be satisfactorily approximated by both lognormal and gamma distributions.

Two two-parameter distributions, the extreme value type-1 and the lognormal, were found by Watt and Nozdryn-Plotnicki (1983) to be reasonable fits to rainfall amounts at three long-term Canadian stations for periods varying from 5 minutes to 6 hours. They also compared four fitting methods by Monte Carlo simulation: moments, modified moments or regression, maximum likelihood, and adjusted maximum likelihood; the modified moments method was poorest with respect to both bias and variance and maximum likelihood the best.

Swift and Schreuder (1981) fitted daily precipitation amounts from the high precipitation zone of the southern Appalachian Mts. for each calendar month for a 38-year period using lognormal, gamma, Weibull, Johnson's S_B, and beta forms. Johnson's S_B, a three-parameter distribution, consistently fitted the data best, but the gamma distribution was the best fit to rainfall amounts accumulated for two or three consecutive wet days. Following the methods of R. Katz, they found that higher-order Markov chains, through the fifth order, described the data for successive days better than lower-order chains. Thus the probability of precipitation on any day depends on what happened on at least the five previous days. Bivariate S_B distributions were used to study precipitation amounts on neighboring days. The cumulative amounts over n days could be approximated by a normal distribution for large n, with mean $0.53n$ cm and standard deviation $1.55n^{1/2}$ cm.

Eriksson (1980) is a very extensive and thorough statistical analysis of monthly precipitation amounts at 262 Swedish and 26 Norwegian stations over the period 1931-1978. Two– or three–parameter gamma and lognormal distributions and the Weibull distribution (three parameters) were tested. He concluded that the gamma distribution provided the best description, with no significant difference between observed and fitted frequencies in 86 percent of the cases. The lognormal distribution had a "much lower capacity" for description of the observed distributions. Eriksson studied measurement errors of precipitation observations, the three most important being aerodynamic, evaporation, and adhesion losses. He concluded that, on the average, for all Swedish stations in his investigation, the mean annual amounts of precipitation for 1931–1978 should be increased by 18 percent to compensate for the systematic errors.

Sneyers (1979) analyzed the precipitation amounts in 1–, 10–, and 60–minute and daily periods at Uccle, Belgium, and then found that the monthly maxima of these amounts followed a lognormal distribution. He also found that the monthly maxima of the duration of precipitation also were lognormally distributed. He derived the distribution of the annual maximum duration from that of the monthly maximum assuming independence among monthly values.

Segal (1980) used a large data base to examine the mathematical form of the distribution of rainfall *rates*, as measured by tipping bucket instruments. "In general, the lognormal function appears to provide a good approximation." However, "a better is provided by piece-wise power-law approximations to different portions of the distributions."

An example of the delta-lognormal distribution (Sec. 3.1 in Chapter 1) is discussed by Koziel (1975)—the daily precipitation amounts at a point, the rainless days providing a probability mass at the origin and the rainy days providing amounts having a lognormal distribution. He tested the form with a chi-squared test.

5. WEATHER MODIFICATION

The scientific study of weather modification began with Vincent Schaefer's (1946) discovery that pellets of dry ice would nucleate ice crystals in supercooled clouds. Since then many field experiments have been analyzed with a variety of statistical methods, among them parametric ones such as fitting gamma distributions to rainfall amounts and transformation to approximate normality by taking square roots, cube roots, and logarithms and nonparametric ones such as Wilcoxon-Mann-Whitney tests and permutation tests (e.g., re-randomization). The use of transformations raised the problem of bias in estimates after inverse transformation (Neyman and Scott, 1960). Parametric tests were advocated because they were more powerful than nonparametric tests (Neyman and Scott, 1967), but their validity was questioned because of lack of certainty of the assumed form (Gabriel, 1979). The relative usage of parametric and nonparametric tests is indicated by the fact that a bibliographic index of statistical techniques used that lists 19 techniques in all has 45 papers mentioning the rank sum test and 15 the t test (Wiorkowski and Odell, 1979, p. 1153). Here some brief background on statistical methods used in testing for effects in weather modification is presented, some uses of the lognormal distribution are cited, and then some details of a particular application in the National Hail Research Experiment are presented.

There have been several valuable discussions of the role of statistics in weather modification experiments: Le Cam and Neyman (1967), Brier (1974), Statistical Task Force of the Weather Modification Advisory Board (1978), Braham (1979), and Wiorkowski and Odell (1979). Most experiments have been directed toward increase in rainfall. The Task Force (1978) concluded in part as follows (p. C-1): "Today we do not look at weather modification of rainfall as a source of major %'s of improvement. Rather wer look in the range 20%, 10%, 0% ... Why are we in this purgatory, between the heaven of conclusive success and the hell of apparent uselessness? Basically, we believe, because the inherent difficulties of the situation and the well-founded need for completely anchored conclusions have not been taken seriously enough." Two large-scale tests of the effectiveness of cloud seeding in reducing hail have likewise failed to obtain statistically significant results (Crow et al., 1979; Federer et al., 1986).

We shall discuss the use of the lognormal distribution appreciably only for the National Hail Research Experiment (NHRE), but it (or the log transformation) has also been used, along with other methods, by, for example, Decker and Schickedanz (1967) for Project Whitetop, a five-year randomized cloud seeding experiment to test increase in rain in Missouri, by Biondini (1976) in a Florida cumulus modification experiment, and by Federer et al. (1986) in a five-year test of a widely discussed "Soviet" hail prevention method.

NHRE was aimed at testing the feasibility of diminishing hail by seeding clouds with silver iodide from aircraft (Crow et al., 1979). The experimental unit was the "declared hail day" determined from real time radar reflectivity data. A random 50/50 choice to seed or not to seed was applied to the first day of a sequence of one or more adjacent declared hail days, the choice alternating on any subsequent such days. The experimental area was an approximate 1600-sq-km rectangle in northeast Colorado, in which over 200 instruments were deployed to measure rain and hail mass and hail sizes and numbers. In the three years 1972–74 there were 27 seed days and 30 control days.

By eye there was little apparent difference between the hail mass distributions (Fig. 1) and this was confirmed by Kolmogorov-Smirnov two-sample tests, so the seed and control data were combined to test for consistency with both lognormal and gamma distributions. No significant deviation from either form was found using the chi-squared test, and the Shapiro-Wilk and Anderson-Darling tests of normality gave the same result (for the lognormal).

It was assumed in the NHRE parametric analyses that there was no change in the shape parameters (σ for the lognormal) as the result of seed-

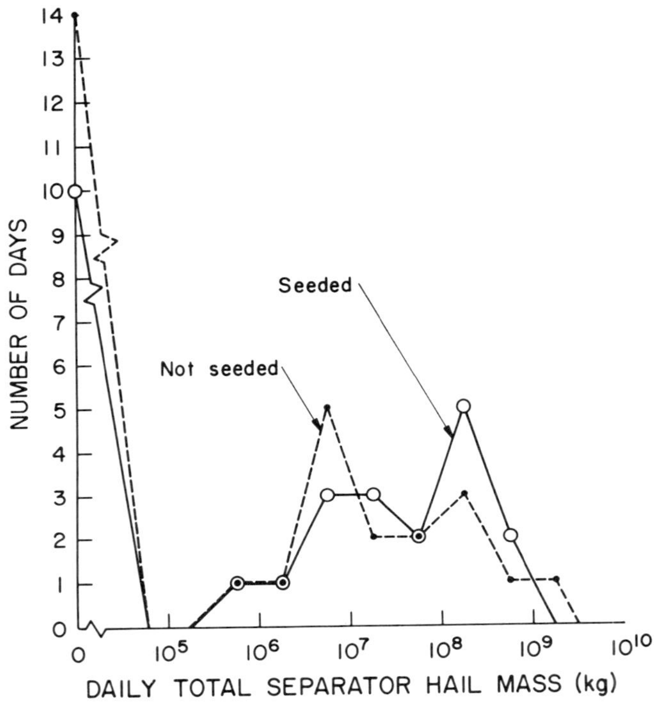

Figure 1 Distributions of 1972–74 seeded and not-seeded daily hail masses measured by hail/rain separators and integrated over the experimental area. Data are classified into half-power intervals, e.g., 10^8 to $10^{8.5}$ kg. (Reproduced from Crow et al. (1979), *Journal of Applied Meteorology* by permission.)

ing, only possible change in scale. This assumption was confirmed by test (F test for the lognormal). The mean of positive daily hail masses for control days is denoted in the lognormal case by

$$E(X) = \exp\left(\mu + \frac{\sigma^2}{2}\right)$$

and that for seed days by

$$E(Y) = \exp\left(\mu' + \frac{\sigma^2}{2}\right) = \rho E(X)$$

so that $\ln \rho = \mu' - \mu$. The test for seeding effect is thus a Student t test, and an equal-tail confidence interval for ρ follows from it also. Both the median-unbiased estimate of ρ, $\tilde{\rho} = \exp(\overline{\ln y} - \overline{\ln x})$, and the uniformly minimum variance unbiased estimate, $\hat{\rho}$ (now more elegantly represented by (3.7) of Chapter 2), were obtained. For example, for hail masses measured by rain-hail separator instruments, these estimates and 90 percent confidence limits for ρ for the two distributions are:

	$\tilde{\rho}$	$\hat{\rho}$	90% limits	
Lognormal	1.41	1.08	0.41	4.88
Gamma	0.99	0.84	0.38	2.54

Thus there was no significant seeding effect for this response variable (or for any of the eight other variables studied).

Since 10 of the 27 declared seed days and 14 of the 30 declared control days resulted in no hail, the experiment was in effect dealing with the delta-lognormal (Sec. 3.1 of Chapter 2) or delta-gamma distribution rather than the lognormal or gamma. The two types of responses were dealt with in combination by an asymptotic likelihood ratio test and confidence interval derived therefrom. There was no change in the significance of the results, however.

As the result of review by K. Ruben Gabriel, Crow et al. (1979) supplemented the above parametric tests by the nonparametric re-randomization tests. These are based on the fact that, under the null hypothesis, all days belong to the same population. The days can be divided into two samples by a coin toss (with alternation of adjacent days) and the difference in the samples characterized by the ratio of mean values (for example). The distribution of this sample ratio can be well approximated by calculating it for 2000 random permutations of the "control" and "seed" days, and the sample significance level of the actual sample thus found. The very interesting result was that this level was close to that for the gamma distribution, for which the ratio of mean values is the optimal measure. When the difference of mean logarithms (optimal for the lognormal) was used to characterize the difference between "seed" and "control" sample, the significance level of the actual sample was close to that obtained from the lognormal assumption. On the other hand, the two re-randomization significance levels (with the different measures, ratio of mean values and difference of means of logs of values) differed substantially. Thus it makes more difference what measure of seeding effect is used than whether a re-randomization or parametric test is used.

As an addendum to Crow et al. (1979), the method of Land (1975) (summarized in Chapter 3 of this volume) is applied to obtain equal-tail

90 percent confidence limits for the mean "integrated" separator hail mass for control days. There are $n = 16$ days with non-zero mass. The unknown mean is $\exp(\mu + \frac{1}{2}\sigma^2)$, where μ is estimated by $\hat{\mu} = \overline{\ln x} = 0.860$ and σ^2 by $\hat{\sigma}^2 = 4.279$. The Land (1975) table is for general situations, but we easily evaluate several constants required in the table: For the lognormal mean his λ is $\frac{1}{2}$, and here

$$\nu = n - 1 = 15 \qquad \gamma^2 = n = 16 \qquad k = \tfrac{1}{2}(\nu + 1)/(\lambda\gamma^2) = 1$$

$$S = (2\lambda\hat{\sigma}^2/k)^{1/2} = 2.069$$

$$\{\text{Upper } 95\% \text{ limit on } \mu + \lambda\sigma^2\} = \hat{\mu} + \lambda\hat{\sigma}^2 + kS\nu^{-1/2}C(S;\nu,.95) = 5.457$$

where C (1.069; 15, .95) is obtained by interpolation (4-point Lagrangian preferably) in Land's table as 4.601. Similarly the lower 95% limit is obtained as 1.824, so the equal-tail 90 percent confidence limits on the mean are 6.20 and 234 (10^7 kg). The minimum variance unbiased estimate of the mean ((2.6) of Chapter 2) is 14.55, whereas the minimum variance unbiased estimate of the median (or "geometric mean"), ((2.4) of Chapter 2), is 2.06, with equal-tail 90 percent confidence limits from the Student t distribution of 0.0629 and 88.9, much different from those for the mean.

In summary, nonparametric methods of analysis now tend to be preferred in weather modification statistical analysis to any parametric method such as lognormal because of their wider ranges of validity, though parametric methods could have somewhat greater power if their assumptions were known to apply.

6. HYDROLOGY

Hydrology is a large-scale user of statistical methods, including estimating and modeling with lognormal distributions, particularly to predict precipitation, flood frequencies, water supply, and flow through porous media. A basic parameter is the *return period*, which is the reciprocal of the probability that the flow of a river equals or exceeds a given flood level in any year. This section provides a sampling of recent hydrology literature that makes significant use of lognormal distributions. The main sources of papers using lognormal distributions are *Water Resources Research*, *Journal of Hydrology*, and *Nordic Hydrology*.

The *Handbook on the Principles of Hydrology* (Gray, 1970) has a substantial section on statistical methods, especially fitting frequency curves, and discusses the normal distribution and transformations thereto, a modified lognormal distribution ($Y = a + b \ln X$), the gamma distribution, and Gumbel's extreme value distribution. Graphical and least squares methods of estimation are most prominent, and goodness of fit is tested with chi squared and the Kolmogorov-Smirnov statistic.

Hydrology for Engineers (Linsley et al., 1975), following a U.S. Government recommendation for flood analysis, shows how to fit by moments a log-Pearson Type III distribution, of which the two-parameter lognormal distribution is the special case with zero skewness of the logs of the measurements. This recommendation is followed in a number of papers, but the lognormal with two or three parameters is also used in many papers. According to a review, the book by Kite (1977) is a "convenient compendium of formulas for method-of-moments and maximum likelihood estimation of most of the commonly used distributions of flood frequency analyses," including two– and three–parameter lognormal.

Condie and Lee (1982) enlarge the information for predicting floods by adjoining the historic information on large floods (above a certain level only) in the past to the recent continuous flow records that give the maximum in each year, whether large or not. They treat these data as a censored sample from a three-parameter lognormal distribution and fit it by maximum likelihood as well as by the previously used method of historically weighted moments. They concluded from a series of 600 Monte Carlo samples that the maximum likelihood method always provided estimates and that they had substantially smaller bias and smaller mean square error than the moment method. Later Condie (1986) evaluated the asymptotic standard of the estimate of the 100-year flood at Adelphi Weir on River Irwell in England from the information matrix. He thus was able to conclude that the historical information of 10 overbank flows (floods) increased the effective sample size from 31 years to about 57 years, worth over 100,000 Canadian dollars in the cost of operating a flow gauging station. (Condie had treated the standard error of the T-year flood for the log-Pearson Type III distribution in 1977.)

Maller and Sharma (1984) consider the distributions of two variables important in determining surface runoff resulting from a given rain rate: infiltration flux and time to ponding t_p. They are functions of two soil properties, sorptivity and hydraulic conductivity, which can be measured and are assumed to vary spatially lognormally with correlation. The theoretical distributions of infiltration flux and t_p were derived analytically and evaluated by numerical integration. An interesting feature is that t_p has a positive probability of being infinite, and even the remaining part of the distribution has infinite mean. Hence Maller and Sharma consider estimation of the median.

Kottegoda and Katuuk (1983) study the behavior of an aquifer (in particular the head of water available) as a function of the spatial variation in hydraulic conductivity, which is assumed to be lognormally distributed. Two methods of solving the partial differential equations involved are used: Monte Carlo using random walks of water particles, and finite-differences. The latter took less computer time for the two-dimensional study undertaken, but the former may be more efficient if a three-dimensional study were to be made. Similar problems using randomly variable infiltration had been studied by previous authors, such as in a Monte Carlo analysis by Smith and Hebbert (1979).

Reinius (1982) compared graphical and analytical methods of estimating flood flows (quantiles) assuming the lognormal distribution (log-Pearson Type III and Weibull being mentioned also). He proposed a new plotting position formula.

Burges et al. (1975) compared two methods of estimating the threshold parameter of the three-parameter lognormal distribution, one using the first three moments, the other substituting the median for the third moment. They concluded from 1000 Monte Carlo samples of size 50 that the three-moment method gave a less biased and less variable estimator of the threshold parameter for samples encountered in operational hydrology. They provided graphical solutions. However, the maximum likelihood solution of Chapter 4 should be considered.

Some flow volumes are found to be fitted by normal distributions rather than lognormal, but for consistency when both occur over several sites Burges and Hoshi (1978) found it desirable to approximate a normal distribution by a three-parameter lognormal with an arbitrary small skewness. This approximation is used also in Hoshi and Burges (1980) to predict runoff from a snowpack in each of several adjacent months, so that multivariate lognormal distribution is involved.

Stedinger (1980) reports a substantial Monte Carlo comparison of five different methods of estimating the parametrs of the two-parameter lognormal and four different methods of estimating those of the three-parameter lognormal. The criteria adopted were the root-mean-square error of the 99% quantile, pertinent to estimating the 100-year flood, the same for the 1% quantile, pertinent to reliable yields of reservoirs, and the integrated mean-square error of the whole distribution, also of interest for reservoir yields. The superiority of maximum likelihood for the two-parameter case was confirmed except there was little difference for sample size as low as 10. For the three-parameter case the results were not as clearcut, but the use of three symmetric quantiles to estimate the threshold parameter only is recommended. Cohen's maximum likelihood method was not included because of the iteration required. Robustness was studied using a log-Pearson Type III distribution.

Boughton (1980) derived an empirical three-parameter form of distribution for annual floods from data at 78 stations in eastern Australia. It is easier to use than the log-Pearson Type III and fits better than the two-parameter distributions lognormal, Gumbel, and log-Gumbel.

The papers of Kuczera (1982a, 1982b) consider the general problem of predicting floods and provide an excellent introduction to several earlier papers. The five-parameter Wakeby distribution (a member of the Tukey lambda family) had been introduced by Houghton to fit high-quality flood records better than two- or three-parameter distributions, but Kuczera suggested that it may be too general; a more parsimonious distribution may suffice, especially in view of the small sample sizes often available. He considered empirical Bayes estimators which incorporate information from other sites in a general way and thus include regionalized estimators (based on data from several sites after normalization) as a special case. A limited range of sampling experiments suggested that the two-parameter lognormal estimated by maximum likelihood is the most "resistant" of the site-specific procedures considered, but more robust procedures may be obtained by using regional flood information.

Simmons (1982) studied dispersion of a solute in a porous medium mathematically assuming two different distributions of particle travel time, one of them the two-parameter lognormal.

Gilliom and Helsel (1986) conducted an extensive Monte Carlo comparison of eight methods of estimating the mean, median, standard deviation, and interquartile range of censored samples from four types of distributions, with four different values of coefficient of variation. The four types of distributions were (two-parameter) lognormal, mixture of two lognormals, delta-lognormal, and gamma. Sample sizes were 10, 25, and 50, and censoring was performed at the 20th, 40th, 60th, and 80th percentiles of the parent distribution. The study was inspired by the increasing need for estimating levels of toxic substances in surface and ground waters, many of the measurements of which are below detection limits. The smallest root-mean-square error for estimating either mean or standard deviation was provided by the log-probability regression method, in which the censored observations are assumed to follow the zero-to-censoring level portion of a lognormal distribution obtained by a least-squares regression of logarithms of the uncensored observations on their z scores. The smallest root-mean-square error for estimating either median or interquartile range was provided by estimating parameters from the uncensored observations by Cohen's 1959 maximum likelihood method.

Helsel and Gilliom (1986) verified these conclusions on actual water-quality data.

7. OTHER PHENOMENA

There are almost unnumerable further meterological phenomena that have been modeled by lognormal distributions, usually empirically. Among the most common are cloud sizes, especially as measured by radar, radio and optical signal and noise magnitudes, and turbulence variables.

A review by Houze and Betts (1981) of the Global Atmospheric Research Program's Atlantic Tropical Experiment (GATE) characterized the distribution of convective features accompanying a deep cumulonimbus cluster as lognormal, "ranging from moderate comulonimbus down to tiny, nonprecipitating cumulus."

Lopez (1977; see also Lopez et al., 1984) assembled a variety of cloud data sets, most of them radar measurements, to test whether their heights, horizontal sizes, and durations are lognormally distributed. They were all consistent with the lognormal according to the chi-squared test at the 5 percent significance level. Lopez formulated two hypotheses to explain this distribution of cloud sizes, one comprising a *growth* process like Kapteyn's law of proportionate effects, the other a process of *merging* of rising smaller clouds, also a proportionate process. He noted a departure from lognormality of some of the physical distributions corresponding to fewer, or an absence of, large values and fitted truncated lognormals to them. He attributed this phenomenon to physical limits to growth. Raghavan et al. (1983) tested Lopez's hypotheses on monsoon clouds in and near India. They found deviations over land areas usually corresponding to those of Lopez but deviations in the other direction—more large radar echoes than predicted by the lognormal—over sea in the monsoon season, and they offered explanations for the deviations.

The lognormal has been used to fit the distribution of the variable "irradiance" resulting when electromagnetic waves are propagated through a turbulent medium. However, Bissonnette and Wizinowich (1979) concluded that the lognormal is theoretically incorrect because it has the irradiance variance diverging with propagation distance or turbulence strength. They derived an alternative distribution, an exponential-Bessel function, and showed good fits of experimental data to it as well as to the lognormal. Majumdar and Gamo (1982) compare laser irradiance distributions with the lognormal, gamma, Rice-Nakagami, and a new "K" distribution by calculating standardized moments of orders 3 to 6. The lognormal fit was close for short and medium propagation paths but deviated considerably for long paths.

Crawford (1982) measured atmospheric velocity and temperature fluctuations over the Pacific Ocean with a free-fall vehicle and computed two measures of turbulence from them, a rate of energy dissipation and a heat diffusivity; both have close to lognormal distributions.

LeMone and Zipser (1980) collected and analyzed a large amount of data on vertical air velocity (updrafts and downdrafts) in cumulonimbus clouds gathered on aircraft flights in GATE. They concluded that the draft diameter, average vertical velocity, maximum vertical velocity magnitude, and mass flux were all approximately lognormally distributed within each altitude group. Melling and List (1980) measured updrafts and downdrafts in a very different setting, over Toronto, Canada, with a very different instrument, an acoustic radar, and showed approximately lognormal distributions of alongstream dimensions of the drafts.

Chadwick and Moran (1980) conducted a year-long program of measuring the refractive index structure constant C_n^2 of air with Doppler radar to determine the feasibility of radar detection of hazardous wind shear at airports. They found that the 1-min values of C_n^2 have a lognormal distribution for each hour, with a diurnal and seasonal variation in mean but approximately constant standard deviation of the logarithms of 6 dB.

Other examples of lognormal variation occur in the attenuation of satellite radio signals (Capsoni et al., 1981), sea surface fluxes (Mollo-Christensen, 1978), times between lightning flashes (Le Vine et al., 1980), and spatial variation of snow depths (Gottschalk and Jutman, 1979).

Still further examples could be cited, but it is evident that examples of the lognormal distribution are ubiquitous. Statisticians should find it of interest to look into applications in the atmospheric sciences, and atmospheric scientists may find the recent developments in lognormal statistical methods in Chapters 1–6 to be useful.

REFERENCES

Aitchison, J., and Brown, J. A. C. (1957). *The Lognormal Distribution*, Cambridge University Press, Cambridge, Massachusetts.

Allen, T. (1981). *Particle Size Measurement*, 3rd ed., Chapman and Hall, New York.

Biondini, R. (1976). Cloud motion and rainfall statistics, *J. Appl. Meteor.*, 205–224.

Bissonnette, L. R., and Wizinowich, P. L. (1979). Probability distribution of turbulent irradiance in a saturation regime, *Appl. Optics.*, *18*, 1590–1599.

Boughton, W. C. (1980). A frequency distribution for annual floods, *Water Resources Res.*, *16*, 347–354.

Braham, R. R., Jr. (1979). Field experimentation in weather modification, *J. Amer. Statist. Assoc.*, *74*, 57–68, with Comments by many statisticians and two meteorologists, 68–97, and Rejoinder 97–104.

Brier, G. W. (1974). Design and evaluation of weather modification tests, *Weather and Climate Modification*, W. N. Hess, ed., John Wiley & Sons, New York, 206–225.

Burges, S. J., and Hoshi, K. (1978). Approximation of a normal distribution by a three-parameter log normal distribution, *Water Resourses Res.*, *14*, 620–622.

Burges, S. J., Lettenmaier, D. P., and Bates, C. L. (1975). Properties of the three-parameter log normal probability distribution, *Water Resources Res.*, *11*, 229–235.

Capsoni, C., Mattricciani, E., and Paraboni, A. (1981). First attempts at modeling Earth-to-space radio propagation using SIRIO measurements in the 11- and 18-GHz bands, *Annales des Telecommunciations*, *36*, 60–64.

Chadwick, R. B., and Moran, K. P. (1980). Long-term measurements of C_n^2 in the boundary layer, *Radio Sci.*, *15*, 355–361.

Condie, R. (1986). Flood samples from a three-parameter lognormal population with historic information: The asymptotic standard of estimate of the T-year flood, *J. Hydrology*, *85*, 139–150.

Condie, R., and Lee, K. A. (1982). Flood frequency analysis with historic information, *J. Hydrology*, *58*, 47–61.

Council on Environmental Quality (1984). *Environmental Quality.* (15th Annual Report of CEQ), Government Printing Office, Washington, D.C.

Crawford, W. R. (1982). Pacific equatorial turbulence, *J. Phys. Oceanography*, *12*, 1137–1149.

Crow, E. L., Long, A. B., Dye, J. E., and Heymsfield, A. J. (1979). Results of a randomized hail suppression experiment in northeast Colorado, Part II: Surface data base and primary statistical analysis, *J. Appl. Meteor.*, *18*, 1538–1558.

Decker, W. L., and Schickedanz, P. T. (1967). The evaluation of rainfall records from a five year cloud seeding experiment in Missouri, *Proceedings, 5th Berkeley Symposium on Mathematical Statistics and Probability, Vol. V: Weather Modification*, Le Cam, L. M., and Neyman, J., eds. 55–63, University of California Press, Berkeley, California.

Epstein, B. (1947). The mathematical description of certain breakage mechanisms leading to the logarithmic-normal distribution, *J. Franklin Inst.*, *244*, 471–477.

Eriksson, B. (1980). Statistisk analys av nederbord, Del 2, Frekvensanalys av manadsnederbord (statistical analysis of precipitation data, Pt.2, Frequency analysis of monthly precipitation data), Norrkoping, Sweden: Meteorologiska och Hydrologiska Institut (Swedish, with English summary) (Meteor/ Geoastrophysical Abstracts Database, ID No. MGA32120031).

Federer, B., Waldfogel, A., Schmid, W., Schiesser, H. H., Hampel, F., Schweingruber, M., Stahel, W., Bader, J., Mezeix, J. F., Doras, N., d'Aubigny, G., Der Megredichian, G., and Vento, D. (1986). Main results of Grossversuch IV, *J. Climate & Appl. Meteor.*, *25*, 917–957.

Feingold, G., and Levin, Z. (1986). The lognormal fit to raindrop spectra from frontal convective clouds in Israel, *J. Climate & Appl. Meteor.*, *25*, 1346–1363.

Fitch, B. W. (1983). Spatial and temporal variations of tropospheric aerosol volume distributions, *J. Climate & Appl. Meteor.*, *22*, 1262–1269.

Gabriel, K. R. (1979). Some statistical issues in weather experimentation, *Commun. Statist.*, *A8*, 975–1015.

Gilliom, R. J., and Helsel, D. R. (1986). Estimation of distributional parameters for censored trace level water quality data, 1. Estimation techniques, *Water Resources Res.*, *22*, 135–146.

Giorgi, F. (1986). Development of an atmospheric aerosol model for studies of global budgets and effects of airborne particulate material, Ph.D. thesis, Cooperative thesis No. 102, Georgia Institute of Technology, Atlanta, GA, and National Center for Atmospheric Research, Boulder, CO.

Gottschalk, L., and Jutman, T. (1979). Statistical analysis of snow survey data. *Hydrologi och Oceanografi*, No. RHO 20, Norrkoping, Sweden: Meteorologiska och Hydrologiska Institut (Meteor/Geoastrophysical Abstracts Database, ID No. MGA31090341).

Gras, J. L., and Laby, J. E. (1981). Southern Hemisphere stratospheric aerosol measurements: 3. Size distribution 1974–1979, *J. Geophy. Res.*, *86*, 9767–9775.

Gray, D. M. (1970). *Handbook on the Principles of Hydrology*, Water Information Center, Inc., Port Washington, NY.

Harrison, R. M., and Pio, C. A. (1983). Size-differentiated composition of inorganic atmospheric aerosols of both marine and polluted continental origin, *Atmospheric Environment*, *17*, 1733–1738.

Helsel, D. R., and Gilliom, R. J. (1986). Estimation of distributional parameters for censored trace level water quality data, 2. Verification and applications, *Water Resources Res.*, *22*, 147–155.

Herdan, G. (1960). *Small Particle Statistics*. 2nd rev. ed., Butterworths, London.

Hoshi, K., and Burges, S. J. (1980). Seasonal runoff volumes conditioned on forecasted total runoff volume, *Water Resources Res.*, *16*, 1079–1084.

Houze, R. A., Jr., and Betts, A. K. (1981). Convection in GATE, *Revs. of Geophysics & Space Physics*, *19*, 541–576.

Kelkar, D. N., and Joshi, P. V. (1977). A note on the size distribution of aerosols in urban atmospheres, *Atmospheric Environment*, *11*, 531–534.

Kite, G. W. (1977). *Frequency and Risk Analyses in Hydrology*, Water Resources Publications, Ft. Collins, CO. (Meteor/Geoastrophysical Abstracts Data Base, ID No. MGA31100364).

Kottegoda, N. T., and Katuuk, G. C. (1983). Effect of spatial variation in hydraulic conductivity on groundwater flow by alternate solution techniques, *J. Hydrology*, *65*, 349–362.

Koziel, S. (1975). Normalizacja zmiennej losowej o rozkladzie Jyskvetno-ciaglym przykladzie sum dobowych opadu atmosferyeznego [Normalization of the discrete-continuous distribution in application to daily total amount of precipitation] *Przeglad Geofizyczny*, *20*, 9–13 (Polish) (Meteor/ Geoastrophysical Abstracts Database, ID No. MGA27004039).

Kuczera, G. (1982a). Combining site-specific and regional information: An empirical Bayes approach, *Water Resources Res.*, *18*, 306–314.

Kuczera, G. (1982b). Robust flood frequency models, *Water Resources Res.*, *18*, 315–324.

Land, C. E. (1975). Tables of confidence limits for linear functions of the normal mean and variance, *Selected Tables in Mathematical Statistics*, *Vol. III*, Harter, H. L., and Owen, D. B., eds., 385–419, American Mathematical Society, Providence, RI.

Larsen, R. I. (1969). A new mathematical model of air pollutant concentration averaging time and frequency, *J. Air Pollution Control Assoc.*, *19*, 24–30.

Larsen, R. I. (1977). An air quality data analysis system for inter-relating effects, standards, and needed source reduction: Part 4. A three-parameter averaging-time model, *J. Air Pollution Control Assoc.*, *27*, 454–459.

Le Cam, L. M., and Neyman, J., eds. (1967). *Proceeding of the Fifth Berkeley Symposium on Mathematical Statistics and Probability*, *Vol. V.*, *Weather Modification*, University of California Press, Berkeley.

Legrand, M. (1974). Statistical studies of urban air pollution—sulfur dioxide and smoke, 25–37, *Statistical and Mathematical Aspects of Pollution Problems*, Marcel Dekker, Inc., New York.

LeMone, M. A., and Zipser, E. J. (1980). Cumulonimbus vertical velocity events in GATE. Part 1: Diameter, intensity, and flux mass. *J. Atmos. Sci.*, *37*, 2444–2457.

Le Vine, D. M., Meneghini, R., and Tretter, S. A. (1980). Rate statistics for radio noise from lightning, NASA Technical Paper 1665.

Linsley, R. K., Jr., Kohler, M. A., and Paulhus, J. L. H. (1975). *Hydrology for Engineers*, 2nd ed. McGraw-Hill Book Co., New York.

López, R. E. (1977). The lognormal distribution and cumulus cloud populations, *Monthly Weather Rev.*, *105*, 865–872.

López, R. E., Blanchard, D.O., Rosenfeld, D., Hiscox, W. L., and Casey, M. J. (1984). Population characteristics, development processes and structure of radar echoes in south Florida, *Monthly Weather Rev.*, *112*, 56–75.

Low, R. D. H. (1979). Theoretical investigation of cloud-fog extinction coefficients and their spectral correlations, *Contributions to Atmospheric Physics*, *52*, 44–57.

Machta, L., and Telegadas, K. (1974). Inadvertent large-scale weather modification, *Weather and Climate Modification*, W. N. Hess, ed., 687–725.

Mage, D. T., and Ott, W. R. (1978). Refinements of the lognormal probability model for analysis of aerometric data, *J. Air Pollution Control Assoc.*, *28*, 796–798.

Majumdar, A. K., and Gamo, H. (1982). Statistical measurements of irradiance fluctuations of a multipass laser beam propagated through laboratory-simulated atmospheric turbulence, *Appl. Optics*, *21*, 2229–2235.

Maller, R. A., and Sharma, M. L. (1984). Aspects of rainfall excess from spatially varying hydrological parameters, *J. Hydrology*, *67*, 115–127.

Melling, H., and List, R. (1980). Characteristics of vertical velocity fluctuations in a convective urban boundary layer, *J. Appl. Meteor.*, *19*, 1184–1195.

Mollo-Christensen, E. (1978). Introductory remark to the session on sea surface fluxes. *Turbulent Fluxes Through the Sea Surface, Wave Dynamics, and Prediction*, 1–4, Favre, A., and Hasselmann, K., eds., Plenum Press, New York.

Monin, A. S. (1986). *Introduction to the Theory of Climate*, D. Reidel Publishing Co., Boston, MA.

Neiburger, M., and Weickmann, H. K. (1974). The meteorological background for weather modification, *Weather and Climate Modification*, W. N. Hess, ed., 93–135, John Wiley & Sons, New York.

Neyman, J., and Scott, E. L. (1960). Correction for bias introduced by a transformation of variables, *Ann. Math. Statist.*, *31*, 643–655.

Neyman, J., and Scott, E. L. (1967). Some outstanding problems relating to rain modification. *Proceedings of the Fifth Berkeley Symposium on Mathematical Statistics and Probability*, Le Cam, L. M., and Neyman, J., eds., 293–326. University of California Press, Berkeley.

Ott, W. R. (1978). *Environmental Indices: Theory and Practice*, Ann Arbor Science Publishers, Ann Arbor, MI.

Patterson, E. M., and Gillette, D. A. (1977). Commonalities in measured size distributions for aerosols having a soil-derived component, *J. Geophys. Res.*, *82*, 2074–2082.

Raghavan, S., Sivaramakvishnan, T. R., and Ramakrishnan, B. (1983). Size distribution of radar echoes as an indicator of growth mechanisms in monsoon clouds around Madras, *J. Atmos. Sci.*, *40*, 428–434.

Reinius, E. (1982). Statistical flood flow estimation, *Nordic Hydrology 13*, 49–64.

Renoux, A., Tymen, G., and Madelaine, G. (1982). Granulometric de l'aerosol marin. *Proc. 2nd European Symp. Physico-Chemical Behaviour of Atmospheric Pollutants*, 249–256. D. Reidel Publishing Co., Dordrecht, Holland.

Schaefer, V. J. (1946). The production of ice crystals in a cloud of supercooled water droplets, *Science*, *104*, 457–459.

Segal, B. (1980). Analytical examination of mathematical models for the rainfall rate distribution function. *Annales des Telecommunications*, *35*, 434–438 (Meteor/Geophysical Abstracts Database, ID No. MGA32110404).

Simmons, C. S. (1982). A stochastic-convective transport representation of dispersion in one-dimensional porous media systems, *Water Resources Res. 18*, 1193–1214.

Smith, R. E., and Hebbert, R. H. B. (1979). A Monte Carlo analysis of the hydrologic effects of spatial variability of infiltration, *Water Resources Res.*, *15*, 419–429.

Sneyers, R. (1979). L'intensité et la durée maximales des précipitations en Belgique, Brussels Publications, Series B., 99, Belgium: Institut Royal Météorologique (Meteor/Geoastrophysical Abstracts Database, ID No. MGA31030362).

Statistical Task Force of the Weather Modification Advisory Board (1978). *The Management of Weather Resources, Vol. II: The Role of Statistics in Weather Resources Management.* U. S. Government Printing Office, Washington, D. C.

Stedinger, J. R. (1980). Fitting log normal distributions to hydrologic data, *Water Resources Res.*, *16*, 481–490.

Swift, L. W., and Schreuder, H. T. (1981). Fitting daily precipitation amounts, using the S_B distribution, *Monthly Weather Rev.*, *109*, 2535–2541.

Tanaka, M., Takamura, T., and Nakajima, T. (1983). Refractive index and size distribution of aerosols as estimated from light scattering measurements, *J. Climate & Appl. Meteor.*, *22*, 1253–1261.

Tsukatani, T. (1981). Statistical aspects of air pollution quality, Annual Report, Res. Reactor Inst., Kyoto, Univ., *14*, 129–152.

Volynets, L. M. (1982). Statisticheskie kharakteristiki osadkov iz individual'nykh kuchevo-dozhdevykh oblakov (Statistical characteristics of precipitation from individual cumulonimbus clouds, Trudy No. 187, 32–43 (Russian), Kiev, Ukraine: Ukrainskiy Regional Nyy Nauchno-Issledovatelskiy Institut (Meteor/Geoastrophysical Abstracts Database, ID No. MGA36020472).

Watt, W. E., and Nozdryn-Plotnicki, M. J. (1983). Frequency analysis of short-duration rainfall, *Atmosphere-Ocean*, *21*, 387–396.

Whitby, K. T. (1978). The physical effects of sulfur aerosols, *Atmospheric Environment*, *12*, 135–159.

Wiorkowski, J. J., and Odell, P. L. (1979). Statistical analysis of weather modification experiments, Special Issues Nos. 10 and 11, *Commun. Statist.*, *A8*, 953–1047, 1049–1153.

14

Applications in Geology

JEAN-MICHEL M. RENDU Technical and Scientific Systems Group, Newmont Mining Corporation, Danbury, Connecticut

1. INTRODUCTION

The lognormal distribution has long been recognized as a useful model in the evaluation of geological variables whose value distribution is often extremely skew. The metal content of rock chips in a mineral deposit, the dimension of sand and gravel particles in a river bed, and the size of oil or gas pools in a geographic region are examples of geological variables whose values have been modeled using the lognormal distribution. Correct evaluation of these variables is often critical. Many significant economic decisions are made and millions of dollars are invested on the assumption that the properties of critical geological variables are well understood.

Statisticians have long been strongly motivated to develop improved methods to analyze geological variables. The particular nature of these variables, including the ubiquitous relation between the location where a measurement is made and the value of this measurement, has required the original development of appropriate statistical methods. The various applications of the lognormal distribution to geological variables can be classified according to whether or not spatial correlations are taken into account.

When studying statistical applications in geology, one should be aware of the difference between the statistical definition of a sample and the corresponding geological definition. In statistics, a sample is a group of observations. In geology a sample is a representative fraction of a body of material, whose properties are analyzed to obtain a single value. The geological definition has been used in this chapter.

2. LOGNORMAL MODELING WITHOUT SPATIAL CORRELATION

The significance of the lognormal distribution in the evaluation of geologic data was brought to light by Sichel (1947). In this original study, the frequency distribution of sample values obtained from South African gold deposits was shown to well approximate the lognormal model. Methods were proposed to eliminate biases in parameter estimations, most significantly in estimation of the mean gold value of the deposit being sampled. The approach to gold evaluation proposed by Sichel was soon used by others (Krige, 1952), and later refined by the introduction of the three-parameter lognormal distribution (Krige, 1960). The same model was subsequently shown to apply to a very large number and variety of deposits (Matheron, 1957; Sichel, 1973; Oosterveld, 1973; David, 1977; Rendu, 1984).

Theoretical investigations completed by geostatisticians resulted in the development of mean and confidence interval estimators for small numbers of sample values (Sichel, 1952, 1966; Wainstein, 1975). Genetic models have also been developed to explain the process which may have resulted in a lognormal distribution of sample values (de Wijs, 1951, 1953). When analyzing samples taken from mineral deposits, truncated and censored distributions are commonly observed, as well as mixed populations which may be indicative of complex geologic environments or multistage orebody genesis.

Analysis of mine production statistics resulted in some instances in the development of empirical laws (Lasky, 1950) which were later recognized as consequences of the lognormal distribution (Matheron, 1959). Lasky observed that when variable cutoff grades are applied to porphyry copper deposits, the average grade above cutoff grade and the corresponding tonnage are linearly related. Matheron showed that a sufficient condition for this estimation to be well approximated is that the sample values are lognormally distributed. The lognormal model has been used to develop mathematical formulae and graphs which relate tonnage and average grade above cutoff grade in deposits whose sample values are lognormally distributed (Krige, 1978; Rendu, 1981). These formulae and graphs are used in the early evaluation of mineral deposits and are preferred to Lasky's law because of their less empirical nature. In the same context, the lognormal distribution has been used to apply statistical decision theory to mineral exploration (Matheron and Formery, 1963; Rendu, 1971).

The lognormal distribution is also used to model the size distribution of mineral deposits in large areas. The first application was by Allais (1957) in an effort to model the economics of mineral exploration in the Sahara. The model was also used by the petroleum industry to represent the size distribution of oil and gas pools (Harbaugh et al., 1977; Lee and Wang, 1985).

Geochemical data, as are obtained from stream samples or grab samples in regional exploration, are often analyzed using graphical or numerical methods developed for lognormal distributions. The problems faced by the decision maker include the recognition of extreme values, the separation of mixed populations (Bridges and McCammon, 1980; Clark and Garnett, 1974) and the determination of structures within multivariate datasets using such techniques as cluster analysis and discriminate analysis.

Particle size distributions are also commonly analyzed using lognormal models. If lognormality of size distribution is accepted, experimental techniques can be conceived to obtain distributions based on weight percent, from which distributions indicative of the number of particles can be derived (Aitchison and Brown, 1957, p. 101). A model of breakage was proposed by Dacey and Krumbein (1979) to explain the lognormality of particle size distributions.

3. UNEXPLAINED STATISTICAL RELATIONSHIPS IN MINERAL DEPOSITS WITH LOGNORMAL PROPERTIES

The most complex applications of lognormal theory in geological sciences can arguably be found in the evaluation of mineral deposits. Evaluation of a deposit typically requires geologic sampling at specified locations. The samples taken may represent as little as one millionth of the total volume of the mineralized body under investigation. Using this very small dataset, the value of mineralized blocks, of parts of the deposit, or of the entire deposit must be inferred. In some instances the sample values can be considered spatially uncorrelated and the relatively simple methods mentioned earlier are applicable. In other more common situations, correlation between sample values must be recognized, and more complex statistical models had to be developed accordingly.

Typically, the development of statistical methods to analyze spatially correlated lognormally distributed variates originated from the need to solve practical problems. In most mineral deposits, some degree of selectivity is imposed by economic constraints. The deposit is divided into parts or blocks, whose values are estimated from the value of surrounding or contained samples. Blocks with high estimated value may be mined at a profit while leaner blocks may be treated as waste. It has long been observed however that samples with high values tend to overestimate the blocks containing or surrounded by the samples, while samples with low values underestimate the corresponding blocks.

This conditional bias is most significant in deposits with lognormal distributions. In these deposits overestimation of blocks near high grade samples can be extremely high, sometimes reaching one or two orders of magnitude, with catastrophic economic consequences. Early attempts were made to explain the bias by sampling and assaying errors. Work completed on data from the South African gold mines showed however that the difference between the distribution of sample values and that of block values could be explained using a combination of lognormal theory and empirical laws.

Dr. D. G. Krige (1951, 1952, 1959, 1966a) observed that in the gold deposits, two fundamental relationships were verified:

—If the sample values were lognormally distributed, the block values were also lognormally distributed with an equal mean but a smaller variance.

—The logarithmic variance of the block values within the deposit was equal to the difference between the logarithmic variance of the sample values within the deposits and that of the sample values within the blocks.

These relationships are empirical and a satisfactory theoretical explanation has as yet not been offered. The variance of the block values being less than that of the sample values however, the regression line which relates the block values to the sample values in the logarithmic space has a slope less than unity. The conditional estimation bias can therefore be explained by statistical theory and statistical methods can be used to reduce or eliminate this bias (Krige 1959, 1962a, 1962b, 1978).

A number of methods have been devised to improve the economic evaluation and exploitation of mineral deposits, which are based on the lognormal laws relating block distributions to sample distributions (Parker, 1980). One of these methods is known as the "lognormal shortcut" (David et al., 1977) because of the lack of a complete theoretical foundation.

4. LOGNORMAL MODELING WITH SPATIAL CORRELATIONS

The correlation between sample values and block values is indicative of a more basic autocorrelation between sample values. It is generally expected that geological samples located next to each other are more likely to have similar values than samples located farther apart. The correlation between sample values is a decreasing function of the distance between samples.

When dealing with lognormal variates this autocorrelation is usually measured using the logarithm of the sample values (Krige, 1978).

Rather than calculating covariance functions, the common practice in geostatistics consists in comparing sample values by calculating the semivariogram. The semivariogram is defined as one half the mean squared difference between sample values and is typically an increasing function of the distance between samples. In large deposits with a lognormal sample value distribution, the semivariogram is often observed to be a linear function of the logarithm of distance (Krige, 1978; Matheron, 1962, 1963). Other semivariogram models have been proposed and are now commonly used (Journel and Huijbregts, 1978).

A general theory of estimation of blocks from sample values has been developed, based on a prior analysis of the sample autocorrelation function (Matheron, 1971; David, 1977; Journel and Huijbregts, 1978; Rendu, 1981). The estimation method was originally developed empirically as a multivariate linear regression technique (Krige, 1966b). An integrated theoretical foundation was later built, and the method was generalized and redefined as "kriging" (Matheron, 1962).

In deposits with lognormally distributed values, the estimation procedure was originally based on calculating logarithmic autocorrelation functions, but estimating blocks from sample values using linear regression (Krige, 1978; Rendu, 1976). This mixed approach presents the advantage of reducing the influence of extreme values in the study of autocorrelations, while avoiding the systematic biases which may be introduced by a lognormal approach when variances are poorly estimated. The disadvantage is that linear regression between lognormal variates is likely to give conditionally biased estimates.

A complete and theoretically more desirable lognormal approach to multivariate lognormal regression was proposed by Marechal (1974), Lallement (1975) and Rendu (1979a). This approach, known as "lognormal kriging" was shown to give improved results in the South African gold fields (Rendu, 1979b) and has been successfully applied to other deposits. Lognormal kriging has been used to find optimal theoretical solutions to a variety of decision making problems (Rendu, 1980a).

Lognormal kriging is generally used to estimate the value of blocks of ore from samples. A related problem encountered by many economic geologists is that of estimating the *proportion* of a deposit (or part of a deposit) which can be mined selectively, and the average value of this proportion, when the sample information is not sufficient to determine the *location* of this mineable material. What must then be estimated is the frequency distribution of blocks of a given size (defined by mining constraints), conditional on the location and value of the available sample information. An answer to this "conditional recovery" estimation problem is the "lognormal shortcut" mentioned earlier. Another answer is based on the assumption of a lognormal distribution of block values, the parameters of this distribution being estimated from the sample locations, sample values, and sample autocorrelation function (Parker and Switzer, 1976; Journel and Huijbregts, 1978; Parker et al., 1979; Journel, 1980).

5. CONCLUSIONS

The lognormal distribution has continuously played a very significant role in the study of geological data. The need to find better answers to specific questions of economic significance, has pushed statistical geologists towards new applications of the lognormal theory. A practical problem remains that the lognormal model is often only an approximation of the true distribution of geological data. For this reason, statistical methods are always sought to analyze geological data which are independent of the lognormal assumption (Marechal, 1976; Rendu, 1980b; Journel, 1983; Verly, 1983). The relative ease with which the lognormal model can be used, however, has made it a very powerful tool which is widely and successfully applied. It is likely that geological statisticians will continue to contribute to the development of the lognormal theory and its applications.

REFERENCES

Aitchison, J., and Brown, J. A. C. (1957). *The Lognormal Distribution*, Cambridge University Press, Cambridge, 176p.

Allais, M. (1957). Method of appraising economic aspects of mining exploration over large territories, *Management Science*, *3*, 285–347.

Bridges, N. J., and McCammon, R. B. (1980). DISCRIM: A computer program using an interactive approach to dissect a mixture of normal or lognormal distributions, *Computers and Geosciences*, *6*, 361–396.

Clark, I., and Garnett, R. H. T. (1974). Identification of multiple mineralization phases by statistical methods, *Transactions, Inst. Min. Metall., Sec. A*, *83*, 43–52.

Dacey, M. G., and Krumbein, W. C. (1979). Models of breakage and selection for particle size distributions, *J. Internat. Assoc. Math. Geol.*, *11*, 193–222.

David, M. (1977). *Geostatistical Ore Reserve Estimation*, Elsevier, Amsterdam, 364p.

David, M., Dagbert, M., and Belisle, J. M. (1977). The practice of porphyry copper deposit estimation for grade and ore-waste tonnages demonstrated by several case studies, in *Application of Computers and Operations Research in the Mineral Industry*, Australian Inst. Min. Metall., Brisbane, Australia, 243–254.

Harbaugh, J. W., Doveton, J. H., and Davis, J. C. (1977). *Probability Methods in Oil Exploration*, John Wiley & Sons, New York, 269p.

Journel, A. G. (1980). The lognormal approach to predicting local distributions of selective mining unit grades, *J. Internat. Assoc. Math. Geol.*, *12*, 285–303.

Journel, A. G. (1983). Nonparametric estimation of spatial distributions, *J. Internat. Assoc. Math Geol.*, *15*, 445–467.

Journel, A. G., and Huijbregts, C. J. (1978). *Mining Geostatistics*, Academic Press, London, 600p.

Krige, D. G. (1951). A statistical approach to some basic mine valuation problems on the Witwatersrand, *J. Chem. Metall. Min. Soc. S. Afr.*, *52*, 119–139.

Krige, D. G. (1952). A statistical analysis of some of the borehole values in the Orange Free State goldfield, *J. Chem. Metall. Min. Soc. S. Afr.*, September, 47–70.

Krige, D. G. (1959). A study of the relationship between development values and recovery grades on the South African goldfields; *J. S. Afr. Inst. Min. Metall.*, January, 317–331.

Krige, D. G. (1960). On the departure of ore value distributions from the lognormal model in South African gold mines. *J. S. Afr. Inst. Min. Metall.*, November, 231–244.

Krige, D. G. (1962a). Effective pay limits for selective mining, *J. S. Afr. Inst. Min. Metall.*, January, 345–363.

Krige, D. G. (1962b). *Statistical Applications in Mine Valuation*, J. Inst. Mine Surveyors S. Afr., 82p.

Krige, D. G. (1966a). A study of gold and uranium distribution patterns in the Klerksdorp gold field, *Geoexploration*, *4*, 43–53.

Krige, D. G. (1966b). Two-dimensional weighted moving average trend surfaces for ore valuation, in *Symposium on Mathematical Statistics and Computer Applications in Ore Valuation*, S. Afr. Inst. Min. Metall., 13–79.

Krige, D. G. (1978). *Lognormal de wijsian geostatistics for ore evaluation*, S. Afr. Inst. Min. Metall., Monograph Series, Johannesburg, 50p.

Lallement, B. (1975). Geostatistical evaluation of a gold deposit, in *Proceedings 13th International APCOM Symposium*, University Clausthal, Federal Republic of Germany, 1-IV, 1–15.

Lasky, S. G. (1950). How tonnage and grade relations help predict ore reserves, *Eng. Min. J.*, *151*, 81–85.

Lee, P. J., and Wang, P. C. C. (1985). Prediction of oil or gas pool sizes when discovery record is available, *J. Internat. Assoc. Math. Geol.* *17*, 95–113.

Marechal, A. (1974). *Krigeage Normal et Lognormal*, Centre Morphologie Mathématique Fontainebleau, École Nat. Sup. Mines Paris, N376, 10p.

Marechal, A. (1976). The practice of transfer functions: numerical methods and their application, in *Advanced Geostatistics in the Mining Industry*, (M. Guarascio et al., eds.), D. Reidel, Boston, 253–276.

Matheron, G. (1957). Théorie lognormale de l'échantillonage systématique des gisements, *Annales des Mines*, September, 566–584.

Matheron, G. (1959). Remarques sur la loi de Lasky, *Chronique Mines Outre Mer et Recherche Miniére*, *282*, 463–466.

Matheron, G. (1962). *Traité de Géostatistique Appliquée, Tome 1*, Memoires Bureau Recherches Géologique Minière, *14*, Technip, Paris, 333 p.

Matheron, G. (1963). *Traité de Géostatistique Appliquée, Tome 2, le Krigeage*, Memoires Bureau Recherches Géologique Minière, *24*, Technip, Paris, 171 p.

Matheron, G. (1971). *The Theory of Regionalized Variables and its Applications*, Cahiers Centre Morphologie Mathématique Fontainebleau, École Nat. Sup. Mines Paris.

Matheron, G., and Formery, Ph. (1963). Recherche d'optimum dans la reconnaissance et la mise en exploitation des gisements miniers, *Annale des Mines*, May, 23–42, June, 2–30.

Oosterveld, M. M. (1973) Ore reserve estimation and depletion planning for a beach diamond deposit, in *Application of Computer Methods in the Mineral Industry*, (M. D. G. Salamon et al., eds.), S. Afr. Inst. Min. Metall., 65–71.

Parker, H. M. (1980). The volume-variance relationship: a useful tool for mine planning, *Geostatistics*, McGraw Hill, New York, 61–91.

Parker, H. M., Journel, A. G., Dixon, W. C. (1979). The use of conditional lognormal probability distribution for the estimation of open-pit ore reserves in stratabound uranium deposits—A case study, in *Application of Computers and Operations Research in the Mineral Industry*, (T. J. O'Neil, ed.), Soc. Min. Eng., New York, 133–148.

Parker, H. M., and Switzer, P. (1976). The problem of ore versus waste discrimination for individual blocks: The lognormal model, *Advanced Geostatistics in the Mining Industry*, (M. Guarascio et al., eds.), D. Reidel, Boston, 203–218.

Rendu, J. M. (1971). Some applications of geostatistics to decision making in exploration, in *Decision Making in the Mineral Industry*, Can. Inst. Min. Metall., *12*, 175–184.

Rendu, J. M. (1976). The optimization of sample spacing in South African gold mines, *J. S. Afr. Inst. Min. Metall.*, *76*, 392–399.

Rendu, J. M. (1979a). Normal and lognormal estimation, *Mathematical Geology*, *11*, 407–422.

Rendu, J. M. (1979b). Kriging, logarithmic kriging and conditional expectation: Comparison of theory with actual results, in *Application of Computers and Operations Research in the Mineral Industry*, (T. J. O'Neil, ed.), Soc. Min. Eng., New York, 199–212.

Rendu, J. M. (1980a) Optimization of sampling policies: A geostatistical approach, in *Proceedings of Fourth Joint Meeting MMIJ-AIME*, Min. Metall. Inst. Japan, *A-1-3*, 27–48.

Rendu, J. M. (1980b). Disjunctive Kriging: comparison of theory with actual results, *J. Internat. Assoc. Math. Geol.*, *12*, 305–320.

Rendu, J. M. (1981). *An Introduction to Geostatistical Methods of Mineral Evaluations*, S. Afr. Inst. Min. Metall., Monograph Series, 2nd edition, 84p.

Rendu, J. M. (1984). Geostatistical methods of ore reserve estimation, *Mining Geology*, Soc. Min. Geol. Japan, *34*, 197–224.

Sichel, H. S. (1947). An experimental and theoretical investigation of bias error in mine sampling with special reference to narrow gold reefs, *Trans. Inst. Min. Metall.*, *56*, 403–443.

Sichel, H. S. (1952). New method in the statistical evaluation of mine sampling data, *Trans. Inst. Min. Metall.*, *61*, 261–288.

Sichel, H. S. (1966). The estimation of means and associated confidence limits for small samples from lognormal population, in *Symposium on Mathematical Statistics and Computer Applications in Ore Valuation*, S. Afr. Inst. Min. Metall., 106–123.

Sichel, H. S. (1973). Statistical valuation of diamondiferous deposits, in *Application of Computer Methods in the Mineral Industry*, (M. D. G. Salamon et al., eds.), S. Afr. Inst. Min. Metall., 17–33.

Verly, G. (1983). The multigaussian approach and its applications to the estimation of local reserves, *J. Internat. Assoc. Math. Geol.*, *15*, 259–286.

Wainstein, B. M. (1975). An extension of lognormal theory and its application to risk analysis models for new mining ventures, *J. S. Afr. Inst. Min. Metall.*, *75*, 221–238.

Wijs, H. J. de (1951). Statistics of ore distribution, part 1, frequency distribution of assay values, *Geologie en Mijnbouw*, November, 365–375.

Wijs, H. J. de (1953). Statistics of ore distribution, part 2, theory of binomial distribution applied to sampling and engineering problems, *Geologie en Mijnbouw*, January, 12–24.

Author Index

Italic numbers give the page on which the complete reference is listed.

Subject Index